Water Gas Shift Reaction

Research Developments and Applications

Water Gas Shift Reaction

Research Developments and Applications

By

Gunugunuri K. Reddy
Research Scientist, Lab M-3
Materials Research Department
Toyota Research Institute of North America
Toyota Motor Engineering &
Manufacturing North America, Inc.
Ann Arbor, MI, USA

Panagiotis G. Smirniotis
Department of Biomedical, Chemical, and Environmental Engineering
Engineering Research Center,
University of Cincinnati
Cincinnati OH, USA

ELSEVIER
AMSTERDAM • BOSTON • HEIDELBERG • LONDON • NEW YORK • OXFORD
PARIS • SAN DIEGO • SAN FRANCISCO • SINGAPORE • SYDNEY • TOKYO

Elsevier
Radarweg 29, PO Box 211, 1000 AE Amsterdam, Netherlands
The Boulevard, Langford Lane, Kidlington, Oxford OX5 1GB, UK
225 Wyman Street, Waltham, MA 02451, USA

Library of Congress Cataloging-in-Publication Data
A catalog record for this book is available from the Library of Congress

British Library Cataloguing in Publication Data
A catalogue record for this book is available from the British Library

ISBN: 978-0-12-420154-5

Dedication

To my wife, Sireesha, and new born baby, who has given me love, understanding, focus and a new vision of the wonders of life. I also dedicate this book to my parents, Rosi Reddy and Padmavathi, and my brother, Konda Reddy, for their endless love throughout my life.

I also take this occasion to thank all my research supervisors Dr. B. M. Reddy, S. Loridant and P. Smirniotis for their support in my research career.

Gunugunuri K. Reddy

To my wife, Vasiliki, whose friendship, support, understanding and love made this possible, my loving daughter, Lydia, and son, George.

Panagiotis G. Smirniotis

Contents

Chapter 1

Introduction About WGS Reaction

1.1 HISTORY AND BACKGROUND

1.1.1 Water Gas

Water gas is an equimolar mixture of carbon monoxide and hydrogen. It can be synthesized by passing steam through coke. During late nineteenth century town gas developed the manufacturing process for the water gas.

$$C + H_2O \rightarrow CO + H_2 \quad \Delta H = 131.2 kJ/mol \tag{1.1}$$

The reaction is endothermic, so the fuel must be continually re-heated to keep the reaction going. In order to do this, an air stream, which alternates with the vapour stream, is introduced for the combustion of carbon.

$$2C + O_2 \rightarrow 2CO \quad \Delta H = -220.8 kJ/mol \tag{1.2}$$

These two reactions take place in cycle basis, as the temperature of the second reaction reaches sufficiently high the steam cycle restarts. Because of the wide temperature range in reality a small amount of carbon dioxide is always present in the water gas. Because of contamination in the air blow cycle a small amount of nitrogen is also present in the water gas.

In the early 1990s, production of water gas using steam reforming of methane received tremendous importance if the ultimate objective is generation of pure hydrogen since it provides highest molar ratio of H_2/CO of Equation (1.3).

$$CH_4 + H_2O \rightarrow CO + 3H_2 \quad \Delta H = 206.3 kJ/mol \tag{1.3}$$

Partial oxidation of methane is another way to produce water gas (Equation 1.3). This process is mainly used when we need lesser H_2/CO ratio and if there are difficulties in external heat supply, internal heat generation is needed as in the case of fuel processors for fuel cell applications. Partial oxidation of methane produces H_2/CO in a ratio of 2. If we need H_2/CO in a ratio of 1, dry reforming of methane can be done (Equation 1.4).

$$CH_4 + 0:5\,O_2 \rightarrow CO_2 + 2H_2 \quad \Delta H = 35.6 kJ/mol \tag{1.4}$$

$$CH_4 + CO_2 \rightarrow 2CO_2 + 2H_2 \quad \Delta H = 247.4 kJ/mol \tag{1.5}$$

Water Gas Shift Reaction. http://dx.doi.org/10.1016/B978-0-12-420154-5.00001-2

Water gas is used extensively in the industry for the manufacture of ammonia, methanol, hydrogen (for hydrotreating, hydrocracking of petroleum fractions and other hydrogenations in the petroleum refining and petrochemical industry), hydrocarbons (by the Fischer-Tropsch process) and metals (by the reduction of the oxide ore).

1.1.1.1 Types of Water Gas

1.1.1.1.1 Carburetted Water Gas

Water gas had a lower calorific value than coal gas, so the calorific value was often boosted by passing the gas through a heated retort into which oil was sprayed. The resulting mixed gas was called carburetted water gas.

1.1.1.1.2 Semi-Water Gas

Semi-water gas is a mixture of water gas and producer gas made by passing a mixture of air and steam through heated coke. The heat generated when producer gas is formed keeps the temperature of the coke high enough to allow water gas to be formed.

1.1.2 Water-Gas Shift Reaction

The water-gas shift reaction (WGSR) was discovered by Italian physicist Felice Fontana in 1780 [1,2]. However, the reaction was first patented by the British scientists Ludwig Mond and Langer C in 1888 for fuel cell application in coal gasification [3]. Ludwig Mond, one of the greatest chemist-industrialists of all time, focused part of his industrial chemical technology developments on the synthesis of ammonia from coal. Mond developed the process for producing the so-called Mond gas (the product of the reaction of air and steam passed through coal/coke – CO_2, CO, H_2, N_2, etc.), which became the basis for future coal gasification processes [4]. Mond and his assistant Carl Langer were the first to use the term 'fuel cells' while performing experiments with the world's first working fuel cell using coal-derived Mond gas [4]. One of the hardest tasks was to feed pure hydrogen to the 'Mond battery' due to the large quantities of carbon monoxide present in Mond gas, which poisoned the Pt electrode. Therefore, Mond solved this problem by passing the Mond gas mixture and steam over finely divided nickel at 400 °C, reacting the carbon monoxide and steam to give carbon dioxide and more hydrogen. This reaction is termed as 'Water-Gas Shift Reaction'. After CO_2 removal by a simple alkaline wash, the H_2-rich stream obtained could be successfully fed to the hydrogen cell [5].

The WGSR is a reversible chemical reaction between carbon monoxide and steam to form carbon dioxide and hydrogen.

$$CO_{(g)} + H_2O \leftrightarrow CO_{2(g)} + H_{2(g)} \quad \Delta H_{298K} = -41.16 kJ/mol \qquad (1.6)$$

In the past, ammonia synthesis plants used water-gas process to produce hydrogen since it provides an economical way to produce hydrogen in the quantity required by the Haber ammonia synthesis process [6,7]. In the ammonia synthesis plant, first CO was removed from water gas by liquefication and scrubbing with hot caustic soda solution. Very soon they realized that the carbon monoxide liquefication process was unsuitable for large-scale plants. Then they used WGSR to convert CO into CO_2 by passing steam into the water-gas mixture. In this way, in 1913, the WGSR found industrial application in the production of synthesis gas as a part of the Haber-Bosch process of ammonia manufacture. Industrially, the process integration of the WGS reaction is dependent upon the origin of the synthesis gas. By the beginning of the twentieth century, and because the major source of synthesis gas was from coal and coke, the WGS reaction was used as a standalone process. By that time, the most common and economical design was to conduct the reaction in a single stage, at temperatures around 450-600 °C, and employing Fe oxide stabilized in Cr oxide as catalyst [5]. The next evolution of the process was the introduction of a second-stage converter at temperatures around 320-360 °C using the same catalyst. The two-stage converter systems reduced the CO level to 3000-4000 ppm compared to the single-stage converters that could not reduce the CO content to much less than 10,000 ppm (1%) [8]. With the discovery, in the 1960s, of Cu-based low-temperature (LT) shift catalysts and improvements in high-temperature (HT) Fe-based shift catalysts, a CO content of $<0.5\%$ in the reformate stream was achieved [9].

In 1914, BASF scientists Bosch and Wild [10] screened a series of metallic oxides for their ability to accelerate the attainment of the water-gas shift equilibrium and discovered iron oxide/chromium oxide type catalyst which can convert carbon monoxide to carbon dioxide by reacting with steam. In this way, Bosch and Wild reported the first industrial catalyst for the WGSR. The iron based discovered by BASF still forms the basis for today's HT WGS catalyst.

In terms of coal gasification process to produce hydrogen for the ammonia synthesis via Haber process, coal/coke was first blasted with air in a water-gas generator, exothermically producing carbon dioxide and carbon monoxide, thereby maintaining the HT of the coal (around 1000 °C) and raising it into an incandescent form. The water gas ($CO+H_2$) was then produced via an endothermic reaction ($C+H_2O \rightarrow CO+H_2$), by passing steam through the fuel bed. After the removal of dust particles, the mixed gas was added to steam, and CO was catalytically shifted in a converter to produce more hydrogen and carbon dioxide. Then, the gas was dried and compressed and passed through caustic scrubbers for CO_2 removal. By passing the gas mixture counter currently over an ammoniacal cuprous solution, the CO was posteriorly eliminated by absorption [5].

As a fundamental part of ammonia synthesis and coal gasification to produce hydrogen, the WGSR continues to maintain important position in chemical industry. From the beginning of the twentieth century until today, the use of the WGS reaction followed the increased industrial demand for hydrogen production. This has been accomplished using natural gas as feedstock instead of coal and employing better catalysts that improve the yields and permit the adjustment of the H_2 to CO ratio of the product, mainly for ammonia and methanol synthesis but also for the Fisher-Tropsch process and in refining operations (desulphuration, hydrogenation processes, etc.). The WGS process facilities are economically dependent on the feedstock used for the syngas generation, both in terms of equipment and catalysts. Coal, the most abundant fossil fuel on the planet, is being looked at as the possible future major source of H_2, due to the development of the integrated gasification combined cycle (IGCC) and integrated gasification fuel cell (IGFC) technologies. The gasification of coal produces syngas consisting of predominately CO and H_2 with some remaining hydrocarbons, CO_2 and water. The syngas then can be used directly to produce electricity or further processed and purified to give pure hydrogen product for such end uses, for ammonia production or hydrocracking of petroleum, or as a fuel for fuel cells to power vehicles and for stationary electricity production.

1.2 THERMODYNAMIC CONSIDERATIONS

WGSR is a reversible exothermic reaction. The equilibrium constant and equilibrium CO conversion decrease with increasing temperature. The CO conversion and H_2 production favour at lower temperatures. Since the reaction is reversible, the rate of forward reaction is strongly inhibited by the reaction products. The equilibrium constant as a function of temperature is shown in Figure 1.1.

$$K_p = e^{\left(\frac{4577.8}{T} - 4.33\right)} \quad (1.7)$$

Equation (1.7) is widely seen in the literature to describe the equilibrium constant (K_p) as a function of temperature [11]: From Figure 1.1 one can say that the CO content at equilibrium can be 20 times lesser at 200 °C than at 400 °C. In the WGS reaction, the effect of reaction pressure does not affect the equilibrium of the WGS reaction because there is no variation in the number of moles during the course of the reaction.

Nevertheless, up to that point (i.e., up to the equilibrium), total pressure positively affects the CO conversion because it increases the reaction rate. However, the addition of steam effects the rate of forward reaction. Addition of more amount of steam than the stoichiometric ratio improves the CO conversion by pushing the equilibrium towards right. Figure 1.2 shows the effect of steam to CO ratio (R) and temperature on the equilibrium CO conversion for the WGSR. In the adiabatic single bed reactor, the CO conversion is thermodynamically limited – as the reaction proceeds the heat of reaction increases the

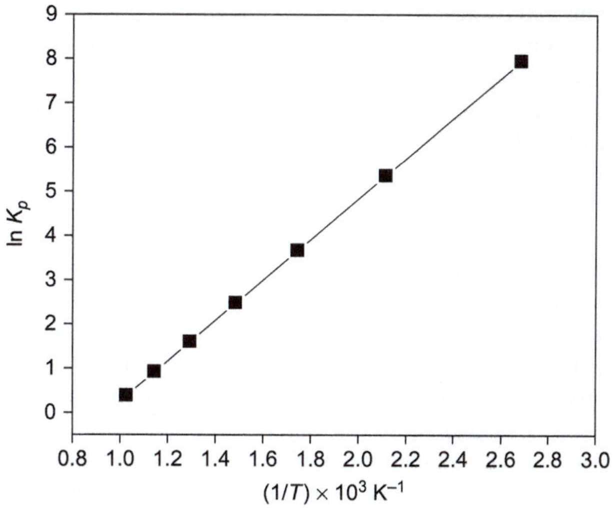

FIGURE 1.1 Effect of temperature on equilibrium constant.

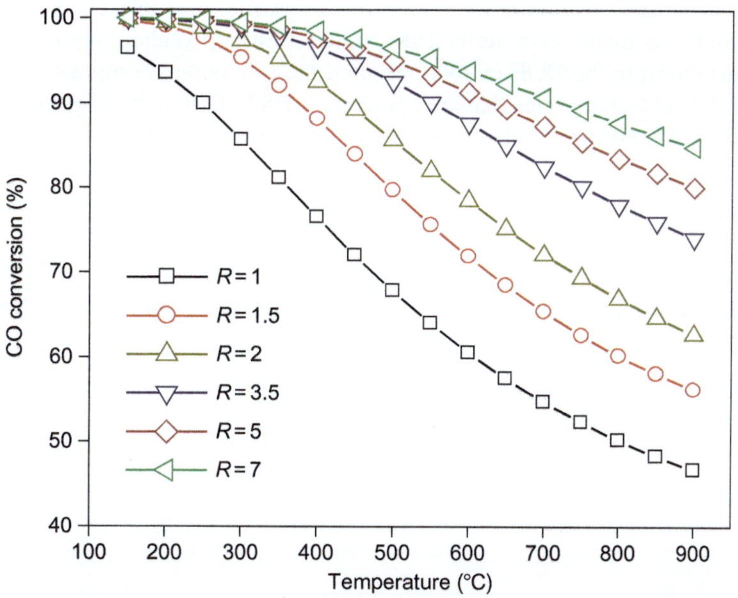

FIGURE 1.2 Effect of steam to CO ratio (R) on equilibrium CO conversion as a function of temperature.

operating temperature, and so restricts the conversion possible. The maximum CO conversion that can be achieved in the single bed adiabatic reactor over iron-chromia catalysts is 96-98% in the ammonia plant. Because it is necessary to operate these catalysts at relatively high inlet temperature (370-420 °C) they are known as HT shift catalysts.

The method of producing the syngas will also affect the WGS equilibrium compositions. For example auto-thermal reforming produces a syngas with lower H_2 concentration (due to the dilution with nitrogen) compared to steam reforming. The lower H_2 concentration increases the equilibrium CO conversion, whereas the high H_2 concentration expected with steam reforming lowers the equilibrium CO conversion. The typical outlet syngas composition of the ammonia synthesis plant and outlet syngas composition for the auto-thermal reforming process after the WGS reaction are presented in Table 1.1. The CO, H_2 and CO_2 contents in the outlet of auto-thermal reformate depend on several factors such as operation temperature, pressure, steam to CO ratio and oxygen to carbon ratio. The effect of temperature and water concentration on the equilibrium CO conversion for the WGS reaction is shown in Figure 1.3 for a typical dry reformate gas used in a large-scale hydrogen or syngas production plant, excluding any residual hydrocarbons. It is clear from Figure 1.3 that an increase in the molar steam to dry gas (*S/G*) ratio improves the equilibrium CO conversion, especially above 150 °C.

Similarly, the syngas composition from the coal gasifier depends on several factors such as type of coal used, operating temperature, steam addition, etc. The syngas composition at the outlet of coal gasifier also influences the equilibrium CO conversion in the WGSR. National Energy Technology Laboratory in their report to the DOE in 2008 reported the raw syngas composition of different IGCC technologies that are operated in USA. Different syngas compositions of different IGCC plants in USA are presented in Table 1.2.

TABLE 1.1 Typical exit syngas compositions (%) of the ammonia synthesis plant and auto-thermal reforming for H_2 production

Component	Ammonia synthesis plant	Auto-thermal reforming plant
CO	0.2	8
CO_2	18	12
H_2	61.2	73
N_2	20	3
CH_4	0	4

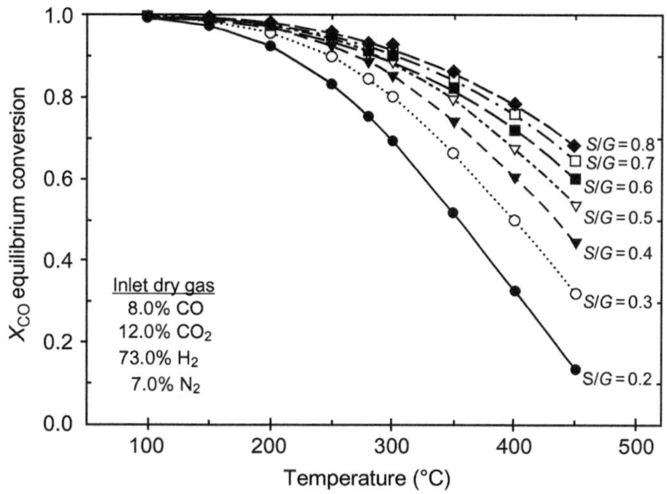

FIGURE 1.3 Equilibrium CO conversions of a typical reformate stream from a methane steam reforming process at various steam to dry gas (*S/G*) ratios. *(Taken from D. Mendes, A. Mendes, L.M. Maderia, A. Lulianelli, J.M. Sousa, A. Basile, Asia-Pac. J. Chem. Eng. 5 (2010) 111.)*

Mendes *et al.* [12] reported the equation to determine the equilibrium constant in the WGSR for syngas as a reactant based on the feed composition and assuming ideal gas behaviour

$$K_p = \frac{\left(Y_{CO_2,in} + Y_{CO,in}X_{CO,eq}\right)\left(Y_{H_2,in} + Y_{CO,in}X_{CO,eq}\right)}{\left(Y_{CO,in}\left(1 - X_{CO,eq}\right)\right)\left(Y_{H_2O,in} - Y_{CO,in}X_{CO,eq}\right)} \tag{1.8}$$

where y_i refers to the molar fraction of species i at the reactor inlet.

As stated above, complete CO conversion is possible only at lower temperatures, i.e., around 200 °C. However, the Fe-Cr catalyst is inactive at these temperatures. Hence, the thermodynamic equilibrium limitation can be overcome by using multi-stage water-gas shift reactor or by using two or more beds of HT shift catalyst with inter bed cooling. By late 1950s, researchers successfully reduced the CO concentration to 1% at the exit of the WGS reactor and decreased the thermodynamic limitation [13]. Even though the CO concentration was reduced to 1%, still the remaining CO has to be removed by absorption in cooper liquor. Some plants also used methanator because of the simplicity of process and accepted the attendant hydrogen loss. Hence, attempts are made to improve the CO conversion and that led to the development of cooper-based catalyst in early 1960s which can operate at much lower temperatures compared to the Fe-Cr catalyst.

Then, the significant improvement in the CO conversion was obtained by introducing multi-stage water-gas shift reactor in the ammonia synthesis plant. The first stage is characterized by working at higher temperatures, favouring

TABLE 1.2 Different syngas compositions in different IGCC plants

	Vendor/gasifier							
	GE energy radiant		*ConocoPhilips E-Gas*		*KBR transport gasifier (TRIG)*		*Shell*	
Syngas pressure (psia)	800		614		400+		565	
Syngas temperature (F)	410		1700		500-700		500	
Syngas composition (mol%)	H_2	25.9	H_2	26	H_2	29.2	H_2	15.6
	CO	26.7	CO	37	CO	34.3	CO	30.7
	CO_2	11.6	CO_2	14	CO_2	13.6	CO_2	1.13
	H_2O	33.6	H_2O	15	H_2O	18.9	H_2O	48.3
	CH_4	0.08	CH_4	4	CH_4	2.5	CH_4	0.02
	H_2S	0.56	H_2S	0.51	H_2S	0.056	H_2S	0.43
	CoS	0.01	CoS	0.00	HCN	0.032	CoS	0.04
	NH_3	0.13	NH_3	0.19	NH_3	0.28	NH_3	0.18

fast CO consumption and minimizing catalyst bed volume. In the following stages, the reaction takes place at progressively lower temperatures for obtaining higher conversions, which are limited by the reaction equilibrium. The lower limit of the operating temperature in the LTS reactor is the dew point of water at the operating pressure (190-200 °C at 30 bars). Condensed steam affects, adversely, the catalytic activity of the Cu-based LTS catalysts. The catalyst for the first stage reactor is Fe-Cr and for the second-stage reactor is Cu-based catalyst. These copper-based catalysts are known as LT shift catalysts. During the first stage CO concentration levels reduced to 3-5% and during the second stage the CO concentration levels further reduced to 0.1-0.3% which is acceptable for subsequent methanation in an economic way. In the multi-stage water-gas shift reactor setup, the outlet gas temperature after HT water-gas shift converter will be more than 400 °C, whereas we would conduct LT shift at 200 °C. Hence, it is important to lower the temperature of the syngas at the exit of the HT shift converter. Inter-stage cooling system is used to decrease the syngas temperature before LT shift. Inter-stage cooling is usually achieved by heat exchange. In some cases the temperature may also be decreased by injecting steam or condensate into the syngas.

1.3 WAYS TO CONDUCT WGSR

There are three different ways to conduct WGS reaction with respect to the reactors.

1.3.1. WGS reaction in traditional catalytic reactor
1.3.2. WGS reaction in membrane catalytic reactor
1.3.3. Photo-catalytic WGS reaction

1.3.1 WGS Reaction in Traditional Reactor

In traditional catalytic reactor, we can conduct WGS reaction in two different ways: homogeneous catalytic WGS reaction and heterogeneous catalytic WGS reaction.

1.3.1.1 Homogeneous Catalytic WGS Reaction

If the catalyst and reactant are in the same state (in this case both are in liquid state) then the reaction is called homogeneous WGS reaction. In 1953, Reppe was the first researcher who discussed the concept of homogeneous WGS reaction [14]. Although he published a number of catalytic reactions of CO and H_2O with organic substrates, Fenton was the first person who patented the homogeneous catalysis of the WGSR. He filed three patents on homogeneous WGS reaction by using group VIII metals in conjunction with phosphine, arsine or stibine ligands and amine or inorganic bases [15–17]. In 1977, Cheng *et al.* first ever reported homogeneous WGS reaction using Rh carbonyl iodide in open literature in aqueous solution [18]. In the same year, Laine *et al.* reported Rh carbonyl for the WGS reaction in the alkaline solution [19]. Kang *et al.* also reported the variety of group VIII metals for the homogeneous WGS catalysts as active catalysts in base medium [20]. On the other hand, Baker *et al.* reported the active WGS catalysts in acidic solutions over Rh carbonyl complexes in the same year [21].

1.3.1.1.1 Heterogeneous Catalytic WGS Reaction

If the catalyst and reactant are in different states (in this case catalyst was in solid state and reactant was in gaseous state) then the reaction is called heterogeneous WGS reaction. All the industries use heterogeneous WGS reaction to conduct WGS reaction. Two different catalyst beds are utilized, one at higher temperature and the other at lower temperature to achieve 100% CO conversion. The different catalysts used for heterogeneous WGS reaction are presented in the following section.

1.3.2 WGSR in Membrane Reactors

1.3.2.1 Membrane Reactor

Membrane reactor is a catalytic reactor that additionally contains cylinder of some porous material within it, the tube within the shell of a shell-and-tube heat exchanger. This porous inner cylinder is the membrane that gives the membrane reactor its name.

A simple example of catalytic ceramic membrane reactor is shown in Figure 1.4. The membrane is a barrier that allows only certain components

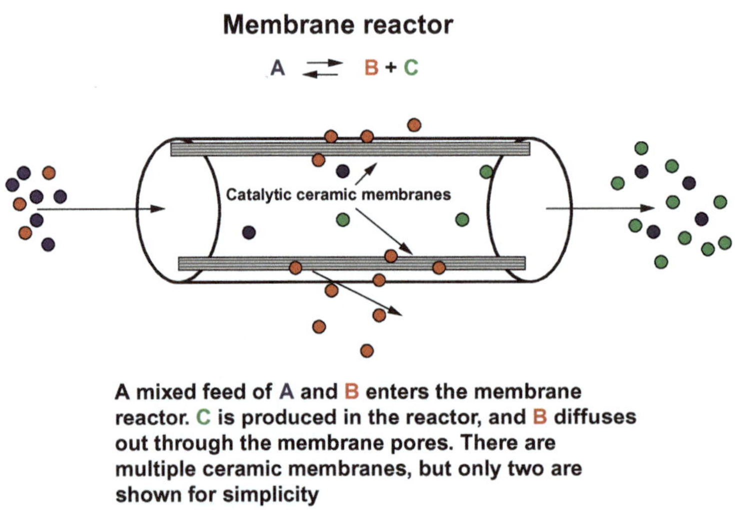

FIGURE 1.4 Schematic diagram of the membrane reactor.

to pass through it. The selectivity of the membrane is controlled by its pore diameter, which can be on the order of Angstroms, for micro-porous layers, or on the order of microns for macro-porous layers. Membrane reactors combine reaction with separation to increase conversion. One of the products of a given reaction is removed from the reactor through the membrane, thus forcing the equilibrium of the reaction 'to the right' (according to Le Chatelier's principle), so that more of that product is generated. A catalytic membrane reactor has a membrane that has either been coated with or is made of a material that contains catalyst, which means that the membrane itself participates in the reaction. Some of the reaction products (those that are small enough) pass through the membrane and exit the reactor on the permeate side.

We know that due to thermodynamic limitations it is difficult to achieve 100% CO conversion. The CO conversion depends on the nature of the catalyst and temperature range, and therefore the reaction is clearly more effective in membrane reactors. In the WGS membrane reactor it is possible to achieve 100% CO conversion in single stage and H_2/CO_2 separation. Based on the membrane properties WGS membrane reactors are classified into two categories, namely, CO_2 selective membrane reactors and H_2 selective membrane reactors. In the CO_2 selective membrane reactors, CO_2 was removed from the catalytic membrane reactor and the reaction mixture becomes H_2 rich steam. This may cause over reduction of Fe- or Cu-based catalysts. However, in the H_2 selective membrane, CO_2 will be present at a higher concentration in the reaction medium, affecting the reaction rate. The membrane reactor offers many potential advantages: reduced capital and downstream separation costs, as well as enhanced yields and selectivity. From the viewpoint of the WGS process in an membrane reactor, a reaction product moves to the permeate side, enabling the WGS reaction to proceed towards completion and so making it possible to achieve the following: (1) higher conversion than a TR working under the same operating conditions or (2) the same conversion as a TR, but working under milder operative conditions. Among the various membrane reactors, both Pd and zeolites membrane reactors received much attention because of their thermal stability, sulphur tolerance and better permeation properties. Among the various catalysts, Cu-, Pt-, Au-based catalysts are suitable for membrane reactor applications. More description about the type of the membranes, catalysts used, mechanism and kinetics of WGS membrane reactors will be presented in future chapters.

1.3.3 Photo-Catalytic WGS Reaction

In the photo-catalytic reaction we use visible or UV-light to irradiate the catalyst instead of light. The reaction is conducted at room temperature. Sato and White first reported the first photo-catalytic WGS reaction on Pt/TiO_2 catalyst [22]. Then, they prepared NaOH-coated Pt/TiO_2 catalysts for the photo-WGS reaction. When Pt/TiO_2 was coated with NaOH, the photo-catalytic activity for WGS reaction increased significantly [23]. Yixuan *et al.* [24] proved that HT hydrogen treatment increases the photo-catalytic activity of Pt-TiO_2 for the WGS reaction. They explained the new concept of strong metal-semiconductor interaction (SMScI) in connection with a proposed novel charge transfer mechanism for the WGS reaction. Tsai *et al.* [25] prepared Pt/TiO_2 (100) catalyst and investigated for the photo-catalytic WGS reaction. Cole-Hamilton *et al.* has evaluated the [RuCl(CO)(bipy)$_2$]C complex for the photochemical homogeneous WGS reaction [26]. However, not much research effort has been focused on photo-assisted WGS reaction because of its disadvantages for the commercialization.

1.4 TYPES OF HETEROGENEOUS WATER-GAS SHIFT CATALYSTS

Since pioneering work from Bosch and Wild [10], many researchers focused on developing different catalysts for HT WGSR. Based on the temperature of the reaction and reaction environment we can classify the water-gas shift catalysts into three categories.

1.4.1. HT WGS catalysts
1.4.2. LT-WGS catalysts
1.4.3. Co-Mo sulphided catalysts

1.4.1 HT WGS Catalysts

The catalysts that operate between 310 and 450 °C temperatures are called HT shift catalysts. Industries are using Fe-Cr for more than 70 years for commercial use. Until the invention of Cu-based catalyst which operates at lower temperature ammonia synthesis plants use two-stage high water-gas shift reactors where the first stage reactor operates at 420 °C and the second stage operates at 320 °C. Both stages used Fe-Cr as catalyst. Conventional Fe_2O_3-Cr_2O_3 catalyst contains 80-90% of Fe and 8-10% of Cr_2O_3. In the Fe-Cr catalysts iron oxide is the active phase for the HT WGS reaction and Cr_2O_3 acts as a stabilizer. In the absence of chromium oxide the effective lifetime of the catalyst is severely restricted because of rapid thermal sintering. The extended activity of the iron oxide/chromium oxide catalyst results from the presence of Cr_2O_3 which prevents the sintering of neighbouring iron oxide crystallites. During the initial stages of the catalyst use, rapid catalyst deactivation occurs, but after 1500 h catalyst activity stabilizes. The catalysts typically operate in plants for 2-5 years before relatively slow thermal sintering leads to a sufficiently large decrease in activity to warrant catalyst replacement. In addition to being a textural promoter preventing the sintering of iron oxide crystallites, Cr_2O_3 also functions as a structural promoter to enhance the intrinsic catalytic activity of Fe_2O_3.

Preparation method for Fe-Cr catalyst has a strong influence on their structural and catalytic properties. Fe-Cr catalysts are usually prepared by co-precipitation method by using nitrates as precursors and ammonia as a precipitating agent followed by drying and calcination. Wet-impregnation technique is an alternate method to synthesize Fe-Cr catalyst. In the impregnation method, first iron hydroxide is prepared by precipitation method and chromium is deposited over iron oxide by impregnation method using chromium nitrate solution [27,28]. X-ray photoelectron spectroscopy revealed that there was surface enrichment of Cr ions in the fresh samples prepared by both the co-precipitation and impregnation routes. The surface chromium concentration was higher in the sample prepared by the impregnation method. After synthesizing Fe-Cr the iron oxide was in α-Fe_2O_3 (hematite) form. Before the WGSR the hematite has to be transformed into magnetite (Fe_3O_4) because Fe_3O_4 is the active phase for the WGSR. Process was used to reduce hematite to magnetite in the temperature range 350-450 °C. Process gas is a mixture of CO, CO_2, H_2 and steam. The ratio between oxidant (H_2O, CO_2) and reductant (CO, H_2) is very important to achieve higher CO conversion.

During the last three decades, due to the rising cost of hydrocarbon feedstocks, plants have been forced to keep operating costs low by being as energy efficient as possible. One way to reduce the cost and improve efficiency is to decrease the steam to gas ratio in the water-gas shift plant. However, using of lower steam to CO ratios can cause over reduction of iron oxide and form iron carbide during the HT WGSR. Iron carbides are very active catalysts for Fisher-Tropsch reactions and can form hydrocarbons like methane during WGSR. Attempts are made in the literature to improve the activity of Fe-Cr at low steam to gas ratios by adding promoters. Attempts are also made in the literature to develop Cr-free iron based HT WGS catalyst. The details about Cr-free iron oxide catalysts and other non-iron oxide catalysts will be discussed in Chapter 5.

1.4.2 LT-WGS Catalysts

1.4.2.1 Cu-Based Catalysts

The catalysts which operate between 150 and 250 °C are known as LT-WGS catalysts. The development of highly efficient desulphurization technologies using Co (Ni)-MoO_3-Al_2O_3 catalysts for sulphur removal provided ammonia manufacturers with syngas mixture containing less than 1.0 ppm sulphur. This, in turn, enabled the use of the otherwise sulphur-sensitive Cu-ZnO catalysts at sufficiently LTs (190-200 °C) when the equilibrium CO concentrations, at the exit of the LTS converters, can be below 0.3%. The invention of Cu catalysts for WGSR at lower temperatures made revolution in the WGS process industry. The operators began to appreciate the economics associated with ammonia synthesis with large production units, i.e., up to 1000 ton/day or more. Since, copper catalysts deliver the CO concentration less than 0.5% after the second-stage WGS reactor it was easy to incorporate a methanation stage in place of the very much complicated copper liquor scrubbing system that was used before. LT-WGS catalysts were used virtually in all ammonia and refinery hydrogen plants. Ray *et al.* [29] reported the advantages of using LT catalyst in hydrogen production to decrease the steam and equipment requirements. The first Cu-based catalysts are used in USA in 1963 [30]. This was a mixture of Cu-Zn with 1:2 ratios. Since their first usage several Cu formulations have been employed in the LT-WGS stage. Improvements in the catalyst activity, stability as well as resistance to poisoning and sintering were based on such developments. The copper metal crystallites disperse over zinc oxide when reduced in hydrogen. The method of preparation is extremely important concerning the activity of LT-WGS catalyst. Uchida *et al.* [31] reported several preparative methods for Cu-Zn catalysts and compared the catalyst activity and stability. In their test the addition of Cu to the Zn oxide rapidly increases the activity up to 0.4. They also did X-ray analysis and found after the WGSR the main components in Cu-Zn catalysts is Cu metal particles and ZnO. However, the main problem associated with this catalyst is that the operating time is as low as 6 months. As like HT WGS shift catalyst, chromia was doped into copper to stabilize the catalyst against sintering. The oxides of Zr, Cr and Cu by themselves were not good catalysts for WGSR. The mixtures of various compositions of Cu-Cr-Zn have been found to have excellent activity. Yureva *et al.* [32] investigated the effect of additives Zn, Co, Mg on WGS activity of $CuO-Cr_2O_3$. They found that the addition of additives did not improve the activity; however, they improved the temperature stability of the Cu/Cr catalysts. They also found that WGS activity is not a function of degree of reduction of the catalyst and concluded that metallic Cu was not the active site of the catalyst. They believe that active sites were Cu ions in the lattice of copper chromite. When Zn was present Cu-Zn-Cr solid was formed. Unfortunately, the Cu-Cr catalysts are very susceptible to thermal sintering via surface migration. The formulations containing Cr did not have significantly improved operating lives. Efforts have been made to enhance the activity and stability of Cu-Zn catalysts by addition of various promoters and supports. One of the major advances in achieving enhanced stability in the Cu catalysts was the introduction of promoters acting as structural spacers that decreases the aggregation of Cu crystallites during WGS reaction. One such advancement is the invention of alumina together with ZnO support. The Al_2O_3-ZnO support not only stabilizes the Cu crystallites against thermal sintering but also enhanced the strength of the catalyst and minimized shrinkage during reduction. The composition of Cu/ZnO-Al_2O_3 for better activity, stability and poison resistance depends on the crystallite size of the components, Cu dispersion and method of synthesis. For example, catalyst containing high copper levels may display very high initial activity that quite rapidly decays.

Gines *et al.* [33] reported the detailed study of the effect of synthesis method on structural properties and WGS activity of Cu-Zn-Al_2O_3 catalysts. The catalysts were prepared by co-precipitation method by using nitrates as precursors and sodium carbonate as a precipitating agent at 60 °C and constant pH around 7. The precipitate was filtered and dried at 100 °C and finally calcined between 400 and 700 °C. They found that the turnover frequency remained constant not only when the copper metallic surface area was varied between 3 and 35 m^2/g Cu, but also for the Cu loading of 30-50 wt%, Al/Zn atomic ratios of 0 and 2.5, Cu dispersion between 0.5% and 5%, calcination temperatures between 400 and 700 °C. These results clearly suggest that WGS reaction is structure insensitive over Cu/ZnO-Al_2O_3 catalysts. On the contrary, Chinchen and Spencer [34] reported that turnover frequency vary by an order of magnitude when Cu metallic surface area was changed from 10 to 40 m^2/g for Cu/ZnO-Al_2O_3 system. Even though, in general, the WGS activity of Cu/ZnO-Al_2O_3 increases with increasing Cu metallic surface area, there is no literature report available on the rate of reaction that correlates linearly with the metallic Cu area over the entire Cu-Zn composition range. Along with high Cu surface area, the bulk structural changes in the ZnO and Cu crystallites during the synthesis and pre-reduction like micro-strain, oxygen vacancies in the ZnO cannot be ignored.

Attempts have been made during the past decade to develop superior catalyst than the conventional Cu-ZnO-Al_2O_3. The details about recent developments in the LT-WGSR will be discussed in the future chapters.

1.4.2.2 Nobel Metal-Based Catalysts

As described above thermodynamics of the WGS reaction are well known that at HTs the CO conversion is equilibrium limited and at LTs the reaction is kinetically limited. The Cu-based catalysts are not suitable active catalysts for single-stage WGS applications. For this reason industries first operate at higher temperatures to have 95-97% CO conversion, and then the remaining CO was converted by Cu-based catalysts at LT. Hence, much research effort has been focused on developing noble metal-based catalysts at lower temperatures. Noble metal catalysts are very active catalysts for several applications especially automotive exhaust applications. Nobel metals supported on partially reduced metal oxide supports are the active catalysts for the WGSR. Of the many catalysts that have been studied, precious catalysts (Pt, Rh, Ru, Au and Pd) deposited on partially reducible oxides (CeO_2, ZrO_2, TiO_2, Fe_2O_3 and mixed oxides) have been the most investigated. These catalysts are quite active in 250-400 °C. In 1925 itself Prichard and Hinshelwood studied the adsorption behaviour of CO, H_2O, CO_2 and H_2 on Pt surface [35]. In 1974, Taylor *et al.* reported that Ru, Pt and Pd catalysts are very active for the WGS reaction. In 1974, Japanese researcher Masuda investigated the kinetics of the WGSR over Pt surface [36]. In 1980 Grenoble and Estadt did a broad investigation over several metal supported Al_2O_3 catalysts for WGS reaction [37]. They investigated Ru, Rh, Pd, Os, Ir, Pt, Fe, Co, Ni, Re, Cu and Au catalysts for WGS reaction. In 1985 Mendelovici [38] reported revolutionary Pt/CeO_2 catalysts for methanation and WGS reaction. Even though the high WGS activity of Pt group metals was known for several years, much research has been done in early 1990s. In 1993 Shido and Iwasawa [39] explained the mechanism of the WGS reaction over $Rh\text{-}CeO_2$ catalysts. They proposed that WGS reaction proceeds on Ce-O pair and not on the Rh metallic particles. Rh/SiO_2 catalyst is inactive for WGS reaction under similar conditions. The OH groups of Ce react with the CO and forms bi-dentate formate species. The bi-dentate formate decomposed into hydrogen and uni-dentate carbonate species. Rate constants suggest that the decomposition of bi-dentate formate species is the rate determining step. Water molecule promotes the desorption of the carbonate species as CO_2. Electron donor-acceptor interaction between Ce-O and H_2O is the important factor for reactant-promoted catalysis. In 1996 Xue *et al.* [40] investigated the effect of H_2S and low steam to CO ratios on Cu-Zn, Pt/ZrO_2 and Fe-Cr commercial catalysts. The Cu-Zn catalyst retained its activity when operated at low steam to CO ratios. However, it lost its activity completely when sulphur was present in the feed. When operated at a low steam to CO ratios the commercial Fe-Cr catalyst becomes active only at relatively higher temperatures. The presence of sulphur deactivated the catalyst. The Pt/ZrO_2 catalyst showed higher activity than that of Fe-Cr catalyst. In the presence of sulphur, the Pt/ZrO_2 catalyst deactivated; however, the activity remained higher than any of the other catalysts tested under the same conditions.

From this point onwards a large number of different formulations combining precious metals with partially reducible oxides have been proposed as promising catalysts for the WGS reaction. Some of the examples are $Au\text{-}Fe_2O_3$ [41,42], $Au\text{-}CeO_2$ [42,43], $Au\text{-}TiO_2$ [44], $Ru\text{-}ZrO_2$ [45], $Rh\text{-}CeO_2$ [46], $Pt\text{-}CeO_2$ [41,47–50], $Pt\text{-}ZrO_2$ [40], $Pt\text{-}TiO_2$ [51], $Pt\text{-}Fe_2O_3$ [52] and $Pd\text{-}CeO_2$ [53,54]. Among the various noble metals investigated Pt shows better activity at medium temperature 250-400 °C and Au showed better activity at lower temperatures >250 °C.

1.4.3 Sulphur Tolerant Co-Mo Catalysts

The utilization of hydrogen for various purposes like liquid synthesis, fuel cell applications increased the demand of using the sulphur containing coal for gasification to produce syngas. Although Fe-Cr catalysts are stable to sulphur poisoning, large amount of sulphur deactivates the HT WGS catalysts. The LT-WGS Cu catalysts cannot tolerate the low levels of sulphur. Hence, completely sulphur tolerant catalysts with high activity and selectivity for the WGS reaction would be desirable. Moreover, if such a catalyst was not only sulphur tolerant but if its activity was even enhanced by sulphur, then single-stage WGS reactors with only one subsequent step to remove CO_2 would be sufficient for pure hydrogen production. Then a family of group VI and VIII metals other than Fe and Cr are tested for WGSR. In 1954 German scientists Wustrow, Maedrich and Macura [55] pointed out that not only can the Mo-based catalysts be used as methanation, but also Fisher-Tropsch and hydro-desulphurization (HDS) catalysts have very good WGS activity. Then further research has been done and the following catalysts were studied for the WGS reaction in the presence of sulphur: Ni or Co with Mo [56,57], Ni or Co or Mo sulphides [58,55], alkali metal added Co-Mo-Ni [59,60]. Among the various catalysts investigated K or Cs added Co-Mo catalysts showed higher WGS activity. Interestingly, these catalysts are completely sulphided in its most active form. Later, it was found that potassium added Co-Mo/Al_2O_3 or Co-Ni/Al_2O_3 catalysts in its sulphided form was the most active catalyst for WGS reaction in the presence of sulphur, and K_2CO_3 was the most recommendable precursor for the catalyst preparation.

The main advantage with the sulphided Co-Mo catalysts is that they operate at much lower temperature compared to (250-350 °C) the HT iron-chromium oxide catalysts (350-450 °C). Hence, it favours higher CO conversion and lower outlet CO concentrations. These catalysts also can operate at much lower steam to CO ratios, resulting in same or even higher CO conversions compared to the HT catalysts. However, these catalysts operate at much lower space velocities and we need 20% extra catalyst than the corresponding HT shift catalysts. Moreover, these catalysts need sulphur in syngas since they are only active in sulphided form. The minimum inlet sulphur concentration in the feed is

TABLE 1.3 Process conditions for Co-Mo catalysts based water-gas shift reactor for Texaco partial oxidation process

	Bed 1	Bed 2	Bed 3
Inlet feed composition			
CO	46	16	3.1
CO_2	6.9	26	34.2
H_2	47	57.9	62.6
CH_4	0.1	0.1	0.1
Sulphur	0.25		
Inlet steam/gas ratio	0.96	0.7	0.61
Pressure (bar)	35	34	33
Inlet temperature (°C)	266	288	278
Outlet temperature (°C)	411	367	292
Outlet CO (mol%)	16	3.1	1

300 ppm for an acceptable performance. If the catalysts are pre-sulphided enough before use then the catalysts can show good activity in the syngas steam that contain H_2S concentration as low as 35 ppm. The typical process conditions for a Texaco partial oxidation process that generates syngas which uses sour Co-Mo WGS catalyst are shown in Table 1.3.

REFERENCES

[1] D. T. Burns, G. Piccardi, L. Sabbatini, Some people and places important in the history of analytical chemistry in Italy, Mikrochim. Acta 160 (2008) 57–87.

[2] H. W. Haring, Industrial Gases Processing, Wiley-VCH, Verlag, Germany, 2007.

[3] L. Mond, C. Langer, Improvements in obtaining hydrogen, British Patent 12608, 1888.

[4] G. Hoogers, Fuel Cell Technology Handbook, CRC Press, Boca Raton London New York Washington, D.C., 2002.

[5] R. P. Quirk, Kirk-Othmer Encyclopedia of Chemical Engineering Technology, vol. 2, Wiley-VCH, 1963.

[6] S. A. Topham, The history of catalytic synthesis of ammonia, in: J. R. Anderson, M. Boudart (Eds.), Catalysis Science and Technology, vol. 7, Springer Verlag, Berlin, 1985, pp. 1–50.

[7] K. Holderman, Im Bannne der Chemie Carl Bosch Lebeb und Werk, Econ-Verlag, Dusseldorf, 1953.

[8] J. R. Ladebeck, J. P. Wang, Handbook of Fuel Cells, Fuel Cell Technology and Applications, vol. 3, John Wiley & Sons, England, 2003 190 pp..

[9] W. F. Ruettinger, O. Ilinich, in: S. Lee (Ed.), Encyclopedia of Chemical Processing, Taylor & Francis, Columbia, Missouri, U.S.A., 2006, 3205 pp.

[10] C. Bosch, W. Wild, Canadian Patent 153,608, 1888.

[11] C. Ratnaswamy, J. P. Wagner, Water gas shift catalysis, Catal. Rev. Sci. Eng. 51 (3) (2009) 325–440.

[12] D. Mendes, A. Mendes, L. M. Maderia, A. Lulianelli, J. M. Sousa, A. Basile, The water-gas shift reaction: from conventional catalytic systems to Pd-based membrane reactors – a review, Asia-Pac. J. Chem. Eng. 5 (2010) 111–137.

[13] M. Appl, Nitrogen 100 (1976) 47.

[14] J. W. Reppe, L. Liebigs, Carbonylierung I. Über die Umsetzung von Acetylen mit Kohlenoxyd und Verbindungen mit reaktionsfähigen Wasserstoffatomen Synthesen α,β-ungesättigter Carbonsäuren und ihrer Derivate, Ann. Chem. 582 (1953) 1–37.

[15] D.M. Fenton, U.S. Patent 3,490,872, 1970.

[16] D.M. Fenton, U.S. Patent 3,539,298, 1970.

[17] D.M. Fenton, U.S. Patent 3,781,364, 1973.

[18] C. H. Cheng, D. E. Hendriksen, R. Eisenberg, Homogeneous catalysis of the water gas shift reaction using rhodium carbonyl iodide, J. Am. Chem. Soc. 99 (1977) 2791–2792.

[19] R. M. Laine, R. G. Rinker, P. C. Ford, Homogeneous catalysis by ruthenium carbonyl in alkaline solution: the water gas shift reaction, J. Am. Chem. Soc. 99 (1977) 252–253.

[20] H. C. Kang, C. H. Mauldon, T. Cole, W. Slegeir, K. Cann, R. Pettit, Reductions with carbon monoxide and water in place of hydrogen. 1. Hydroformylation reaction and water gas shift reaction, J. Am. Chem. Soc. 99 (1977) 8323–8324.

[21] E. C. Baker, D. E. Hnedriksen, R. Einsenberg, Mechanistic studies of the homogeneous catalysis of the water gas shift reaction by rhodium carbonyl iodide, J. Am. Chem. Soc. 102 (1980) 1020–1021.

[22] S. Sato, J. M. White, Photoassisted water-gas shift reaction over platinized titanium dioxide catalysts, J. Am. Chem. Soc. 102 (1980) 7206–7210.

[23] S. Sato, J. M. White, Photocatalytic water decomposition and water-gas shift reactions over NaOH-coated, platinized TiO_2, J. Catal. 69 (1981) 128–139.

[24] C. Yixuan, W. Zhaobin, C. Y. Anxin, L. Huaxin, H. Zupei, L. Huiqing, D. Yonglei, Y. Chunying, L. Wenzhao, Metal-semiconductor catalyst: photocatalytic and electrochemical behavior of Pt-TiO_2 for the water-gas shift reaction, J. Mol. Catal. 21 (1983) 275–289.

[25] S. C. Tsai, C. C. Kao, Y. W. Chung, Metal-semiconductor catalyst: photocatalytic and electrochemical behavior of Pt-TiO_2 for the water-gas shift reaction, J. Catal. 79 (1983) 451–461.

[26] D. J. Cole-Hamilton, Photocatalysis of the homogeneous water-gas shift reaction, J. Chem. Soc. Chem. Commun. (24) (1980) 1213–1215.

[27] E. Matijevic, P. Scheider, Ferric hydrous oxide sols: III. Preparation of uniform particles by hydrolysis of Fe(III)-chloride, -nitrate, and -perchlorate solutions, J. Colloid Interface Sci. 63 (1978) 509–524.

[28] M. C. Rangel, M. S. Santos, A. Albornoz, The influence of the preparation method on the catalytic properties of lanthanum-doped hematite in the ethylbenzene dehydrogenation, Stud. Surf. Sci. Catal. 162 (2006) 753.

[29] N. Ray, S. K. Roy, N. C. Ganguli, S. P. Sen, Deactivation of low temperature shift catalysts, J. Res. Inst. Catal. 21 (1973) 187–199.

[30] J. M. Moe, Low temperature CO conversion, Am. Chem. Soc. Div. Fuel Chem. 8 (1963) B29–B34.

[31] H. Uchida, N. Isogai, M. Oba, T. Hasegawa, The zinc oxide-copper catalyst for carbon monoxide-shift conversion. I. The dependency of the catalytic activity on the chemical composition of the catalyst, Bull. Chem. Soc. Jpn. 40 (1967) 1981–1986.

[32] T. M. Yureva, G. K. Boreskov, V. Sh. Gruver, Kinet. Katal. 10 (1969) 294.

[33] M. J. L. Gines, N. Amadeo, M. Laborde, C. R. Apesteguia, Activity and structure-sensitivity of the water-gas shift reaction over CuZnAl mixed oxide catalysts, Appl. Catal. A Gen. 131 (1995) 283–296.

[34] G. C. Chinchen, M. S. Spencer, Sensitive and insensitive reactions on copper catalysts: the water-gas shift reaction and methanol synthesis from carbon dioxide, Catal. Today 10 (1991) 293–301.

[35] C. R. Prichard, N. Hinshelwood, The interaction of hydrogen and carbon dioxide on the surface of platinum, J. Chem. Soc. 127 (1925) 806–811.

[36] M. Masuda, K. Migahrua, Isotopic Study of Water Gas Reaction Catalyzed by Platinum, Bull. Chem. Soc. Jpn. 47 (1974) 1058–1063.

[37] D. C. Grenoble, M. M. Estadt, The chemistry and catalysis of the water gas shift reaction: 1. The kinetics over supported metal catalysts, J. Catal. 67 (1981) 90–102.

[38] L. Mendelovici, M. Steinberg, Methanation and water-gas shift reactions over $PtCeO_2$, J. Catal. 96 (1985) 285–287.

[39] T. Shido, Y. Iwasawa, Reactant-promoted reaction mechanism for water-gas shift reaction on Rh-doped CeO_2, J. Catal. 141 (1993) 71–81.

[40] E. Xue, M. O. Keeffe, J. R. H. Ross, Water-gas shift conversion using a feed with a low steam to carbon monoxide ratio and containing sulphur, Catal. Today 30 (1996) 107–118.

[41] D. Andreeva, V. Idakiev, T. Tabakova, A. Andreeva, Low-temperature water-gas shift reaction over Au/α-Fe_2O_3, J. Catal. 158 (1998) 354–362.

[42] A. Luengnaruemitchai, S. Osuwan, E. Gulati, Comparative studies of low-temperature water-gas shift reaction over Pt/CeO_2, Au/CeO_2, and Au/Fe_2O_3 catalysts, Catal. Commun. 4 (2003) 215–221.

[43] D. Andreeva, V. Idakiev, T. Tabakova, L. Ilieva, P. Falaras, A. Bourlinosand, A. Travlos, Low-temperature water-gas shift reaction over Au/CeO_2 catalysts, Catal. Today 72 (2002) 51–57.

[44] D. F. Boccuzzi, A. Chiorino, M. Manzoli, D. Andreeva, T. Tabakova, L. Lievab, L. Ladakiev, Gold, silver and copper catalysts supported on TiO_2 for pure hydrogen production, Catal. Today 75 (2002) 169–175.

[45] O. Goerke, P. Pfeifer, K. Schubert, Water gas shift reaction and selective oxidation of CO in microreactors, Appl. Catal. A Gen. 263 (2004) 11–18.

[46] C.L. Aldridge, T. Kalina, U.K. Patent 1,325,172, 1969.

[47] W.J. Lywood, M.V. Twigg, U.S. Patent 5,030,440, 1991.

[48] P. Panagiotopoulou, J. Papavasiliou, G. Avgouropoulos, T. Ionnides, D. I. Kondarides, Water-gas shift activity of doped Pt/CeO_2 catalysts, Chem. Eng. J. 134 (2007) 16–22.

[49] A. Goguet, F. C. Meunier, J. P. Breen, R. Burch, M. I. Petch, A. F. Ghenciu, Effect of morphological characteristics of TiO_2-supported noble metal catalysts on their activity for the water-gas shift reaction, J. Catal. 226 (2004) 382–392.

[50] G. Germani, Y. Schuurman, WGS reaction kinetics over μ-structured Pt/CeO_2/Al_2O_3 catalysts, AIChE J. 52 (2006) 1806–1813.

[51] P. Panagiotopoulou, D. I. Kondarides, Effect of morphological characteristics of TiO_2-supported noble metal catalysts on their activity for the water-gas shift reaction, J. Catal. 225 (2004) 327.

[52] A. Basinskaa, T. P. Manieckib, W. K. Jozwiakb, Catalytic activity in water-gas shift reaction of platinum group metals supported on iron oxides, React. Kinet. Catal. Lett. 89 (2006) 319–324.

[53] K. Wang, R. J. Gorte, J. P. Wagner, Deactivation Mechanisms for Pd/Ceria during the Water—Gas-Shift Reaction, J. Catal. 212 (2002) 225.

[54] W. Ruettinger, X. Liu, R. J. Farrauto, Mechanism of aging for a Pt/CeO_2-ZrO_2 water gas shift catalyst, Appl. Catal. B Environ. 65 (2006) 135–141.

[55] W. Wustrow, O. Maedrich, H. Macura, German Patent 1,085,287, 1954 (Ruhrgas A.-G).

[56] G.S. Levinson, J. Ramsbotham, S.T. Sie, German Patent 2,416,351, 1974.

[57] L. Harris, German Patent 1,162,505, 1974.

[58] O. Reita, E. Lorenz, British Patent 940,960, 1974.

[59] K.L. Riley, C.L. Aldridge, German Patent 2,453,784, Exxon Research & Engineering Co., 1975.

[60] M.A. Segura, C.L. Aldridge, U.S. Patent 4,054,644, Exxon Research & Engineering Co., 1977.

Chapter 2

High-Temperature WGS Reaction

2.1 Fe-Cr CATALYSTS

The catalysts operate between 310 and 450 °C temperatures are called high-temperature shift catalysts. Mond [1], one of the greatest chemist-industrialists, first time discovered a high-temperature WGS catalyst, i.e., finely divided nickel for converting CO into CO_2 in the coal gasification process. In 1914, BASF scientists Bosch and Wild tested various catalysts for high-temperature WGS reaction and found out that iron oxide stabilized with Cr is the best catalyst [2]. Until the invention of Cu-based catalyst which operates at lower temperatures, industries were using Fe-Cr for more than 70 years for commercial use. Conventional Fe_2O_3-Cr_2O_3 catalyst contains 80-90% of Fe and 8-10% of Cr_2O_3. In Fe-Cr catalysts iron oxide is the active phase for the high-temperature WGS reaction and Cr_2O_3 acts as a stabilizer. In the absence of chromium oxide the effective lifetime of the catalyst is severely restricted because of rapid thermal sintering. The extended activity of the iron oxide/chromium oxide catalyst results from the presence of Cr_2O_3 which prevents the sintering of neighbouring iron oxide crystallites. During the initial stages of the catalyst use, rapid catalyst deactivation occurs, but after 1500 h on steam the catalyst activity stabilizes. The catalysts typically operate in plants for 2-5 years before relatively slow thermal sintering leads to a sufficiently large decrease in activity to warrant catalyst replacement. In addition to being a textural promoter preventing the sintering of iron oxide crystallites, Cr_2O_3 also functions as a structural promoter to enhance the intrinsic catalytic activity of Fe_2O_3.

Chinchen *et al.* [3] suggest that as the reaction progresses, discrete Cr_2O_3 grains grow and become dispersed over Fe_3O_4 domains, thereby blocking the thermal agglomeration of Fe_3O_4 particles. A different opinion is that Cr^{3+} ions enter into the inverse-spinel lattice of Fe_3O_4 and form a solid solution. Robbins *et al.* [4] found that Cr^{3+} ions dissolve into the Fe_3O_4 lattice and occupy the octahedral site and the displaced Fe^{2+} and Fe^{3+} ions (from the octahedral sites) are transferred to the tetrahedral sites. Edwards *et al.* [5] claimed that the dissolved Cr^{3+} is enriched at the surface region of the catalyst and that the Cr-enriched surface shell, being more thermodynamically stable than the Fe-rich core, reduces ion diffusion and sintering effects. Natesakhawat *et al.* [6]

Water Gas Shift Reaction. http://dx.doi.org/10.1016/B978-0-12-420154-5.00002-4

reported that the Cr^{3+} in Fe/Cr was oxidized to Cr^{6+} during WGS catalysis. The $Cr^{3+} \leftrightarrow Cr^{6+}$ oxidation-reduction cycle was expected to enhance the redox rate of magnetite and promote the WGS activity of the catalyst. Today, people believe that Cr acts as both a stabilizer (resist the sintering of magnetite particles) and a promoter (improve the redox efficiency of $Fe^{2+} \leftrightarrow Fe^{3+}$ redox couple).

2.1.1 Fe-Cr Activation

The synthesis of iron oxide catalysts leads to the hematite (Fe_2O_3) phase after calcination. Before the WGS reaction the hematite phase is reduced to magnetite phase (Fe_3O_4), since magnetite is the active phase for the WGS reaction. Usually, the reduction is taking place in the presence of process gas between 350 and 450 °C. Process gas is a mixture of CO, CO_2, H_2 and water vapour. The representative reactions are shown as follows:

$$3Fe_2O_3 + H_2 \rightarrow 2Fe_3O_4 + H_2O \quad \Delta H = -16.3\text{kJ/mol} \tag{2.1}$$

$$3Fe_2O_3 + CO \rightarrow 2Fe_3O_4 + CO_2 \quad \Delta H = +24.8\text{kJ/mol} \tag{2.2}$$

The reduction is typically performed during the HTS reactor startup, and it should be carefully controlled because of a significant heat release resulting from the exothermic nature of the reactions involved, which may damage the catalyst. The ratios of water vapour to hydrogen and CO_2 to CO for the activation step determine the equilibrium of the Fe^{2+} and Fe^{3+} ions in octahedral sites. With common used process gases the catalyst is more reduced than the reduction to reach the state of equilibrium. In this manner more Fe_3O_4 is created, which forms the stable state. It is important to avoid over-reduction of the magnetite active material during the process to lower oxides, carbides or metallic iron species. In such a case, a damaging hot spot can form during the reaction due to exothermic methanation for which metallic iron is known to be a good catalyst.

$$Fe_3O_4 + H_2 \rightarrow FeO + H_2O \quad \Delta H = -63.8\text{kJ/mol} \tag{2.3}$$

$$Fe_3O_4 + CO \rightarrow FeO + H_2O \quad \Delta H = -22.6\text{kJ/mol} \tag{2.4}$$

$$FeO + H_2 \rightarrow Fe + H_2O \quad \Delta H = -24.5\text{kJ/mol} \tag{2.5}$$

$$FeO + CO \rightarrow Fe + CO_2 \quad \Delta H = -12.6\text{kJ/mol} \tag{2.6}$$

$$Fe_3O_4 + 4H_2 \rightarrow 3Fe + 4H_2O \quad \Delta H = -149.4\text{kJ/mol} \tag{2.7}$$

The Cr phase presented in the catalysts is also reduced during the pre-treatment.

$$2CrO_3 + 3H_2 \rightarrow Cr_2O_3 + 3H_2O \quad \Delta H = -684.7\text{kJ/mol} \tag{2.8}$$

$$2CrO_3 + 3CO \rightarrow Cr_2O_3 + 3CO_2 \quad \Delta H = -808.2\text{kJ/mol} \tag{2.9}$$

The metallic iron species especially are active catalysts for the Fischer-Tropsch process, thus leading to methanation Boudouard reactions in the WGS reaction.

$$CO + 3H_2 \rightarrow CH_4 + H_2O \quad \Delta H = -206.2 \text{kJ/mol} \tag{2.10}$$

$$2CO \rightarrow C + CO_2 \quad \Delta H = -172.5 \text{kJ/mol} \tag{2.11}$$

It is generally known from industrial experience that if the reduction factor (R) for the reformed gas is maintained at less than 1.2, then over-reduction of Fe_3O_4 does not occur, while over-reduction consistently occurs when R is greater than 1.6. If the ratio is too low (in other words the H_2O to H_2 ratio is too high) the magnetite reacts with the water and forms hematite again.

$$2Fe_3O_4 + H_2O \rightarrow 3Fe_2O_3 + H_2 \tag{2.12}$$

In 1985 Rethwisch *et al.* [7] reported the effect of pre-treatment condition on the WGS activity of iron catalysts supported on graphite. Initially, when the catalysts reduced in the presence of CO/CO_2 (15/85) mixture, a rapid initial deactivation followed by stable activity was observed. When they changed CO/CO_2 ratio to 4/96 no change in the activity was observed. Treatment in H_2/H_2O (40/60) for 22 h had very little effect on the water-gas shift activity. Treatment in CO/CO_2 15/85 mixture for 32 h produced a greater extent of catalyst activation. However, this effect was small compared to the effects of treatments in the CO/CO_2 gas mixtures. It was found that treatment for 22 h in CO/CO_2 (15/85) mixture followed directly by a 22 h treatment in H_2/H_2O (80/20) had a similar effect on water-gas shift activity to a treatment for 22 h in H_2/H_2O (40/60). This indicates that treatments in H_2/H_2O not only fail to activate the samples to a significant extent but also that these treatments deactivate the catalyst following activation in CO/CO_2. They also investigated the effect of oxygen pre-treatment on the WGS activity. A 0.5 h treatment in oxygen followed by 23 h in CO/CO_2 15/85 mixture led to an increase in catalytic activity, which was comparable to the extent of catalyst activation caused by treatment in CO/CO_2 15/85 mixture alone. Treatment in oxygen for 0.33 h followed by treatment in CO/CO_2 15/85 mixture for 20 h also caused an increase in catalytic activity which was comparable to the effect of treatment in CO/CO_2.

In 1986 Gonzalez *et al.* [8] studied the effect of thermal treatments and reduction process on the activity of Fe/Cr catalysts. It was shown that the reduction of Fe/Cr catalysts with hydrogen does not lead necessarily to Fe formation, unless they are submitted to a previous thermal treatment or an activation at temperatures higher than 500 °C is carried out. The addition of vapour in standard conditions of hydrogen reduction is not justified due to the fact that Fe formation is not possible. Carbon monoxide is a more powerful reducing agent than hydrogen, so it may lead to Fe formation at temperatures lower than 500 °C. In this case, it is convenient to add steam to the reducing mixture.

Process and hydrogen gas are the best reducing agents in order to obtain higher activities without having an over-reduction at temperatures lower than 500 °C.

Li and Chang [9] investigated the mechanical strength of Fe-based catalysts in a reduction process. They found that the process parameters in the heating and reduction such as the heating rate, the reduction temperature, and the steam to gas ratio all have strong effects on the mechanical properties of the catalyst. The reduction temperature and the steam to gas ratio have a stronger effect on the strength than the heating rate. The optimum condition of reduction for the catalyst depends also strongly on the properties of the materials.

2.1.2 Influence of Various Parameters

During the last three decades, due to the rising cost of hydrocarbon feedstocks, plants have been forced to keep operating costs low by being as energy efficient as possible. One way to reduce the cost and improve efficiency is to decrease the steam to gas ratio in the water-gas shift plant. However, using of lower steam to CO ratios can cause over-reduction of iron oxide and form iron carbide during the high-temperature water-gas shift reaction. Iron carbides are very active catalysts for Fisher-Tropsch reactions and can form hydrocarbons like methane during water-gas shift reaction. Attempts are made in the literature to improve the activity of Fe-Cr at low steam to gas ratios by adding promoters. Topsoe *et al.* [10] and Natesakhawat *et al.* [6] among other authors suggested that the addition of other cations (metal promoters) into the metal lattice might change the electronic properties, improving the catalyst activity. In late 1980s Sud-Chemie developed Fe-Cr-Cu catalyst formulation that suppressed the Fisher-Tropsch reactions in WGS plant [11]. The Fe-chrome catalyst was promoted by adding Cu to increase the catalytic activity and selectivity, even at lower H_2O/C ratios where secondary reactions might happen, leading to energy reduction costs. On the other hand, Andreev *et al.* [12] studied the effect of adding CuO, CoO, MnO and ZnO on the activity of Fe-Cr catalysts, where Cu- and Co-oxides proved to be catalytic promoters of interesting relevance. More recently, Rhodes *et al.* [13] studied the promotion of Fe-Cr catalysts with B, Cu, Ba, Pb, Hg and Ag. In this case, by adding Hg, Ag or Ba and then Cu, higher catalytic activity was observed at temperatures between 350 and 440 °C. Attempts are also made in the literature to develop Cr-free iron-based high-temperature WGS catalyst.

Preparation method for Fe-Cr catalyst has a strong influence on their structural and catalytic properties. Fe-Cr catalysts were usually prepared by co-precipitation method by using nitrates as precursors and ammonia as a precipitating agent followed by drying and calcination. Wet-impregnation technique is an alternate method to synthesize Fe-Cr catalyst. In the impregnation method, first iron hydroxide was prepared by precipitation method and chromium was deposited over iron oxide by impregnation method using chromium nitrate solution. X-ray photoelectron spectroscopy revealed that there was surface enrichment of Cr ions in the fresh samples prepared by both the co-precipitation and impregnation routes. The surface chromium concentration was higher in the sample prepared by the impregnation method.

Chinchen *et al.* [14] investigated the effect of mass transport limitations on the WGS activity of Fe-Cr catalysts. They investigated in two reactors, namely, semi-technical plant and micro-reactor. They found that in the semi-technical plant the film diffusion limitations are unlikely to happen. Inspection of the experimental results reveals that the two differently sized catalysts give similar performance at temperatures up to about 360 °C, but above this temperature the smaller pellets give higher conversions. The reaction over 5.4 × 3.6 mm pellets (15-5) is not pore diffusion limited at temperatures of 370 °C and below and at pressures up to 31 bar. Over the larger 8.5 × 10.5 mm pellets (15-4) the reaction becomes pore diffusion limited at temperatures above 350 °C and at 31 bars pressure. The results obtained with the micro-reactor unit are in good agreement with this analysis of catalyst performance in the semi-technical unit.

Chinchen *et al.* investigated [3] the stability of the Fe-Cr catalyst in both the reactors. There are two stages of deactivation. First it was observed in the initial stage which is a fast decay while the other one is a slow decay which occurs in the later stages of the reaction. The Arrhenius plots show that the activation energy for the water-gas shift reaction had not changed significantly between the new catalyst and a catalyst which had been operated for 9 months at about 470 °C. However, it is clear that there is a 30% decrease in the activity. The main reason for the deactivation is the loss of surface area.

2.1.3 Effect of Other Metals on the Activity of Iron-Chromia Catalysts

Andreev *et al.* [12] synthesized first row transition metal doped Fe-Cr catalysts. They synthesized 5% Zn, Mn, Cu and Co Fe-Cr catalysts by co-precipitation method. Cu- and Co-oxides proved to be catalytic promoters of interesting relevance. In the late 1980s Sud-Chemie developed Fe-Cr-Cu catalysts for the low steam to CO ratio applications [11]. Operating iron-chromia catalysts at lower steam to CO ratios leads to the formation of iron carbides and iron carbides are very active catalysts for Fisher-Tropsch reactions. Figure 2.1 compares the methane formation across the conventional Fe-Cr catalysts and Cu-promoted Fe-Cr catalysts. It is seen that Cu promotion suppresses the methane formation. They proposed that Cu suppresses the C-O cleavage and prevents the formation of iron carbides.

Idakiev *et al.* [15] also explained the promotional effect of copper. Initially, they investigated the effect of temperature on the activity of 5% Cu and 15% Cu-doped Fe-Cr catalysts. The CO conversion increases with increasing Cu loading. Also, the degree of conversion of carbon monoxide on the unpromoted iron-chromia catalyst sharply decreases with the increase of the space velocity. On the copper oxide promoted samples, it remains almost unchanged. However,

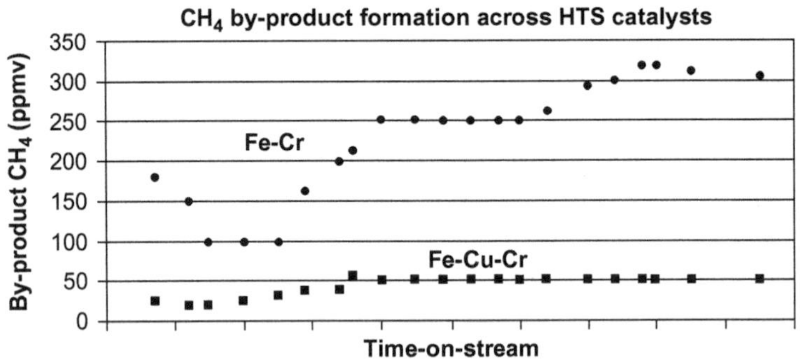

FIGURE 2.1 Comparison of CH_4 formation over standard iron oxide-chromium oxide and copper-promoted iron oxide-chromium oxide catalysts. *(Taken from Figure 7 of Catal. Rev. Sci. Eng. 51 (2009) 325–440.)*

there is a sharp decrease in the catalytic activity of Cu-promoted samples after high-temperature thermal treatment. On the other hand the activity of unpromoted Fe-Cr catalysts remained unchanged after high-temperature thermal treatment. They concluded that the high catalytic activity of the copper oxide promoted samples can be ascribed to the formation of copper aggregates. The high-temperature treatment in the conditions of the reaction leads to an enlargement of the copper aggregates and to the formation of larger copper crystallites, which have lower catalytic activity.

Popaa *et al.* [16] investigated the influence of Al addition on the WGS activity of Fe-Cr-Cu catalysts. They showed that 15 wt% alumina addition to conventional high-temperature water-gas shift catalysts increases CO conversion and increases thermal stability. If the alumina replaces the chromia content of the catalyst, the stability of the catalyst decreases. Hence, alumina alone cannot effectively stabilize Fe_3O_4 without chromia. The alumina added catalysts exhibit much higher surface area compared to pure Fe-Cr-Cu catalysts. After the WGS reaction, the Al-based catalysts retain their higher surface areas compared to the other catalysts. Thus, alumina in Fe-Al-Cr-Cu catalysts significantly increases surface areas, especially for the used catalysts, indicating that thermal stability is also improved.

Dofour *et al.* [17] investigated the influence of synthesis method, precursor and effect of Cu addition on the WGS activity of Fe-Cr-Co catalysts. They prepared FeCr, FeCrCu, FeCrCo and FeCrCuCo formulations by oxidation precipitation method, using chloride (Cl) and sulphate (S) metal precursors. The catalytic activity results of FeCrCo and FeCrCuCo catalysts are presented in Figure 2.2. All the materials prepared from sulphate precursor showed higher carbon monoxide conversion than those synthesized with chloride. As expected Cu-promoted catalysts show better activity than Fe-Cr-Co catalysts. For the catalysts synthesized by chloride precursor, in the case of cobalt, incorporation of this metal into the magnetite lattice could improve the covalency of Fe^{2+} and

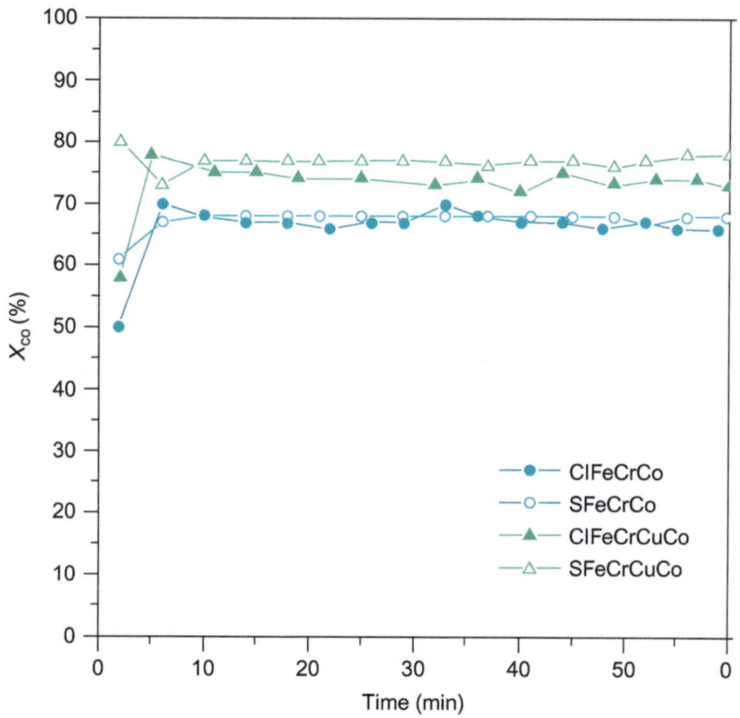

FIGURE 2.2 Catalytic activity results of FeCrCo and FeCrCuCo catalysts. *(Taken from Figure 6 of Int. J. Hydrogen Energy 36 (2011) 8173–8180).)*

Fe^{3+} redox couple, increasing the activity of magnetite. In the case of copper-promoted materials, the higher activity could be related to their higher reducibility as compared to the non-promoted one. On the other hand the addition of promoters for the catalysts synthesized from sulphate precursors did not improve the CO conversion. They proposed that the presence of sulphur poisons the Cu metallic particles and is responsible for the catalyst deactivation.

Meshkani *et al.* [18] synthesized Fe-Cr, Fe-Cr, Fe-Cr-Cu catalysts by facile direct pyrolysis method and compared with commercial catalysts. They found that Fe-Cr-Cu catalysts synthesized by facile direct pyrolysis method exhibit better activity than commercial Fe-Cr-Cu catalysts.

2.2 Cr-FREE Fe-BASED CATALYSTS

In high-temperature Fe/Cr WGS catalysts, Cr is present as Cr^{3+} mostly and to a less extent as Cr^{6+}. Cr^{3+} is less toxic and is a nutrient for the human body. However, hexavalent chromium is a strong carcinogen, threatening human health and the environment. Hence, workers must take precautions when handling the catalyst throughout the span of the operation. Moreover, Cr^{6+} is water-soluble and can be leached from the catalyst by condensed steam or cold water, which could be a threat to the environment, even with minimal disposal. There are several possibilities for producing hexavalent chromium during the manufacturing of the catalyst. Concerns about the environmental hazard and toxicity of hexavalent chromium have been raised since the early 1920s. In 1983 U.S. National Research Council published the general guidelines for chromium compound risk assessments, and the EPA reported many practical guidelines for the identification and assessment of hexavalent chromium [19]. The Occupational Health and Safety Administration (OSHA) under the U.S. Department of Labour enforced strict regulations regarding worker exposure to hexavalent chromium in several industries [20]. Moreover, European Restriction of Hazardous Substances (RoHS) banned the use of six hazardous materials, including hexavalent chromium, in all electronic-electrical equipment.

In 1985 Rethwisch *et al.* [7] first time ever reported the first Cr-free iron oxide catalyst magnetite supported on activated carbon. They prepared magnetite on two graphite supports. Before the reaction the catalysts are reduced in the presence of CO/CO_2 mixture at 487 °C. For a sample containing 14.9 wt% Fe_3O_4 an initial rapid deactivation followed by a long period of relatively constant activity was observed. Then, they investigated the effect of pre-treatment conditions on the WGS activity of magnetite catalysts. Even the pre-treatment in helium gives good catalytic activity.

Basifiska and Domka in 1993 reported [21] Fe-Ru catalysts for the WGS reaction. They took α, β, γ or δ iron hydroxide and impregnated with Rh followed by calcination at 400 or 600 °C. The catalysts were prepared by ruthenium chloride. The catalytic activity results are presented in Table 2.1.

Interestingly, the catalysts without pre-treatment exhibit better activity compared to the catalysts with pre-treatment in CO. They proposed that during the pre-treatment metallic iron was formed.

Then, they prepared the catalysts using ruthenium carbonyl [22]. They found that for two iron oxides (α- and δ-FeOOH) the reaction rate constants are considerably higher than those of reactions with other catalysts based on $RuC1_3$.

Ladebeck and Kochloefl [23] tried to replace Cr with Al-based mixed oxides. They investigated Fe-Al mixed with ZrO_2, La_2O_3, CeO_2 or MnO_2. Among the various catalysts Al combined with Ce exhibits better activity followed by Mn- and Zr-containing systems. These catalysts exhibit better activity than commercial catalyst.

Tsagaroyannis *et al.* [24] investigated Fe-Gd ferrites for WGS reaction. These catalysts exhibit remarkable catalytic activity. This activity is attributed to their p-type semi-conductivity due to a limited number of Fe^{4+} cations, existing in their crystal lattice and associated with the presence of some cation vacancies. Catalytic activity is considerably higher for the ferrites thermally treated in rather lower temperatures (about 800 °C) as they included more defects in their crystal lattices.

TABLE 2.1 Conversion Degree (%) in the Water-Gas Shift Reaction, $RuCl_3$-Impregnated Iron Oxide-Hydroxide Catalysts (α- and δ-FeOOH); 0.5 g of the Catalyst, Reaction Temperature 350 °C

	Conversion degree after 3 h on stream			
	With 30 min CO pre-treatment		*Without CO pre-treatment*	
Catalyst	*(%)*	k *($cm^3\,g^{-1}\,min^{-1}$)*	*(%)*	k *($cm^3\,g^{-1}\,min^{-1}$)*
α-FeOOH-400	13.4	18.2		
α+0.5%Ru-400	5.2	6.7	53.1	99.0
α+2%Ru-400	19.5	27.5	70.0	163.7
α+5%Ru-400	42.4	71.2	73.4	182.5
α-FeOOH-600	20.3	28.8	20.5	29.1
α+0.5%Ru-600	19.0	26.7	33.1	51.4
α+5%Ru-600	62.1	129.0	82.6	257.5
δ+2%Ru-400	36.2	57.7	89.7	396.3
δ+5%Ru-400	54.2	102.3	82.2	253.0
δ+2%Ru-600	58.1	114.7	87.6	336.0
δ+5%Ru-600	64.1	136.8	89.0	372.6

Taken from Table 1 of Catal. Lett. 17 (1993) 327–331.

Silva *et al.* [25] also reported Fe-Gd catalysts for the WGS reaction. The samples were prepared by co-precipitation method. As expected CO conversion increases with increasing reaction temperature. The activity increased with the dopant amount up to the sample with Gd/Fe = 0.10 (FG10), indicating that here is an optimum amount of gadolinium for its promoting action. Rangel Costa *et al.* [26] synthesize Th- and Cu-doped iron oxide catalysts. They found that thorium prevents sintering and the production of metallic iron which can catalyze undesirable reactions. However, thorium damages some active sites and decreases the activity per area. The iron oxide catalyst doped with both Th and Cu exhibits remarkable catalytic activity. It is more active and selective than a chromium- and copper-doped commercial catalyst.

Junior *et al.* [27] substituted Cr with V in the iron oxide catalysts. They synthesized the catalyst by heating vanadium-doped iron(III) hydroxoacetate (IHA) under nitrogen. The vanadium-doped catalysts exhibit better activity compared to pure iron oxide and commercial catalyst. This is due to both textural and structural action. Vanadium resists sintering as well as stabilizes the excess Fe(III) in the structure, thus favouring the successive oxidation and reduction cycles during the reaction and avoiding the formation of metallic iron which causes the catalyst deactivation. Martos *et al.* [28] synthesized Mo-doped iron oxide catalyst by oxidation precipitation and wet-impregnation methods. They found that the properties of Fe-Mo-Cu catalysts are similar to Fe-Cr-Cu catalysts. Pereira *et al.* [29] reported Fe/Co catalysts for the WGS reaction. The addition of small amounts of cobalt decreased the activity, as compared to pure iron oxide, but higher amounts did the opposite.

Lee *et al.* [30] replaced Cr with Ni, Zn, Co doped in the iron oxide. Performance of the catalysts in high-temperature WGS reaction was evaluated under gas mixture conditions (42% H_2, 10% CO, 37% H_2O, 8% CO_2 and 3% CH_4). Among the prepared catalysts, Zn5FeNi20 and Co5FeNi20 catalysts showed good catalytic activity (around 65-75% CO conversion at 400 °C) in the HTS condition of LPG reformate.

Then, they synthesized several Fe/Ni catalysts [31]. The catalytic activity results are presented in Figure 2.3. The incorporation of Ni increased the CO conversion until the Ni content reached 40 wt% but the further increase of Ni reduced the CO conversion. Among the Fe/Ni catalysts, the FN6634, FN6040 and FN5050 catalysts were more active than the Cr-containing,

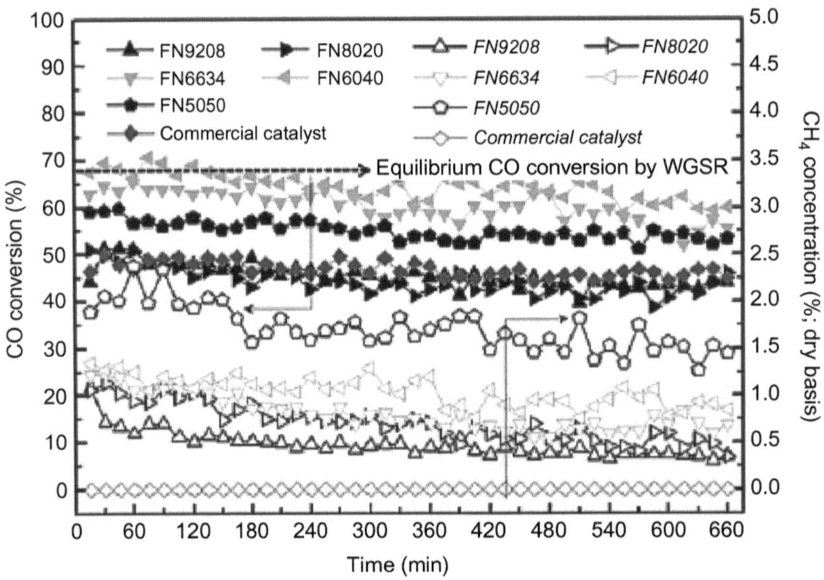

FIGURE 2.3 HTS test results of the Fe/Ni metal oxide catalysts under LNG reformate condition (solid symbols for CO conversion and open ones for CH_4 concentration): $R=2$, H_2 (56.7 vol%), CO (10 vol%), CO_2 (6.7 vol%) and H_2O (26.7 vol%); temperature = 400 °C; WHSV = 25,000 ml $(g_{cat}\ h)^{-1}$. *(Taken from Figure 6 of Int. J. Hydrogen Energy 36 (2011) 8173–8180.)*

commercial catalyst (Fe/Cr/Cu). The FN6040 catalyst exhibited the highest CO conversion, at a level quite close to the equilibrium value. However, the methane production and accompanying hydrogen consumption continuously increased with increasing Ni content, and the increments were largest when the Ni content reached 50 wt%.

Correia *et al.* [32] reported Fe/MCM-41 catalysts. They prepared three concentrations of iron nitrate on the MCM-41 by using impregnation method. It can be noted that the activity increased with the amount of iron and then the sample with the highest amount of iron was the most active. All samples were more active than hematite and this can be assigned both to the higher specific surface area and to the higher activity per gram of iron of the samples. The most active catalyst exhibits better activity than commercial catalyst.

2.3 Fe-Al CATALYSTS

Replacement of Cr with Al has been investigated extensively in recent times. Several groups concluded that Cr can be successfully replaced with Al. In 2000 Araujo and Rangel [33] first time ever reported Fe-Al-Cu catalysts for WGS reaction. They prepared Fe-Al, Fe-Cu, Fe-Al-Cu catalysts by co-precipitation. They maintained the iron to dopant molar ratio of 10. Addition of Al to the iron oxide increases the catalytic activity of iron oxide slightly. However, addition of both Al and Cu to the iron oxide increases the WGS activity tremendously (34×10^{-4} mol g^{-1} h^{-1}). The catalyst with both dopants showed higher activity than a chromium- and copper-doped commercial catalyst (25×10^{-4} mol g^{-1} h^{-1}). This sample produces the active phase more easily than the other catalysts and shows resistance to a further magnetite reduction.

Li *et al.* [34] synthesized Fe-Al-Ce catalysts and compared with Fe-Cr catalysts and evaluated for high-temperature WGS reaction under simulated coal derived syngas at a steam to CO ratio of 1. The Fe-Al-Ce catalyst exhibits better activity compared to Fe-Cr catalyst.

Natesakhawat *et al.* [6] investigated Fe/promoter ratio, pH of precipitation medium, calcination, reduction temperatures and preparation methods on the catalytic activity of Fe-Al and Fe-Al-Cu catalysts. The WGS activity of Fe-Al catalysts increases with decreasing Fe/Al molar ratio and reaches a maximum at Fe/Al = 10. Further addition of Al (Fe/Al = 5) causes a significant drop in the WGS activity. They also proposed that optimum pH value of the precipitation medium is 9 to get the highest activity of Fe-Al catalyst. The catalytic activity of Fe-Al catalysts is seen to go through a maximum at 450 °C calcination temperature and further increase in the calcination temperature results in a decrease in WGS activity possibly due to loss of surface area caused by sintering.

Natesakhawat *et al.* then promoted [6] Fe-Al catalysts with first row transition metals Cu, Co and Zn using two-step co-precipitation – impregnation method and one-step co-precipitation method. Addition of Zn did not provide any higher activity compared to Fe-Al catalyst, promotion with Co and Cu leads to significantly higher CO conversions, with Fe-Al-Cu catalyst giving the highest activity. The catalysts synthesized by one-step co-precipitation method exhibit better activity compared to two-step method. Then they investigated the effect of pre-reducing agent on the WGS activity of Fe-Al-Cu catalysts (one-step method). Figure 2.4 shows the conversion of CO and H_2O versus time-on-stream. The catalyst pre-reduced with 20% H_2/N_2 shows the lowest steady-state conversions of CO and H_2O, while the catalyst pre-reduced with CO/H_2O at 350 °C exhibits the highest steady-state activity. The one which has undergone no pre-reduction gives a steady-state performance in between.

Zhang *et al.* also synthesized [35] the Fe-Al-Cu catalysts using sol-gel method and proved that sol-gel synthesized catalysts exhibit better activity than one-step and two-step precipitation methods. Then, they compared sol-gel synthesized catalyst with commercial catalyst. The results are presented in Figure 2.5. Fe-Al-Cu-SG catalyst yielded much higher activity than the commercial catalyst at every temperature tested. The fact that this superior activity was maintained throughout the entire temperature range indicates that this preparation technique widens the operating temperature range, showing promise to combine high-temperature and low temperature water-gas shift stages together.

Then they studied the effect of [36] Cu loading on the WGS activity of Fe-Al-Cu catalysts. The catalyst with Fe/Al = 10 and Fe/Cu = 5 exhibits better

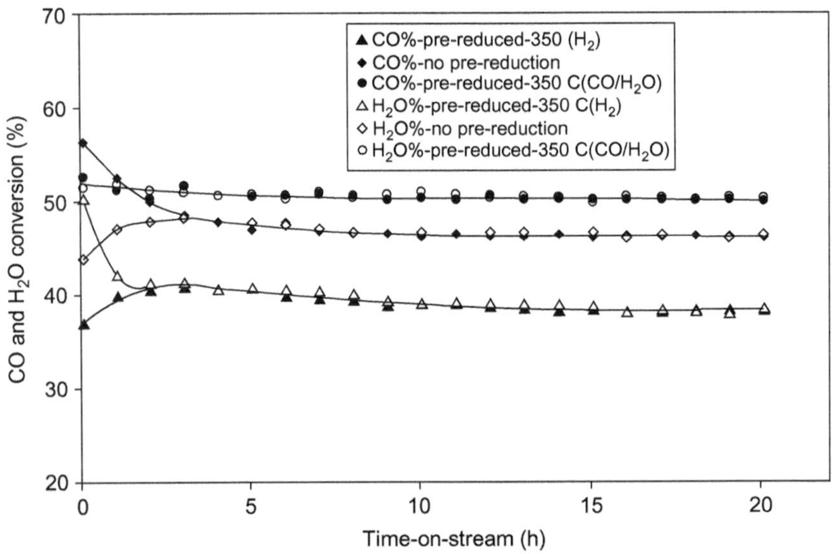

FIGURE 2.4 Effect of reduction steps on catalytic performance of Fe-Al-Cu (one-step) catalyst: conversion of CO and H_2O versus time-on-stream. *(Taken from Figure 11 of J. Mol. Catal. A Chem. 260 (2006) 82–94.)*

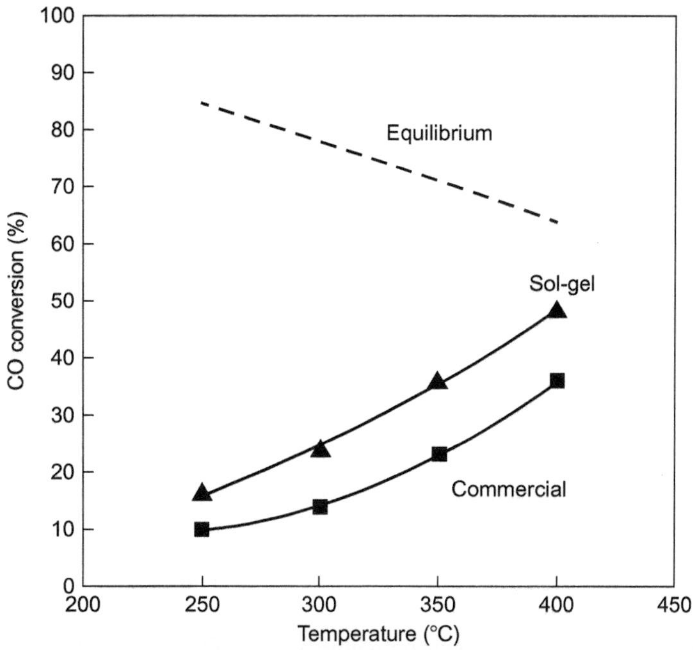

FIGURE 2.5 CO conversion between sol-gel Fe-Al-Cu (▲) catalyst and commercial Fe-Cr-Cu (■) catalyst. *(Taken from Figure 2 of Appl. Catal. A Gen. 351 (2008) 1–8.)*

activity compared to Fe/Cu = 10 and Fe/Cu = 2 catalysts. Increase in the steam to CO ratio from 1 to 2 had no significant effect on the CO conversion at lower reaction temperatures, however, at higher temperatures, increase in the steam content of the feed resulted in a significant increase in CO conversion.

Subramanian *et al.* [37] synthesized Fe-Al-Cu catalysts using citric acid assisted sol-gel method. In this method they directly prepared the magnetite to avoid the reduction process. The catalyst exhibits CO conversion 80% at 350 °C and 84% at 400 and 450 °C. However, the catalyst exhibits continuous deactivation with time.

Jeong *et al.* [38] investigated Fe-Al-Cu catalysts in the presence of simulated waste derived synthesis gas. The simulated waste derived synthesis gas consisted of 17.1 vol% CO, 9.6 vol% CO_2, 1.0 vol% CH_4, 13.0 vol% H_2, 55.4 vol% H_2O and 3.9 vol% N_2, which represents a typical syngas from waste gasifier. They synthesized several catalysts by varying the ratio between Fe and Cu with fixed Al content. Among the various catalysts, the Fe-Al-Cu catalyst with composition 78:7:15 exhibited better activity. Then, they compared this catalyst with commercial catalyst. The catalytic results are presented in Figure 2.6. The Fe-Al-Cu catalyst exhibits much better and stable activity compared to commercial catalyst. The enhanced CO conversion with stability over the Fe-Al-Cu catalyst in severe conditions can be explained as follows. First,

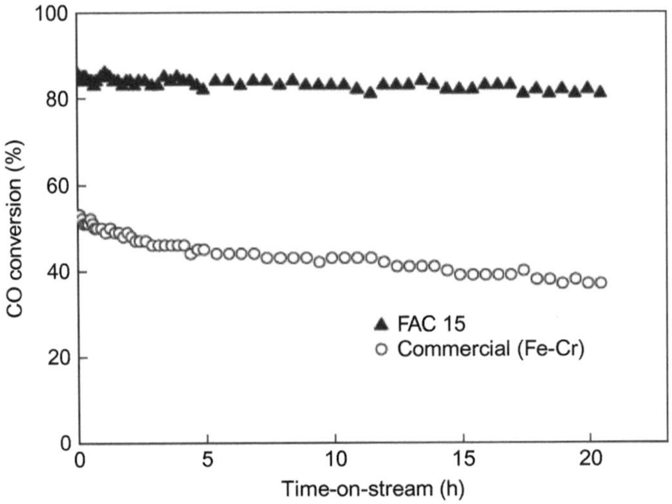

FIGURE 2.6 CO conversion with time-on-stream over FAC 15 and the commercial HTS catalyst ($H_2O/(CH_4+CO+CO_2)=2.0$, $T=400$ °C, GHSV$=40{,}057$ h^{-1}). *(Taken from Figure 5 of Catal. Lett. 143 (2013) 438–444.)*

Fe-Al-Cu catalyst has the easier reducibility of Fe_2O_3 to Fe_3O_4 and higher stability of the active phase (Fe_3O_4). Second, Fe-Al-Cu catalyst has a strong Cu to Fe_2O_3 interaction, resulting in enhanced reducibility and stability of Fe_3O_4. Al_2O_3 has a beneficial effect in preventing sintering of iron.

Meshkani Rezaei [39] synthesized mesoporous Ba promoted Fe-Al-Ni catalysts for the high-temperature WGS reaction. Fe-Al-Ni and Fe-Al-Ni-Ba catalysts exhibit better catalytic activity than commercial catalysts. Also Ba addition suppresses the methanation activity.

2.4 CERIA-BASED CATALYSTS

In 2004 Qi and Stephanopoulos [40] reported Cu-CeO_2 catalysts for the high-temperature WGS reaction. They also investigated the effect of La addition in the Ce lattice on the WGS activity. The catalytic results of various Cu-CeO_2-La_2O_3 catalysts, Fe-Ce catalysts, pure Ce-La and commercial catalyst are presented in Figure 2.7. As expected, pure Ce-La catalyst exhibits lower conversion compared to remaining catalysts. The Fe-Ce(8% La)O_x samples have higher activities than pure Ce(8% La)-O_x, indicating that addition of Fe increases the activity of ceria. However, the commercial catalyst C12-4-02 is more active than the Fe-Ce(8% La)O_x samples. All the Cu-promoted Ce(8% La)-O_x catalysts exhibit better activity than commercial catalyst as well as Fe-Ce(8% La)O_x catalysts. Among the copper-ceria samples, 10% Cu-Ce(8% La)O_x and 15% Cu-Ce(8% La)O_x were the most active. The 13% Fe-Ce(8% La)O_x was the most active among the iron-ceria samples.

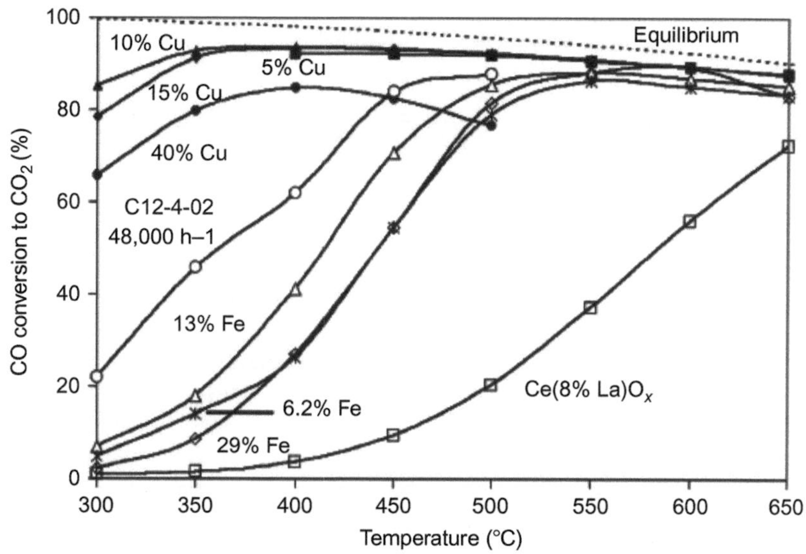

FIGURE 2.7 Steady-state WGS performance of CU- and Fe-Ce-(8% La)O_x in 2% CO-10% H_2O-He, 0.09 g s cm^{-3}, SV = 80,000 h^{-1} (except C12-4-02). *(Taken from Figure 1 of Ind. Eng. Chem. Res. 43 (2004) 3055–3062.)*

Then they tested 10% Cu-Ce(8%La)O_x and 13% Fe-Ce(8% La)O_x catalysts and C12-4-02 in a simulated coal gas mixture with a molar composition 10% CO-34% H_2O-10% CO_2-15% H_2-He and in a CO_2-rich gas containing 2% CO-10% H_2O-35% CO_2-He. Among the three catalysts, Cu-Ce-La exhibits higher activity and reaches equilibrium CO conversions above 450 °C in the simulated coal gas mixture and at high space velocity (80,000 h^{-1}).

Zhi *et al.* [41] investigated the effect of synthesis method on the WGS activity of Cu-Ce-La catalysts. They prepared the catalysts using co-precipitation, homogeneous precipitation and deposition-precipitation techniques. Among the various catalysts, the catalysts synthesized by co-precipitation method exhibit better activity. Then, they investigated the effect of precipitator. The sequence of activities are NaOH (61.10%) > NaOH + $(NH_2)_2CO$ (60.98%) > $(NH_2)_2CO$ (57.4%) ≫ $NH_3 \times H_2O$ (5.35%) at 300 °C.

Djinovic *et al.* synthesized [42] Cu-CeO_2 by hard template method and compared that with co-precipitated catalysts. The activity results show that the catalysts synthesized by hard template method show better activity compared to catalysts synthesized by co-precipitation method regardless of Cu loading.

Kumar and Idem [43] synthesized Cu/Ni/CeO_2-ZrO_2 and compared with commercial low- and high-temperature WGS catalysts. Under the reformate conditions, Cu/Ni/CeO_2-ZrO_2 was found to be more stable with no apparent activity loss compared to both commercial low- and high-temperature WGS catalysts. They proposed that the superior catalytic performance of

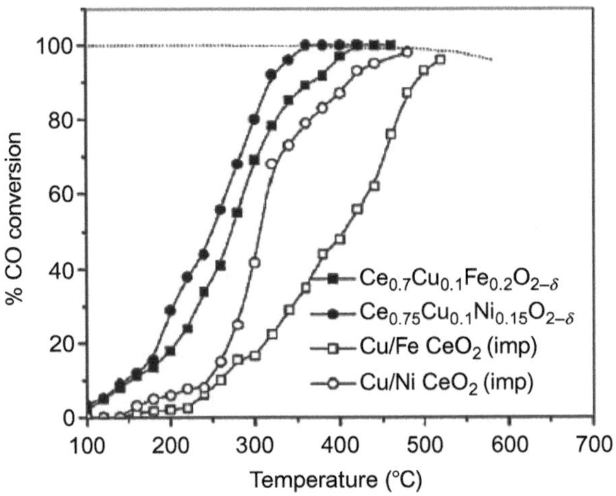

FIGURE 2.8 WGS activity of various ceria-based compounds with temperature. *(Taken from Figure 7 of Appl. Catal. B Environ. 123–124 (2012) 367–378.)*

$Cu/Ni/CeO_2$-ZrO_2 is attributed to the existence of higher amounts of easily reducible well-dispersed copper and nickel oxide species on the catalytic surface strongly interacting with the support material.

Shinde and Madras [44] synthesized Cu-Ni/CeO_2-ZrO_2 and Cu-Fe/CeO_2-ZrO_2 with sono-chemical method and compared with impregnated catalysts. They performed WGS reaction in the temperature region 100-500 °C. The catalytic activity results are presented in Figure 2.8. The catalyst synthesized by sono-chemical method exhibits better activity than impregnated catalyst.

Ni-doped catalysts exhibit better activity compared to Fe-doped catalysts. They proposed that Fe metal is likely to be oxidized under reaction conditions and therefore unlikely to stabilize CO adsorption.

Li *et al.* [45] investigated the effect of lattice oxygen changes on the WGS activity of Cu-CeO_2 catalysts. They investigated the effect of Cu loading and calcination temperature on the WGS activity of Cu-CeO_2 catalysts. CO conversion increased with increasing copper content up to 40 wt% and then decreased with further increasing copper content. They proposed that the lattice oxygen concentration increases with increasing Cu loading and is responsible for the catalytic activity. They also found that catalysts calcined at 500 °C exhibit better activity. However, these catalysts exhibit continuous deactivation with time during the 120 h of operation.

2.5 PEROVSKITES TYPE OF CATALYSTS

In 2011 Maluf *et al.* [46] investigated $La_{(1-y)}Sr_yNi_xCo_{(1-x)}O_3$ perovskite catalysts for WGS reaction. The catalytic activity results of various catalysts are presented in Figure 2.9. The perovskite catalysts exhibit fairly good catalytic

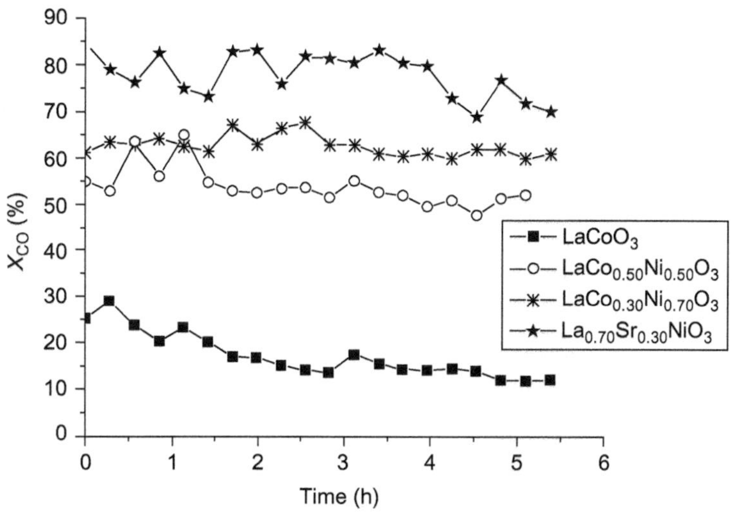

FIGURE 2.9 CO conversion versus time at 400 °C. *(Taken from Figure 6 of Top. Catal. 54 (2011) 210–218.)*

activity. The $LaCoO_3$ perovskite as a precursor showed the lowest conversion (15%). The replacement of Co by Ni caused the precursors to be more active in WGSR, with CO conversions of 50% and 60% for the $LaCo_{0.5}0Ni_{0.50}O_3$ and $LaCo_{0.3}0Ni_{0.70}O_3$ sample.

Sun *et al.* [47] investigated the effect of Ce on the WGS activity of $La_{(0.9-x)}Ce_xFeO_3$ perovskite-like catalysts with $x=0$, 0.2, 0.6 and compared with high-temperature commercial catalyst Fe-Cr-Cu. The activity trend of the catalysts depends on the reaction temperature. At 450 °C, commercial catalyst exhibits higher activity compared to perovskite catalysts. However, at reaction temperatures 550 and 650 °C perovskite catalysts exhibit better activity. They proposed that addition of small amount of Ce has little influence on the formation of the $LaFeO_3$ perovskite structure but enhances catalytic activity. However, addition of larger amount of Ce leads to the formation of segregated phases of Fe_2O_3, CeO_2 and $La_{0.25}Ce_{0.75}O_{1.9}$, which inhibit the formation of the $LaFeO_3$ perovskite structure and decrease catalyst activity.

Then, they investigated [48] the effect of H_2S performance on the WGS activity of $La_{0.7}Ce_{0.2}FeO_3$ perovskite catalyst. They exposed catalyst to various H_2S concentrations and the results are presented in Figure 2.10. The results demonstrate that the activity of $La_{0.7}Ce_{0.2}FeO_3$ perovskite catalyst decreases with increasing concentrations of H_2S at 600 °C. However, the activity is somewhat higher than the original baseline value when H_2S is removed from the feed gas and then eventually stabilizes at the original value. This indicates that the negative effect of H_2S on the activity of the $La_{0.7}Ce_{0.2}FeO_3$ perovskite catalyst is reversible. Also, $La_{0.7}Ce_{0.2}FeO_3$ perovskite catalyst exhibits better activity than Fe-Cr-Cu and Co-Mo catalysts in the presence of H_2S.

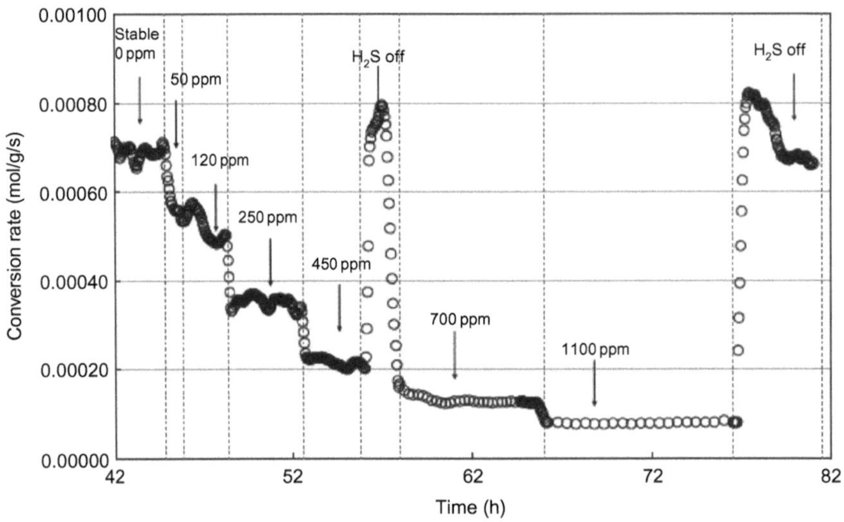

FIGURE 2.10 Performance of $La_{0.7}Ce_{0.2}FeO_3$ perovskite-like catalyst at various concentrations of H_2S (0–1100 ppm wet basis) in the simulated coal-derived syngas at 600 °C and 1 atm. *(Taken from Figure 5 of Int. J. Hydrogen Energy 37 (2012) 1475–1481.)*

They proposed that the activity of $La_{0.7}Ce_{0.2}FeO_3$ perovskite catalyst decreases with increasing concentrations of H_2S at 600 °C. However, the activity is somewhat higher than the original baseline value when H_2S is removed from the feed gas and then eventually stabilizes at the original value. This indicates that the negative effect of H_2S on the activity of the $La_{0.7}Ce_{0.2}FeO_3$ perovskite catalyst is reversible.

Jijil *et al.* [49] investigated $BaCe_{1-x}Pt_xO_{3-d}$ type of catalysts. They found that the activity increases when the catalyst is recycled indicating an enhancement in Pt surface distribution during reaction. Their neutron diffraction data shows that the oxygen vacancies increase with increase in Pt substitution and the oxygen vacancies created are in the O_2 site.

2.6 INTEGRATED GASIFICATION COMBINED CYCLE

Integrated gasification combined cycle (IGCC) technology uses WGS reaction to convert the remaining CO in the syngas to CO_2 and H_2. IGCC produces electricity from a solid or liquid fuel. First, the fuel is converted to syngas which is a mixture of hydrogen and carbon monoxide. Second, the syngas is converted to electricity in a combined cycle power block consisting of a gas turbine process and a steam turbine process which includes a heat recovery steam generator (HRSG). The combined cycle technology is similar to the technology used in modern natural gas fired power plants. Coal utilization as a fuel is anticipated to rise by 30% in the next two decades which will cause coal-fired power plants to have a capacity growth of approximately 40%. IGCC plants have been recognized as an attractive option for coal-based power production because of their high thermal efficiency and low environmental impact. The key benefits of IGCC over the traditional combustion systems are reduced production of solid wastes, higher fuel flexibility, higher efficiency, lower SO_x and NO_x emissions, marketable by-products and higher energy output. However, this technology still requires improvement in terms of process economics to become fully competitive. Also, IGCC technologies are more expensive to build compared to coal-fired power plants. The main disadvantage with the IGCC plant is the capital cost increases by 30% if we install the CO_2 capture unit. Table 2.2 shows some of the current IGCC technologies running in the world. The complete list can be found somewhere else.

In the first generation IGCC plants there is no WGS reactor and CO_2 capture unit was installed. The clean syngas is fed to the combustion turbine with the products going to a HRSG and steam turbine. In the pre-combustion IGCC plant, the clean syngas passes through a shift reactor and an absorption tower to remove the carbon in the form of CO_2. In the conceptual poly-generation IGCC plant, in addition to the shift reactor and CO_2 capture unit the syngas conversion unit will be installed. This unit adjusts the CO to H_2 ratio for the Fisher-Tropsch synthesis. If the unit is used for power generation complete conversion of CO to CO_2 takes place. The schematic of typical conceptual poly-generation IGCC plant is shown in Figure 2.11.

The typical comparison between gasification and combustion units is presented in Table 2.3 [50].

The main difference is that combustion is referred to as full oxidation and gasification as partial oxidation. Also, combustion occurs in an oxidizing (excess oxygen) environment and gasification occurs in a reducing (oxygen depleted) environment. Gasification is more efficient, has lower emissions and competitive capital cost compared to combustion. With respect to the competitiveness of the cost, it is the cost of electricity that is nearly the same for both technologies; the higher capital cost of gasification is offset by the improved efficiency.

In a typical IGCC plant fuel, oxygen and steam are fed to the gasifier to produce raw syngas. Particulates and sulphur are separated and clean syngas is fed to the combustion turbine with the products being sent to a HRSG coupled with a steam turbine. In IGCC plants with CO_2 capture, the syngas is passed through a water-gas shift reactor (water reacts with CO over a catalyst to produce H_2 and CO_2) before the CO_2 capture unit removes the carbon in the form of CO_2. In a typical IGCC power plant without CO_2 capture, the most investment cost is devoted to power generation followed by gasification unit. The addition of CO_2 capture will increase the investment costs by roughly 32-40% resulting in a spread of capital costs from \$2390 to \$2668 kWe^{-1}.

As explained earlier, IGCC operates WGS reaction in two stages: one at higher temperature and another at lower temperature to convert the entire CO into CO_2. Reports had been made in the literature to replace traditional

TABLE 2.2 Major Existing and Planned IGCC Commercial Plants

Plant (location) initial year of operation	Size (MW)	CCS technology	Feedstock	Main features
Buggenum (Netherlands) 1994	253	Pre-combustion (installed in 2011) [9,10]	Coal and biomass	First IGCC commercial plant. Shell entrained-flow gasifier, Siemens V94.2 gas turbine and overall plant efficiency of 43% without CCS
Wabash river (USA) 1995	262	No CCS included [11]	Bituminous coal and petroleum coke	E-GAS technology two stage entrained-flow gasifier, General Electric's MS 7001 gas turbine and overall plant efficiency of 40%
Tampa (USA) 1996	250	No CCS included [11]	Coal and petroleum coke	Texaco entrained-flow oxygen-blown gasifier. GE 7FA gas turbine and overall plant efficiency of 38.2%
Pernis (Netherlands) 1997	155	To be added by 2015 [12]	Heavy oil residue	First IGCC in the world to operate using heavy oil residue gasification, Shell entrained-flow gasifier, GE MS6541B gas turbine and overall plant efficiency of 36.7%
Priolo Gargallo (Italy) 1998	532	Amine based technology [13]	Heavy fuel oil, tar	Siemens V94.2 gas turbine, Texaco entrained-flow gasifier and overall plant efficiency of 38% without CCS
Puertollano (Spain) 1998	335	Amine based technology [14]	Petroleum coke	CO_2 capture process started in 2010, Siemens entrained-flow gasifier. Siemens V 94.3 gas turbine and overall plant efficiency of 40% without CCS
Sarlux (Italy) 2000	548	No CCS included [52]	Residual oil, tar, bitumen	GE entrained-flow gasifier and GE MS9001E gas turbine

Negishi (Japan) 2003	342	No CCS included [1]	Heavy oil, asphalt	GE entrained-flow gasifier, Mitsubishi 701F gas turbine and overall plant efficiency of 36%
Vresova (Czech Republic) 2005	400	No CCS included [14]	Brown coal	Lurgi fixed-bed gasifiers. GE9171E gas turbine and overall plant efficiency of 42.2%
Knox County (USA) 2013	618	Projected CCS plant [15]	Mid-western coal	Replaces 160 MW coal-fired power plant at the site, GE entrained-flow gasifier and two GE 7FB gas turbines
Kemper County (USA) 2014	582	Pre-combustion technology [14]	Lignite coal	Units to be completed in the first half of 2013, non-slagging transport reactor integrated gasifier (TRIG) and two Siemens SGT6-5000F gas turbine
Nakoso (Japan) 2014	500-600	Amine based technology [14]	Bituminous	Demonstration plant completed in 2007, flooded during 2011 tsunami, Mitsubishi Heavy industries (MHI) gasifier. M701G gas turbine and overall plant efficiency of 42% without CCS for the demonstration plant

Taken from Table 1 of Fuel 137 (2014) 245–259.

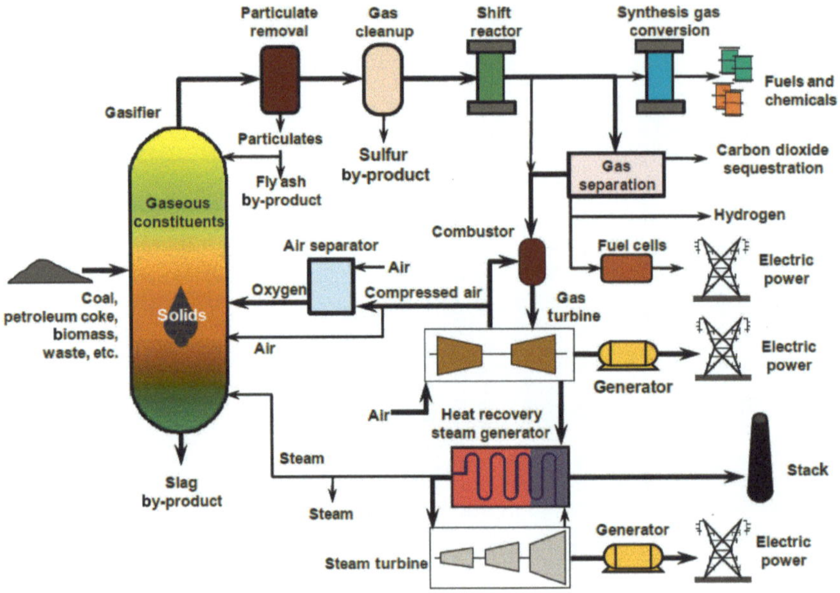

FIGURE 2.11 Poly generation IGCC plant *(Taken from Figure 5 of Energies 3 (2010) 216–240.)*

TABLE 2.3 Contrasts Between Combustion and Gasification [1]

	Combustion	Gasification
Chemical process	Full oxidation	Partial oxidation
Chemical environment	Excess oxygen (air)-oxidizing	Oxygen-starved – reducing
Primary product	Heat (e.g., steam)	Syngas (CO and H_2)
'Downstream' products	Electric power	Electric power, pure H_2, liquid fuels, chemicals
Current application	Dominates coal-fired power generation worldwide	Mostly chemicals and fuels, power generation demonstrated
Efficiency	35-37% (HHV)	39-42% HHV
Emissions	~NSPS	~1/10 NSPS
Capital cost	\$1000-1150 kW^{-1}	Competitive
Maturity/risk	High experience, low risk	Reliability needs improved

Taken from Table 2 of Energies 3 (2010) 216–240.

WGS reactors with the membrane reactors. A combination of membrane designs such as hydrogen-selective sweep membranes for bulk hydrogen recovery and CO_2-selective membranes for the cold end of the process to support the CO_2 liquefaction system has been shown to improve the IGCC's commercial aspects. Future developments on membrane technology are required to improve the short lifetime, selectivity and permeability coefficients of these systems.

REFERENCES

[1] L. Mond, C. Langer, British Patent 12608, 1888.

[2] C.B. Wild, Canadian Patent 153,608, 1888.

[3] G. C. Chinchen, R. H. Logan, M. S. Spencer, Water-gas shift reaction over an iron oxide/chromium oxide catalyst. II: stability of activity, Appl. Catal. 12 (1984) 89–96.

[4] M. Robbins, G. K. Wertheim, R. C. Siterwood, D. N. E. Buchananj, Magnetic properties and distribution in the system $FeCr_2O_4$-Fe_3O_4 ($Fe^{2+}Cr^{2-}$ $x$$Fe_x^{3+}O_4$), Phys. Chem. Solids 32 (1971) 717–721.

[5] M. A. Edwards, D. M. Whittle, C. Rhodes, A. M. Ward, D. Rohan, M. D. Shannon, G. J. Hutchings, C. J. Kiely, Microstructural studies of the copper promoted iron oxide/chromia water-gas shift catalyst, Phys. Chem. Chem. Phys. 4 (2002) 3902–3908.

[6] S. Natesakhawat, X. Wang, L. Zhang, U. S. Ozkan, Development of chromium-free iron-based catalysts for high-temperature water-gas shift reaction, J. Mol. Catal. A Chem. 260 (2006) 82–94.

[7] D. G. Rethwisch, J. Phillips, Y. Chen, T. F. Hayden, J. A. Dumesic, Water-gas shift over magnetite particles supported on graphite: effects of treatments in CO/CO_2 and H_2/H_2O gas mixtures, J. Catal. 91 (1985) 167–180.

[8] J. C. Gonzalez, M. G. Gonzalez, M. A. Laborde, N. Morenoi, Effect of temperature and reduction on the activity of high temperature water gas shift catalysts, Appl. Catal. 20 (1986) 3–13.

[9] Y. Li, L. Chang, Optimizing the mechanical strength of Fe-based commercial high-temperature water-gas shift catalyst in a reduction process, Ind. Eng. Chem. Res. 35 (1996) 4050–4057.

[10] H. Topsoe, J. A. Dumesic, M. Boudart, Alumina as a textural promoter of iron synthetic ammonia catalysts, J. Catal. 28 (1973) 477–488.

[11] C. Ratnasamy, J. P. Wagner, Water gas shift catalysis, Catal. Rev. Sci. Eng. 51 (2009) 325–440.

[12] A. Andreev, V. Idakiev, D. Mihajlova, D. Shopov, Iron-based catalysts for the water-gas shift reaction promoted by first-row transition metal oxides, Appl. Catal. 22 (1986) 385–387.

[13] C. Rhodes, B. P. Williams, F. King, G. J. Hutchings, Promotion of Fe_3O_4/Cr_2O_3 high temperature water gas shift catalyst, Catal. Commun. 3 (8) (2002) 381–384.

[14] G. C. Chinchen, R. H. Logan, M. S. Spencer, Water-gas shift reaction over an iron oxide/chromium oxide catalyst. I: mass transport effects, Appl. Catal. 12 (1984) 69–88.

[15] V. Idakiev, D. Mihajlova, B. Kunev, A. Andreev, Effect of copper oxide on the catalytic activity of iron chromia catalyst for water gas shift, React. Kinet. Catal. Lett. 35 (1987) 119–124.

[16] T. Popaa, G. Xua, T. F. Bartonb, M. D. Argyle, High temperature water gas shift catalysts with alumina, Appl. Catal. A Gen. 379 (2010) 15–23.

[17] J. Dufour, C. Martos, A. Ruiz, F. J. Ayuela, Effect of the precursor on the activity of high temperature water gas shift catalysts, Int. J. Hydrogen Energy 38 (2013) 7647–7653.

[18] F. Meshkani, M. Rezaei, A facile method for preparation of iron based catalysts for high temperature water gas shift reaction, J. Ind. Eng. Chem. 20 (2013) 3297–3303.

[19] P. C. Grevatt, Toxicological Review of Hexavalent Chromium, U.S. Environmental Protection Agency, U.S. Environmental Protection Agency Washington, DC, 1998.
[20] U.S. Department of Labor, Hexavalent Chromium, OSHA 3373-10, 2009.
[21] A. Basifiska, F. Domka, Iron-ruthenium catalyst for the water-gas shift reaction, Catal. Lett. 17 (1993) 327–331.
[22] A. Basifiska, F. Domka, Chlorine-free iron-ruthenium catalyst for the water-gas shift reaction, Catal. Lett. 2 (1993) 327–331.
[23] J. Ladebeck, K. Kochloefl, Cr-free iron-catalysts for water-gas shift reaction, Stud. Surf. Sci. Catal. 91 (1995) 1079–1083.
[24] J. Tsagaroyannis, K.-J. Haralambous, Z. Loizos, G. Petroutsos, N. Spyrellis, Gadolinium-iron ferrites: catalytic effect on the water-gas shift reaction, Mater. Lett. 28 (1996) 393–400.
[25] C. L. S. Silva, S. G. Marchetti, A. C. F. Juniorc, T. F. Silva, J. M. Assafd, M. C. Rangela, Effect of gadolinium on the catalytic properties of iron oxides for WGSR, Catal. Today 213 (2013) 127–134.
[26] J. L. Rangel Costa, G. S. Marchetti, M. C. Rangel, A thorium-doped catalyst for the high temperature shift reaction, Catal. Today 77 (2002) 205–213.
[27] I. L. Junior, J.-M. M. Millet, M. Aouine, M. C. Rangel, The role of vanadium on the properties of iron based catalysts for the water gas shift reaction, Appl. Catal. A Gen. 283 (2005) 91–98.
[28] C. Martos, J. Dufour, A. Ruiz, Synthesis of Fe_3O_4-based catalysts for the high-temperature water gas shift reaction, Int. J. Hydrogen Energy 34 (2009) 4475–4481.
[29] A. L. C. Pereira, N. A. Santos, M. L. O. Ferreira, A. Albornoz, M. C. Rangel, Effect of cobalt on the activity of iron-based catalysts in water gas shift reaction, Stud. Surf. Sci. Catal. 167 (2007) 225–230.
[30] J. Y. Lee, D.-W. Lee, K.-Y. Lee, Y. Wang, Cr-free Fe-based metal oxide catalysts for high temperature water gas shift reaction of fuel processor using LPG, Catal. Today 146 (2009) 260–264.
[31] J. Y. Lee, D.-W. Lee, K.-Y. Lee, Y.-K. Hong, The CO removal performances of Cr-free Fe/Ni catalysts for high temperature WGSR under LNG reformate condition without additional steam, Int. J. Hydrogen Energy 36 (2011) 8173–8180.
[32] S. D. Correia, A. C. Oliveira, P. C. Morais, V. K. Garg, A. C. Oliveira, M. L. S. Correa, M. C. Rangel, Evaluation of Fe/MCM-41 catalysts in the water gas shift reaction, Stud. Surf. Sci. Catal. 154 (2004) 2417–2424.
[33] G. C. Araujo, M. C. Rangel, An environmental friendly catalyst for the high temperature shift reaction, Stud. Surf. Sci. Catal. 130 (2000) 1601–1606.
[34] Q. Li, W. Ma, R. He, Z. Mu, Reaction and characterization studies of an industrial Cr-free iron-based catalyst for high-temperature water gas shift reaction, Catal. Today 106 (2005) 52–56.
[35] L. Zhang, X. Wang, J.-M. M. Millet, P. H. Matter, U. S. Ozkan, Investigation of highly active Fe-Al-Cu catalysts for water-gas shift reaction, Appl. Catal. A Gen. 351 (2008) 1–8.
[36] P. Gawade, B. Mirkelamoglu, B. Tan, U. S. Ozkan, Cr-free Fe-based water-gas shift catalysts prepared through propylene oxide-assisted sol-gel technique, J. Mol. Catal. A Chem. 321 (2010) 61–70.
[37] V. Subramanian, D.-W. Jeong, W.-B. Han, W.-J. Jang, J.-O. Shim, H.-S. Roh, H_2 production from high temperature shift of the simulated waste derived synthesis gas over magnetite catalysts prepared by citric acid assisted direct synthesis method, Int. J. Hydrogen Energy 38 (2013) 8699–8703.
[38] D.-W. Jeong, V. Subramanian, J.-O. Shim, W.-J. Jang, Y.-C. Seo, H.-S. Roh, J. H. Gu, Y. T. Lim, High-temperature water gas shift reaction over Fe/Al/Cu oxide based catalysts using simulated waste-derived synthesis gas, Catal. Lett. 143 (2013) 438–444.

[39] F. Meshkani, M. Rezaei, Mesoporous Ba-promoted chromium free Fe_2O_3-Al_2O_3-NiO catalyst with low methanation activity for high temperature water gas shift reaction, Catal. Commun. 58 (2015) 26–29.

[40] X. Qi, M. F. Stephanopoulos, Activity and stability of Cu-CeO_2 catalysts in high-temperature water-gas shift for fuel-cell applications, Ind. Eng. Chem. Res. 43 (2004) 3055–3062.

[41] K. Zhi, Q. Liu, R. Zhao, R. He, L. Zhang, Preparation and characterization of Cu-Ce-La mixed oxide as water-gas shift catalyst for fuel cells application, J. Rare Earths 26 (2008) 538–543.

[42] P. Djinovic, J. Batista, J. Levec, A. Pintar, Comparison of water-gas shift reaction activity and long-term stability of nanostructured CuO-CeO_2 catalysts prepared by hard template and co-precipitation methods, Appl. Catal. A Gen. 364 (2009) 156–165.

[43] P. Kumar, R. Idem, A comparative study of copper-promoted water-gas-shift (WGS) catalysts, Energy Fuel 21 (2007) 522–529.

[44] V. M. Shinde, G. Madras, Water gas shift reaction over multi-component ceria catalysts, Appl. Catal. B Environ. 123–124 (2012) 367–378.

[45] J. Li, Z. Hongwei, R. Razzaq, C. Li, L. Zengxi, Effect of structural and lattice oxygen changes on the properties of CuO/CeO_2 catalysts for high-temperature water-gas shift of H_2-rich coal-derived synthesis gas, Z. Phys. Chem. 227 (2013) 371–387.

[46] S. S. Maluf, E. Y. Tanabe, P. A. P. Nascente, E. M. Assaf, Study of water-gas-shift reaction over $La_{(12y)}Sr_yNi_xCo_{(12x)}O_3$ perovskite as precursors, Top. Catal. 54 (2011) 210–218.

[47] Y. Sun, S. S. Hla, G. J. Duffy, A. J. Cousins, D. French, L. D. Morpeth, J. H. Edwards, D. G. Roberts, Effect of Ce on the structural features and catalytic properties of $La_{(0.9-x)}Ce_x$-FeO_3 perovskite-like catalysts for the high temperature water-gas shift reaction, Int. J. Hydrogen Energy 36 (2011) 79–86.

[48] L. D. Morpeth, Y. Sun, S. S. Hla, D. French, G. J. Duffy, J. H. Edwards, Effect of H_2S on the performance of $La_{0.7}Ce_{0.2}FeO_3$ perovskite catalyst for high temperature water-gas shift reaction, Int. J. Hydrogen Energy 37 (2012) 1475–1481.

[49] C. P. Jijil, A. K. Rajarajan, R. N. Devi, Noble metal ions incorporated in lattice points of perovskites – water gas shift activityof $BaCe_{1-x}Pt_xO_{3-d}$, AIP Conf. Proc. 1447 (2012) 1277–1278.

[50] R. W. Breault, Gasification processes old and new: a basic review of the major technologies, Energies 3 (2010) 216–240.

[51] M. H. Sahraei, D. McCalden, R. Hughes, L. A. R. Sandoval, A survey on current advanced IGCC power plant technologies, sensorsand control systems, Fuel 137 (2014) 245–259.

[52] M. Moliere, Pre-combustion capture: a powerful concept against long-range, trans-boundary pollution. in: EGTEI-EmTech50-500 Conference Rome, 2011.

Chapter 3

Low-Temperature WGS Reaction

3.1 LOW-TEMPERATURE WGS REACTION

The catalysts which operate between 150 and 250 °C are known as low-temperature water gas shift (WGS) catalysts. The development of highly efficient desulphurization technologies using Co(Ni)-MoO_3-Al_2O_3 catalysts for sulphur removal provided ammonia manufacturers with syngas mixture containing less than 1.0 ppm sulphur. This, in turn, enabled the use of the otherwise sulphur-sensitive Cu-ZnO catalysts at sufficiently low temperatures (190-200 °C) when the equilibrium carbon monoxide (CO) concentrations, at the exit of the LTS converters, can be below 0.3%. The invention of Cu catalysts for water gas shift reaction (WGSR) at lower temperatures made revolution in the WGS process industry. The operators began to appreciate the economics associated with ammonia synthesis with large production units, i.e., up to 1000 tonnes per day or more. Since copper catalysts reduce the CO concentration less than 0.5% after the second stage WGS reactor it was easy to incorporate a methanation stage in place of the very much complicated copper liquor scrubbing system that was used before. Low-temperature WGS catalysts were used virtually in all ammonia and refinery hydrogen plants. Ray *et al.* [1] reported the advantages of using low-temperature catalyst in hydrogen production to decrease the steam and equipment requirements. The first Cu-based catalysts are used in USA in 1963 [2]. This was a mixture of Cu-Zn with 1:2 ratios. Since their first usage several Cu formulations have been employed in the LT-WGS stage. Improvements in the catalyst activity, stability as well as resistance to poisoning and sintering were based on such developments. The copper metal crystallites disperse over zinc oxide when reduced in hydrogen. The method of preparation is extremely important concerning the activity of low-temperature WGS catalyst.

So many metals have been reported for the low-temperature WGSR. The two main categories are non-noble metal catalysts and noble metal catalysts. The following section explains the various metal catalysts investigated for the low-temperature WGSR.

Water Gas Shift Reaction. http://dx.doi.org/10.1016/B978-0-12-420154-5.00003-6

3.2 NON-NOBLE METAL CATALYSTS

The two main non-noble metal catalysts investigated for the low-temperature WGSR are Cu and Ni.

3.2.1 Cu-Based Catalysts

3.2.1.1 Cu-Zn-Al Catalysts

As explained above the first Cu-Zn catalysts were reported in 1963. Uchida *et al.* [3] reported several preparative methods for Cu-Zn catalysts and compared the catalyst activity and stability. In their test the addition of Cu to the Zn oxide rapidly increases the activity up to Cu/Zn 0.4. They also did X-ray analysis and found that after the WGSR the main components in Cu-Zn catalysts are Cu metal particles and ZnO. However, the main problem associated with this catalyst is the operating time is as low as 6 months. As like high-temperature WGS shift catalyst, chromia was doped into copper to stabilize the catalyst against sintering. The oxides of Zr, Cr and Cu by themselves were not good catalysts for WGSR. The mixtures of various compositions of Cu-Cr-Zn have been found to have excellent activity. Boreskov *et al.* [4] investigated the effect of additives Zn, Co, Mg on WGS activity of CuO-Cr_2O_3. They found that the addition of additives did not improve the activity; however, they improved the temperature stability of the Cu/Cr catalysts. They also found that WGS activity is not a function of the degree of reduction of the catalyst and concluded that metallic Cu was not the active site of the catalyst. They believe that active sites were Cu ions in the lattice of copper chromite. When Zn was present Cu-Zn-Cr solid was formed. Unfortunately, the Cu-Cr catalysts are very susceptible to thermal sintering via surface migration. The formulations containing Cr did not have significantly improved operating lives. Efforts have been made to enhance the activity and stability of Cu-Zn catalysts by addition of various promoters and supports. One of the major advances in achieving enhanced stability in the Cu catalysts was the introduction of promoters acting as structural spacers that decreases the aggregation of Cu crystallites during WGSR. One such advancement is the invention of alumina together with ZnO support. The Al_2O_3-ZnO support not only stabilizes the Cu crystallites against thermal sintering but also enhances the strength of the catalyst and minimizes shrinkage during reduction. The composition of Cu/ZnO-Al_2O_3 for better activity, stability and poison resistance depends on the crystallite size of the components, Cu dispersion and method of synthesis. For example, catalyst containing high copper levels may display very high initial activity that quite rapidly decays.

Kuijpers *et al.* [5] synthesized Cu/SiO_2 catalysts using deposition-precipitation method. They proposed that the reaction is structure sensitive and follows reduction-oxidation mechanism. Xue *et al.* [6] compared Cu-Zn catalysts with Pt/ZrO_2 and Fe-Cr catalysts. The catalytic activity results of those catalysts at various temperatures are presented in Figure 3.1. They carried out the reaction

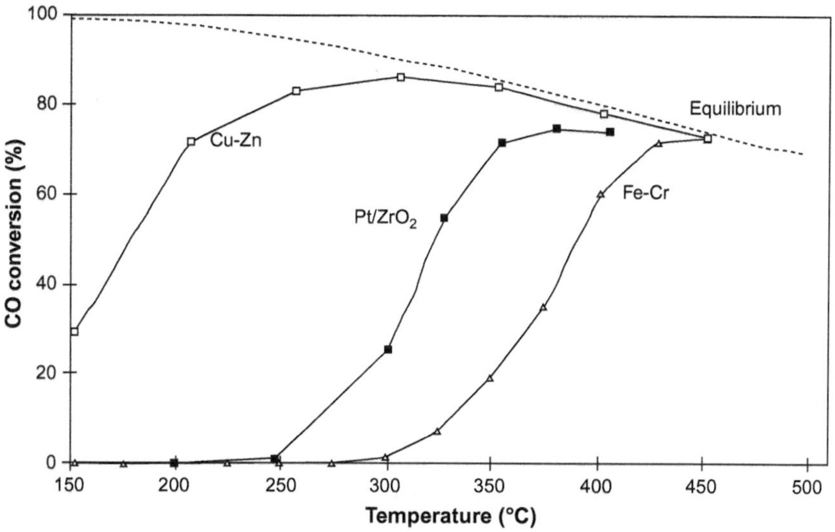

FIGURE 3.1 The conversions of CO as a function of temperature over various catalysts. *(Taken from Figure 5 of Catal. Today 30 (1996) 107-118.)*

at a lower steam to CO ratio of 1.33. The LT shift Cu-Zn catalyst appeared to be very active under the experimental conditions used, a CO conversion of ca. 70% being obtained at about 200 °C and equilibrium conversions were achieved at temperatures above 350 °C. Neither carbon nor CH_4 was observed during the reaction, indicating that the Cu-Zn catalyst is not only very active but also very selective for the WGSR when operated with a low H_2O/CO ratio in the feed.

Xue *et al.* also investigated [6] Cu-Zu catalysts in the presence of sulphur. After the copper catalyst had been stabilized at 200 and 240 °C for about 18 h, a concentration of 50 ppm H_2S was admitted to the feed. The catalyst deactivated completely and removal of H_2S from the feed later did not result in any noticeable return of the catalyst activity. This result indicates that the Cu-Zn catalyst is extremely susceptible to poisoning by sulphur.

Gines *et al.* [7] reported the detailed study of the effect of synthesis method on structural properties and WGS activity of Cu-Zn-Al_2O_3 catalysts. The catalysts were prepared by co-precipitation (CP) method by using nitrates as precursors and sodium carbonate as a precipitating agent at 60 °C and constant pH around 7. The precipitate was filtered and dried at 100 °C and finally calcined between 400 and 700 °C. They found that the turnover frequency was remained constant not only when the copper metallic surface area was varied between 3 and 35 m^2/g Cu, but also for the Cu loading of 30-50 wt%, Al/Zn atomic ratios of 0 and 2.5, Cu dispersion between 0.5% and 5%, calcination temperatures between 400 and 700 °C. These results clearly suggest that WGSR is structure insensitive over Cu/ZnO-Al_2O_3 catalysts. On the contrary, Chinchen and Spencer [8] reported that the turnover frequency varies by an order of magnitude when Cu metallic surface area was changed from 10 to 40 m^2/g for Cu/ZnO-Al_2O_3 system. Even though, in general, the WGS activity of Cu/ZnO-Al_2O_3 increases with increasing the Cu metallic surface area, there is no literature report available on the fact that the rate of reaction correlates linearly with the metallic Cu area over the entire Cu-Zn composition range. Along with high Cu surface area, the bulk structural changes in the ZnO and Cu crystallites during the synthesis and pre-reduction such as micro-strain, oxygen vacancies in the ZnO cannot be ignored.

Mellor *et al.* [9] reported caustic leached Raney copper catalysts with rim compositions of Cu-ZnO-$A1_2O_3$ for low-temperature WGSR. Although initial WGS activities were comparable to the industrial and co-precipitated catalyst, deactivation as a result of copper crystallite sintering took place over long operating times. Catalyst instability resulted from a rim zinc oxide deficiency. Presumably, even greater amounts of zinc for long-term catalyst stability could be provided by the impregnation process. However, the addition of concentrated zincate to the caustic solution also slowed the leach rate, promoting an increase in copper crystallite size with a corresponding decrease in copper surface area and WGS activity. An examination of the relationship between WGS activity and Raney copper surface area showed that catalytic activity was to a certain extent dependent on the magnitude of the copper area.

Andrade *et al.* [10] reported composition effect on the WGS activity of Cu-Zn-Al_2O_3 catalysts. They proposed that the interaction between the catalyst components in the calcined samples led to a synergy between the Cu-rich and the Al-rich Zn-containing compositions and for Zn-rich samples, ZnO may play the role of a copper dispersion agent. There was found a defined correlation between the catalytic activity and the amount of reduced 'free' CuO in the sample. They also proposed that a high Cu, Zn and Al surface interdispersion is necessary in order to attain a high catalytic activity. Li *et al.* [11] investigated the effect of Al content on the WGS activity of Cu/Zn/Al catalysts. 12% Al composition catalyst shows better activity than other catalysts.

Budiman *et al.* [12] synthesized Cu/Zn/Al catalysts using modified CP method. They investigated the effect of pH, temperature and solvent. They found that the WGS activity depends on the metal complexes produced in the precipitated precursor states, which can be adjusted by the conventional CP conditions.

Meland *et al.* [13] synthesized Cu- and Pt-based catalyst by using flame spray pyrolysis method. Among various catalysts, Cu/Zn/Al and Cu/Zn/Ce/Al catalysts are stable and have high activity for 10 h at 310 °C. Figueiredo *et al.* [14] investigated the effect of alkali metal ion doping on the WGS activity of Cu/Zn/Al catalysts. The order of activity is Cu/Zn/Al/K > Cu/Zn/Al > Cu/Zn/Al/Rb > Cu/Zn/Al/Cs. Knapp *et al.* [15] synthesized

1-butyl-2,3-dimethyl-imidazolium trifluoromethane sulphonate ionic liquid-coated Cu/Al catalysts. Catalysts coated with ionic liquid showed a higher turnover frequency for the WGSR at low temperatures compared to uncoated catalysts and to the best commercial systems.

3.2.1.2 Other Cu Catalysts

Tanaka *et al.* [16] synthesized $CuAl_2O_4$ and $CuMn_2O_4$ catalysts and showed that those catalysts exhibit better activity than $Cu/ZnO/Al_2O_3$. Both $CuAl_2O_4$ and $CuMn_2O_4$ exhibited much higher conversion rate than $Cu/ZnO/Al_2O_3$ under the oxygen-containing conditions. In particular, $CuMn_2O_4$ showed an extraordinarily high rate of CO conversion at 200 °C.

Zhi *et al.* [17] investigated the effect of precipitate on the catalytic performance of Cu-Mn catalysts. They prepared Cu-Mn catalysts from copper sulphate and manganese sulphate using NaOH and KOH as precipitates. The sample prepared using NaOH as a precipitator has a stable texture and is well-dispersed in terms of active components, whereas the sample prepared using KOH as a precipitator has an unstable texture and poor dispersion. Therefore, the sample prepared using NaOH as a precipitator has superior catalytic activity and stability than the sample prepared using KOH as a precipitator because of rapid deactivation.

Tabakova *et al.* [18] investigated the effect of synthesis method on the WGS activity of Cu-Mn spinels. They synthesized the catalysts using single step urea combustion method and CP method. The catalysts synthesized by the urea combustion method show better activity compared to co-precipitated catalysts and commercial catalyst in idealized reformate and realistic reformate conditions.

Wang *et al.* [19] investigated the effect of third metal on the WGS activity of Cu-Mn spinels. They synthesized Cu-Mn-M catalysts (M = Ce, Zr, Zn, Fe and Al) using urea-nitrate combustion method. The activity order is Zr > Fe > non-doped > Ce > Al > Zn. In addition, Zr-doped Cu-Mn catalyst with 5 wt% content showed the best catalytic performance and optimal stability exposed to oxygen-stream and on-stream operation.

He *et al.* [20] investigated the influence of La doping on the WGS activity of Cu-Mn catalysts. 0.5% doped catalyst shows better activity. They proposed that the doped La greatly enhanced the synergistic effect of copper-manganese oxide catalysts.

He *et al.* [20] synthesized several $CuZrO_2$ catalysts and proved that compared to a Cu-ZnO-Al_2O_3 catalyst, the Cu-ZrO_2 catalysts showed an excellent performance in the reaction, and the activity of the catalyst increased with the Cu loading. At low Cu loadings below 40 at.%, Cu was present in the form of highly dispersed oxide clusters, while at high loadings, the clusters as well as bulk CuO were present in the catalysts. Based on their TOF results and literature reports they proposed the mechanism. In the mechanism, CO reduces Cu^{2+} to Cu^{+} and produces CO_2, while H_2O oxidizes Cu^{+} to Cu^{2+} and generates H_2.

Aguila *et al.* [21] synthesized Cu-ZrO_2 catalysts by reflux method and compared with impregnation method. The reflux method-synthesized catalysts generate very high active catalysts only for 3 wt% loaded catalysts, whereas the same catalysts synthesized by impregnation method are inactive. They proposed that the activity of the catalysts prepared by refluxing is associated with the formation of small Cu clusters, which would allow the regrouping of the H atoms to generate molecular H_2 in the presence of the crystalline tetragonal ZrO_2.

Sagata *et al.* [22] investigated Cu-Fe/Al_2O_3 catalysts. The WGS activity of various Fe-doped catalysts is presented in Figure 3.2.

The addition of FeOx to Cu/Al catalysts provided the negative effect on the original activity of Cu/Al catalyst below 473 K, whereas it provided the positive effect above 523 K. In particular, Cu Fe/Al (1.0) catalyst showed the highest activity at 573 K. They found that partially reduced Fe_2O_3, probably Fe_3O_4, acts as an active site for WGSR in the addition to CuO, resulting in the improvement of the catalytic activity at higher temperature.

Hossain *et al.* [23] reported Cu-Fe-Cr and Cu-Fe-Mn catalysts and they compared with commercial catalysts. Cu-Fe-Mn catalyst showed significantly higher CO conversion than the commercial Cu-ZnO/Al_2O_3 catalyst, while the activity of Cu-Fe-Cr catalyst is lower than that of the commercial catalyst. Thouchprasitchai *et al.* [24] reported Cu-Fe, Cu-Zn and Cu-Fe-Zn catalysts. The ternary catalyst shows better activity compared to binary catalysts. Takehira *et al.* [25] investigated alkali metal-doped Cu/Zn catalysts for low-temperature WGSR. The highest activity is obtained for Cu/MgO/ZnO catalysts. The Cu/MgO/ZnO catalyst

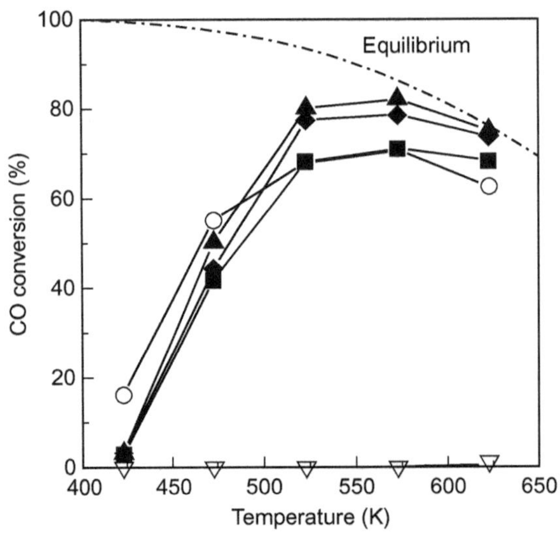

FIGURE 3.2 Catalytic activities of Cu-Fe/Al catalysts with different Fe/Cu ratios. Fe/Cu = (○) 0, (■) 0.5, (▲) 1.0, (◆) 1.5 and (▽) ∞. *(Taken from Figure 8 of Bull. Chem. Soc. Jpn. 85 (2012) 511-516.)*

showed no significant decrease in the activity as well as no detectable sintering in the Cu metal particles during 50 h of the reaction. The formation of Cu^+ was observed on the Cu/MgO/ZnO catalyst where the addition of MgO clearly enhanced the Cu^+ formation.

Yeragi *et al.* [26] synthesized several Mn-doped Cu/Al catalysts. Among various catalysts 8.55% Mn showed higher activity. Yahiro *et al.* [27] investigated the effect of support on the WGS activity of Cu/Zn catalysts. They investigated Al_2O_3, MgO, SiO_2-Al_2O_3, SiO_2-MgO, β-zeolite and CeO_2 supports. Among the various supports, alumina proved to be the best support. Scott *et al.* [28] investigated La doping for the Cu-Zn catalysts. They proved that 2.3% La-doped catalysts exhibit better activity than Cu/Zn/Al catalysts.

3.2.1.3 Cu/CeO_2

In 2000 Li and Stephanopolous [29] reported Cu/CeO_2 catalysts for WGSR. They prepared 2-8 wt% of catalysts using urea CP-gelation method. La was used as a structural stabilizer for ceria. Better than 90% CO conversion to CO_2 was achieved at 250 °C with a gas hourly space velocity (GHSV) = 8000 h^{-1}. At higher temperatures, the conversion of CO to carbon dioxide eventually approaches the equilibrium. Figure 3.3 shows the WGS light-off curves

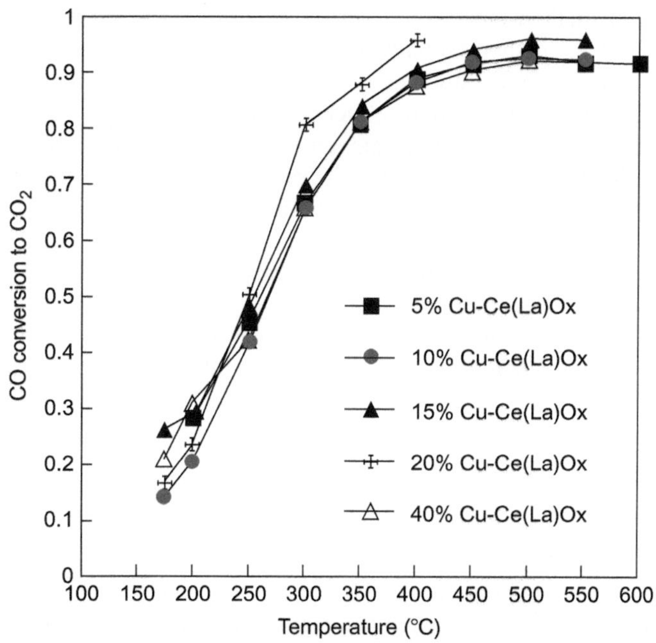

FIGURE 3.3 Effect of copper content on the WGS activity of Cu-Ce(10% La) O_x, 2% CO-10.7% H_2O-He, S.V. = 80,000 h^{-1} (NTP). *(Taken from Figure 5 of Appl. Catal. B Environ. 27 (2000) 179-191.)*

over Cu-Ce(La)Ox catalysts with different copper contents from 5 to 40 at.%. Clearly, there is no appreciable difference in the light-off temperature of the reaction over these ceria-based catalysts, indicating that only a small amount of copper is adequate to change the WGS activity of ceria. Addition of 40% of hydrogen did not significantly suppress the reaction. Above 350 °C, however, we see a decrease in the CO conversion due to the lower equilibrium value at this high concentration of hydrogen.

Assaf and da Silva [30] compared Cu-Ce catalysts with the Cu-Ce-Zr catalysts. The addition of ZrO_2 to the CeO_2 lattice enhances the thermal stability inhibiting the crystal growth of ceria and thus enhances the activity. Then, they synthesized Cu and Cu-Zn supported on CeO_2 and CeO_2-La_2O_3. Interestingly, the Cu/CeO_2 catalysts exhibited better activity than Cu-Zn/CeO_2, Cu-Zn/CeO_2-La_2O_3 and Ce/CeO_2-La_2O_3. They said that the presence of Zn inhibits the interaction between Cu and Ce ions, and lanthanum oxide forms a solid solution with cerium oxide, which will cause a decrease in the surface area of the catalysts.

Zhi *et al.* [31] reported the effect of the precipitator concentration on the activity of mesoporous Cu-Ce-La mixed oxide catalyst. They concluded that the precipitator concentration influences the activity of catalyst via the stability of crystal structure and mesoporous structure.

Park *et al.* [32] synthesized Cu-Ce-Al catalysts by using one plot templating method. The Cu catalysts synthesized by this method exhibit better activity than impregnation method catalysts. Pintar *et al.* [33] compared hard templating method with CPs method on Cu/CeO_2 catalysts. The catalysts synthesized by hard templating method show better activity than that by CP method.

Lin *et al.* [34] investigated the effect of Nb on the WGS activity of Cu/CeO_2 catalysts. It is found that the introduction of Nb^{5+} will result in the substitution of Ce^{4+} with Nb^{5+}, thus creating mobile electronic carriers in the as-prepared catalysts. However, excessive Nb_2O_5 addition prevents the incorporation of Cu^{2+} into CeO_2 lattice and partially covers the surface of CuO and CeO_2, resulting in weakening their reducibility and interaction between them, thus leading to inferior catalytic performance.

Macial *et al.* [35] reported the performance of CuO/CeO_2-TiO_2 catalysts. They showed that CuO/CeO_2-TiO_2 catalysts exhibit better activity compared to Cu-TiO_2 and Cu-CeO_2. Poggio-Fraccari *et al.* [36] reported CuO/CeO_2-Pr_2O_3 catalysts. They varied Pr amount from 0 to 50 wt%. They found 5 wt % Pr is the optimum amount.

3.2.1.4 *Bimetallic Cu/CeO_2 Catalysts*

Shinde and Madras [37] synthesized Ni- and Fe-doped Cu/CeO_2 catalysts. They synthesized catalysts by sonochemical method and impregnation methods. The WGS activity results of various catalysts are presented in Figure 3.4. Among the various catalysts Cu-Ni/CeO_2 catalyst synthesized by sonochemical method shows the best activity.

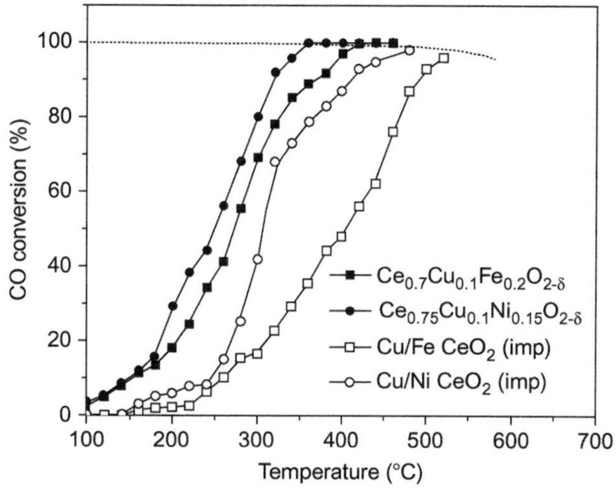

FIGURE 3.4 WGS activity of various ceria-based compounds with temperature. *(Taken from Figure 7 of Appl. Catal. B Environ. 123-124 (2012) 367-378.)*

Jiang *et al.* [38] investigated the effect of other metals (Y, La, Fe, Co and Ni) on the WGS activity of Cu-Ce-Zr catalysts. The WGS activity results of various metal-doped Cu-Ce-Zr catalysts are presented in Figure 3.5. Among the various dopants Fe shows higher activity compared to other dopants. Then they investigated the effect of Fe content. The CO conversion increased with increasing Fe content up to 3 wt% and further increase in Fe content decreases the WGS activity.

3.2.2 Ni Catalysts

In 1995 Andreev *et al.* [39] reported Ni hydroxide catalysts supported on activated char coal for the WGSR for the first time. The catalyst shows better catalytic activity than Cu catalysts. They also proposed the WGSR mechanism over Ni catalysts. The mechanism is presented in Schemes 3.1 and 3.2.

After that no research has been made with the Ni catalysts. In 2009 Haryanto *et al.* [40] investigated the influence of support on the WGS activity of Ni catalysts. The catalytic activity results are presented in Figure 3.6. Among the various catalysts $Ni/CeO_2\text{-}Al_2O_3$, Ni/CeO_2 and $Ni/CeYO_5$ show higher activities. The catalysts were stable at high temperature (450 °C) but unstable at low temperature (250 °C).

Rodriguez *et al.* [41] studied the interaction of Ni with ceria during the WGSR. There is a continuous decrease in the WGS activity, while the rate of formation of methane increases. This trend can be attributed to a decrease in the WGS catalytic activity when the Ni particle size becomes very large. Also, the Ni catalysts show the higher catalytic activity than Cu catalysts at lower metal loadings.

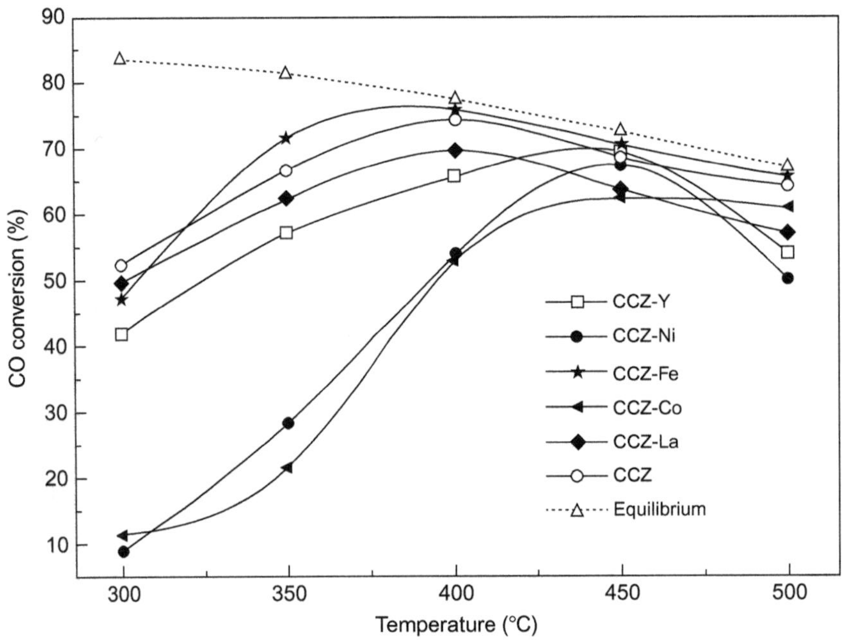

FIGURE 3.5 Catalytic activities of the CCZ-M catalysts with different additives. *(Taken from Figure 9 of Chem. Eng. Technol. 36 (2013) 1891-1898.)*

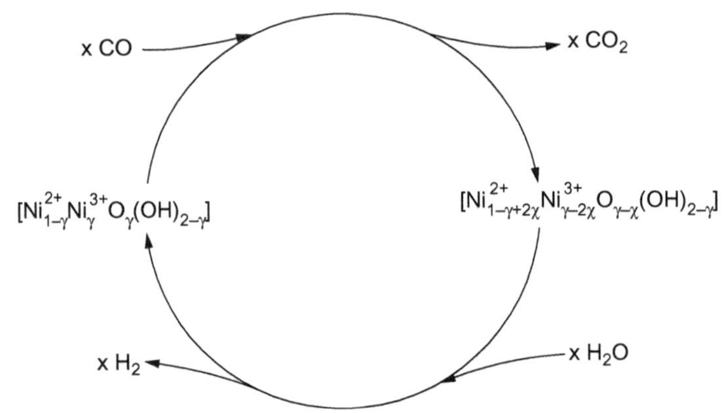

SCHEME 3.1 The mechanism of the WGS reaction over Ni catalysts ***(Taken from Catal. Lett. 31 (1995) 245-252.)***

Shinde and Madras [42] synthesized Pd-modified Ni/CeO_2 catalysts by solution combustion synthesis for the WGSR. The catalyst is highly active and selective towards H_2 production and exhibits a lack of CO methanation activity. Fuentes *et al.* report Ni-Al and Ni-Al-Zn catalysts for the WGSR. Interestingly, Ni-Al catalyst exhibits better activity than Ni-Al-Zn catalysts.

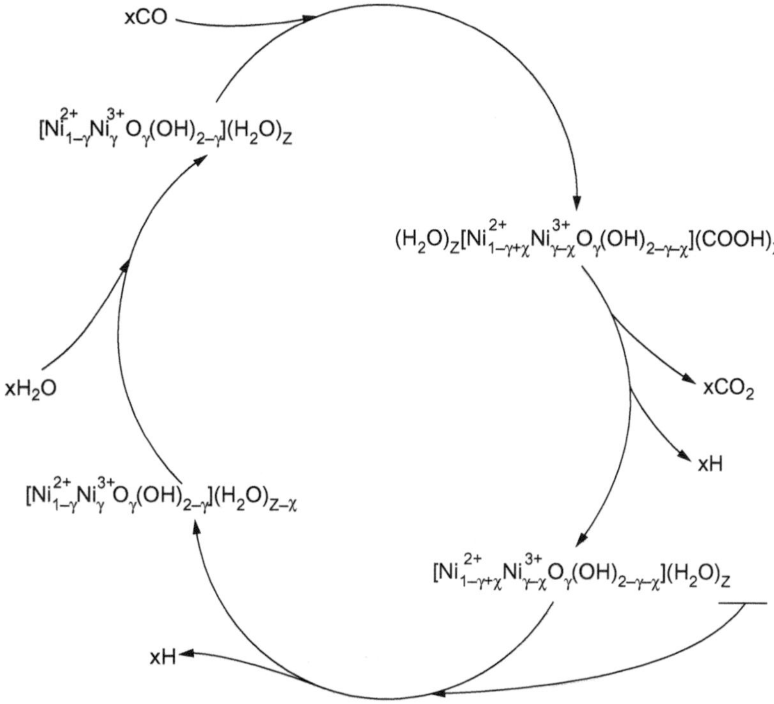

SCHEME 3.2 The mechanism of the WGS reaction over Ni catalysts via carboxylate intermediate formation ***(Taken from Catal. Lett. 31 (1995) 245-252.)***

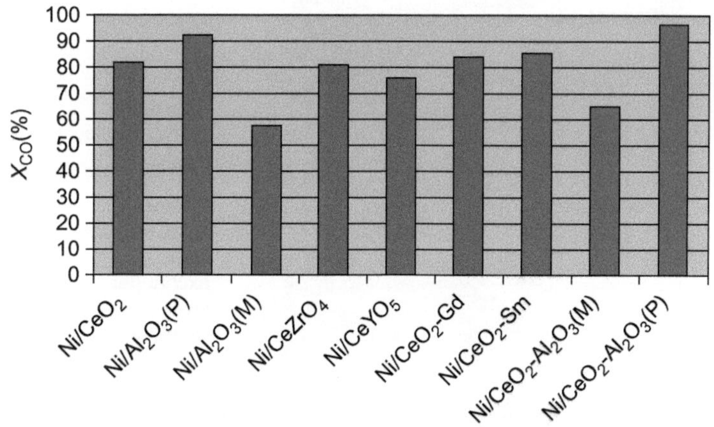

FIGURE 3.6 Initial CO conversion with different nickel catalysts at 250 °C (conditions: catalyst loading, 1 g; CO/steam molar ratio, 1:3; and GHSV, 7500 cm^{-1} h^{-1} g_{cat}^{-1}). *(Taken from Figure 3 of Energy Fuels 23 (2009) 3097-3102.)*

Zhao *et al.* reported Ni-W-Ce catalysts for the WGSR. Ni-W-Ce catalysts exhibited better activity than Ni-Ce catalysts.

Park *et al.* [43] reported a microstructured reactor with catalytic nickel plates, which is newly designed and developed for proper heat management in an exothermic WGSR.

The reactor is designed to increase the reactor capacity simply by numbering-up a set of catalyst layers and heat exchanger layers. The schematic representation of WGS reactor is presented in Figure 3.7. The catalyst exhibits higher CO conversion 96.1% at higher temperatures.

3.3 NOBEL METAL-BASED CATALYSTS

The thermodynamics of the WGSR are well known that at high temperatures the CO conversion is equilibrium limited and at low temperatures the reaction is kinetically limited. The Cu-based catalysts are not suitable active catalysts for single-stage WGS applications. For this reason industries first operate at higher temperatures to have 95-97% CO conversion, and then the remaining

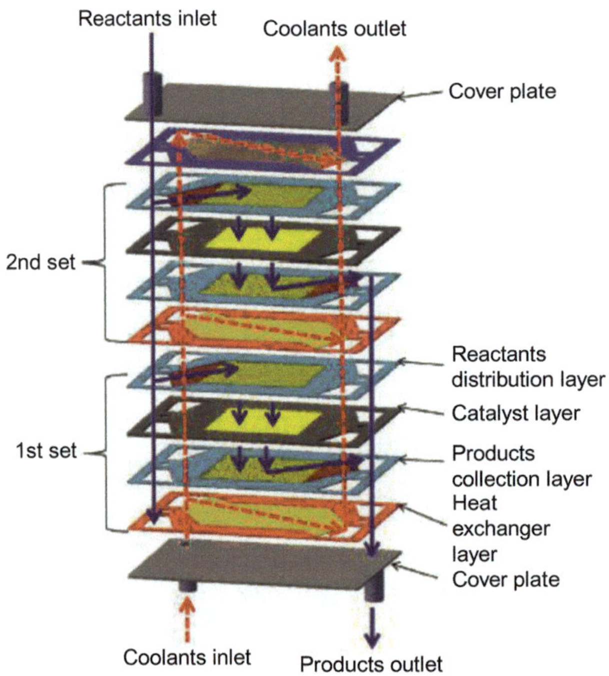

FIGURE 3.7 The schematic representation of the newly designed single-stage temperature-controllable reactor and flow paths. Blue solid line: reactant and product flow path. Red solid line: coolant flow path. *(Taken from Figure 1 of J. Power Sources 247 (2014) 280-285.)*

CO was converted by Cu-based catalysts at low temperature. Hence, much research effort has been focused on developing noble metal-based catalysts at lower temperatures. Noble metal catalysts are very active catalysts for several applications especially automotive exhaust applications. Nobel metals supported on partially reduced metal oxide supports are the active catalysts for the WGSR. Of the many catalysts that have been studied, precious catalysts (Pt, Rh, Ru, Au and Pd) deposited on partially reducible oxides (CeO_2, ZrO_2, TiO_2, Fe_2O_3 and mixed oxides) have been the most investigated. These catalysts are quite active in 250-400 °C. In 1925 Prichard and Hinshelwood studied the adsorption behaviour of CO, H_2O, CO_2 and H_2 on Pt surface [44]. Radhakrishnan *et al.* [45] investigated the WGS activity of noble metals supported on CeO_2-ZrO_2. They conducted the WGSR both at high-temperature shift and low-temperature shift. The feed composition for the high-temperature shift was 4.9% CO, 10.5% CO_2, 33% H_2O and 30.3% H_2 and the low-temperature gas composition was 1.5% CO, 5% CO_2, 45% H_2O and 25% H_2 with the balance of the steam being N_2 in both cases. The activity trend under both feed gases over a broad range of temperatures (200-320 °C) was $Pt > Rh > Ru \sim Pd > Ir > Au$. However, the activity trend for Al_2O_3-SiO_2 supported catalysts observed by Greenoble *et al.* [46] followed the trend $Ru > Pt > Au > Rh \sim Pd$. Gorte and Zhao [47] found that Pt, Pd and Rh supported on ceria to be equally effective for the WGS activity at 1:1 H_2O/CO ratio. The discrepancy between different literature data is due to the differences in the feed compositions and catalyst composition. The work by Stephanopolous and co-workers [48] has determined that non-metallic gold and platinum species strongly associated with cerium-oxygen groups are responsible for the WGS activity in ceria-supported gold and platinum catalysts. Luengnaruemitchai *et al.* [49] investigated the Pt/CeO_2, Au/CeO_2 and Au/Fe_2O_3 catalysts for low-temperature WGSR and found that crystallinity of CeO_2 has negative effect on the activity of the catalyst. They also reported that Pt is the much more active metal than Au. Panagiotopoulou *et al.* [50] also investigated various noble metal-supported TiO_2 catalysts and found that Pt/TiO_2 is the most active one and the order of activity is $Pt > Rh > Ru > Pd$. The catalytic activity depends strongly on the structure and morphology of the titania. The rate per surface Pt atom increases by more than two orders of magnitude with decreasing crystallite size of titania from 16 to 35 nm.

Thinon *et al.* [51] have recently screened about 20 metal-on-oxide catalysts for the WGSR under identical conditions using a model reformate as the reaction mixture. The catalysts based on Pt, Au, Cu, Rh, Pd and Ru supported on ceria, alumina, zirconia, iron oxide and titania were evaluated. The catalysts were synthesized by impregnation of supports with a solution of the corresponding metal precursor. Among the various catalysts Pt/TiO_2 shows the highest activity. The activity results are presented in Figure. 3.8.

Among the various noble metals Pt catalysts are active at medium temperatures 250-350 °C and Au catalysts are active at lower temperatures 150-250 °C.

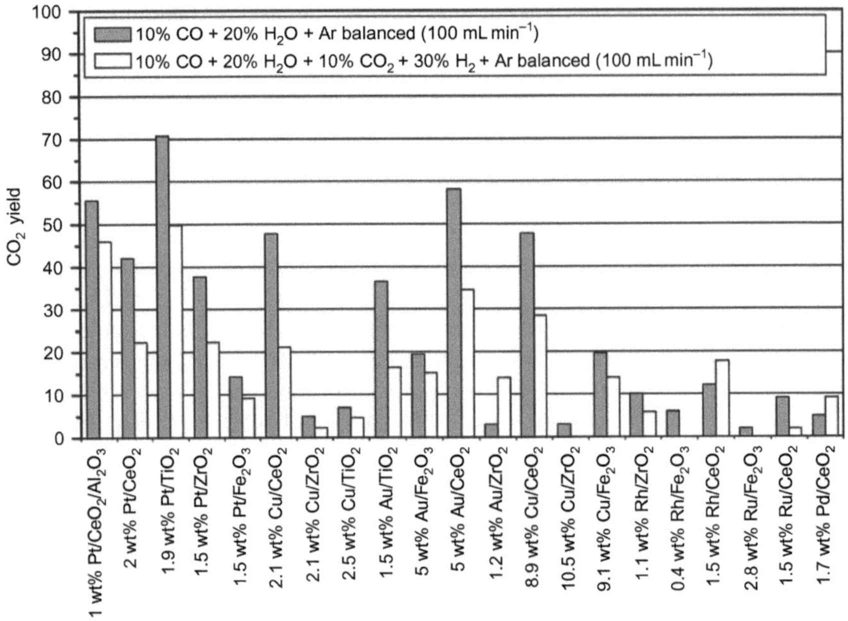

FIGURE 3.8 CO_2 yield at 300 °C with a reformate mixture (100 N mL/min of 10% CO, 10% CO_2, 20% H_2O, 30% H_2 and 30% Ar) and a pure reactant mixture (100 N mL/min of 10% CO, 20% H_2O and 70% Ar) 100 mg of sample. *(Taken from Catal. Today 137 (2008) 29-35.)*

3.3.1 Pt-Based Catalysts

In 1974 Japanese researcher Masuda investigated the kinetics of the WGSR over Pt surface [52]. In 1985 Mendelovici and Steinberg [53] reported revolutionary Pt/CeO_2 catalysts for methanation and WGSR. Even though the high WGS activity of Pt group metals was known for several years, much research has been done in early 1990s.

3.3.1.1 Pt/ZrO_2

In 1996 Xue *et al.* [6] investigated the effect of H_2S and low steam to CO ratios on Cu-Zn, Pt/ZrO_2 and Fe-Cr commercial catalysts. The Cu-Zn catalyst retained its activity when operated at low steam to CO ratios. However, it lost its activity completely when sulphur was present in the feed. When operated at a low steam to CO ratios the commercial Fe-Cr catalyst becomes active only at relatively higher temperatures. The presence of sulphur deactivated the catalyst. The Pt/ZrO_2 catalyst showed higher activity than that of Fe-Cr catalyst. In the presence of sulphur the Pt/ZrO_2 catalyst deactivated, however, the activity remaining higher than any of the other catalysts tested under the same conditions.

Xue *et al.* [54] then modified Pt/ZrO_2 catalysts by $C1^-$ and K^+ ions. Modification of Pt/ZrO_2 with $C1^-$ leads to a reduction of S-uptake by the catalyst and improved its WGS activity in the presence of H_2S; the addition of K^+ to Pt/ZrO_2 catalyst has the opposite effect.

Recently, Park *et al.* [55] investigated the effect of Ti addition on the WGS activity of Pt/ZrO_2 catalysts. Ti impregnation into ZrO_2 increased the surface area of the support and the Pt dispersion. The reducibility of the catalyst was enhanced in the controlled Ti impregnation (20 wt%) over Pt/ZrO_2 by the Pt-catalyzed reduction of supports, particularly, at the interface between ZrO_2 and TiO_2. Ti impregnation over the ZrO_2 support led to a remarkably enhanced CO conversion and the reaction rate of $Pt/Ti/ZrO_2$ increased by a factor of about 3 from the bare Pt/ZrO_2 catalyst.

Xie *et al.* [56] synthesized Na-stabilized non-porous ZrO_2 supports for the Pt catalysts. Na did not promote the WGS activity of monoclinic zirconia-supported catalysts. However, the WGSR rate of Na-promoted tetragonal zirconia catalysts increased by a factor of five times compared to that of pure zirconia and Na-promoted monoclinic zirconia-supported catalysts.

3.3.1.2 *Pt/CeO$_2$*

As explained above Mendelovici and Steinberg [53] reported revolutionary Pt/CeO_2 catalysts for the first time for WGSR. In 2003 Stephanopolous and co-workers [48] reported non-metallic platinum species for the WGSR. Strongly associated cerium-oxygen groups are responsible for the WGS activity in ceria-supported gold and platinum catalysts.

In 2004 Bera *et al.* [57] reported solution combustion-synthesized Pt/CeO_2 catalysts for the WGSR. They prepared the 1% Pt and 2% Pt catalysts using oxalyldihydrazide as fuel. The WGS activity results of Pt catalysts as a function of reaction temperature are presented in Figure 3.9. They conducted WGSR using CO-containing H_2 rich steam reformate exiting the C_2H_5OH reformer. From the figure it is clear that up to 90% of CO is converted at 200 °C over 2% Pt/CeO_2. No methane formation was observed up to 200 °C. But at higher temperature (>225 °C) CH_4 formation was observed due to reaction among CO, CO_2 and H_2. Similarly, reverse WGSR could not be ruled out at higher temperatures. The catalytic activity on combustion-synthesized Pt/CeO_2 catalysts employed in this study is higher than that on supported gold catalysts such as Au/Fe_2O_3, Au/ZrO_2 and Au/ZnO reported by other groups.

Ivanov *et al.* [58] synthesized Pt/CeO_2 catalysts using unique extractive pyrolytic method. They observed that the catalyst sample with the highest platinum loading shows the lowest catalytic activity, while the one with the lowest content of platinum exhibits the highest activity. Interestingly, they observed similar higher activity for the catalysts treated in both hydrogen and air.

Davis *et al.* [59] compared Pt/ceria catalysts with Cu-, Au-, Ni-, Co- and Fe-supported ceria catalysts. Among various catalysts, Pt-supported catalysts

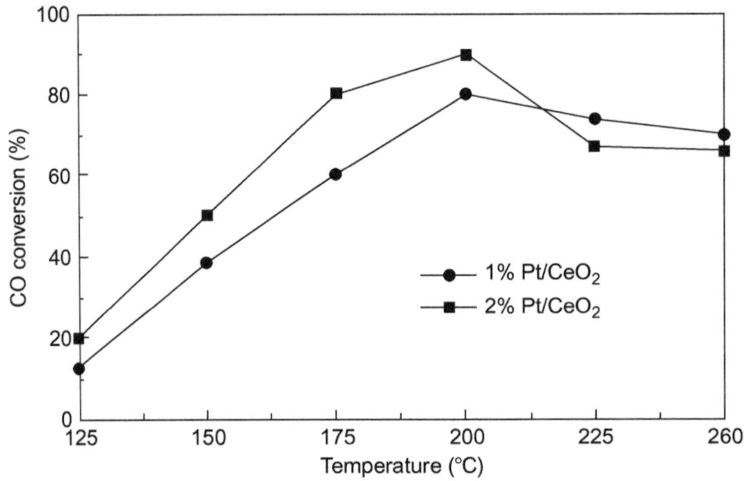

FIGURE 3.9 %CO conversion in WGS reaction over 1% and 2% Pt/CeO_2 as a function of temperature at a GHSV of 2600 h^{-1}. *(Figure 5 of Catal. Lett. 96 (2004) 213-219.)*

exhibit much higher activity compared to all other catalysts in the temperature region 240-360 °C. Then, they compared [60] Pt/ceria catalysts with the Pt/thoria catalysts. Interestingly, Pt/thoria exhibits better catalytic activity compared to Pt/ceria catalysts.

Davis *et al.* then [60] conducted mechanistic studies using X-ray absorption near edge structure and *in situ* diffuse reflectance Fourier transform IR (DRIFT) studies and they made the following conclusions. Increase in Pt loading has positive effect on the catalytic activity. The partial reduction of ceria is necessary for the formation of bridge OH groups which are active sites for the WGSR. During the surface shell reduction process in hydrogen, surface carbonates are liberated, hydrogen is dissociated to produce two bridging OH groups, and two cerium atoms for each dissociated hydrogen molecule change from the Ce^{4+} to the Ce^{3+} oxidation state. Two routes of bridging OH group formation are as follows: (1) once it is reduced, Pt^0 dissociates H_2, which spills over to the ceria surface to generate the bridging OH group active sites directly, accompanied by a change in the oxidation state of the Ce atoms involved with the sites from Ce^{4+} to Ce^{3+} and (2) H_2 or CO removes ceria surface capping oxygen atoms to generate vacancies (and surface Ce^{3+}), followed by H_2O dissociation at the vacancies to generate the bridging OH groups. The results indicate not only that Pt serves to facilitate the generation of the bridging OH groups (active sites) at low temperature, but that it may also be involved in accelerating the decomposition of surface formates, the proposed rate-limiting step of the reaction mechanism.

Davis *et al.* also [61] investigated the influence of zirconium doping into the cerium lattice. The addition of zirconium increases the catalytic activity.

Characterization measurements suggest that zirconium decreases the surface shell reduction temperature of ceria and promotes the WGS activity.

Trovarelli [62] group synthesized ex nitrate and ex chloride $Pt/Ce_{0.56}Zr_{0.44}O_2$ catalysts using solution combustion synthesis method. They reported that the high dispersion of platinum coupled with the high crystallinity of the combustion-synthesized catalyst makes it a potential candidate in parallel with the conventional impregnated catalysts for the WGSR at high space velocities.

Jeong *et al.* [63] also synthesized various Ce-Zr mixed oxides as supports for Pt. They prepared supports using CP and impregnated Pt. Interestingly, Pt/CeO_2 exhibits higher activity compared to Pt/Ce-Zr catalysts. Within the reaction temperature range from 240 to 360 °C, the CO conversion decreased in the order: $Pt/CeO_2 > Pt/Ce_{0.8}Zr_{0.2}O_2 > Pt/Ce_{0.6}Zr_{0.4}O_2 > Pt/Ce_{0.4}Zr_{0.6}O_2 > Pt/Ce_{0.2}Zr_{0.8}O_2 > Pt/ZrO_2$.

Panagiotopoulou *et al.* [50] synthesized Pt-promoted metal-doped ceria catalysts by urea-nitrate combustion method and investigated for the low-temperature WGSR. It has been found that the WGS activity of Pt/Ce-Me-O catalysts depends on the nature of the dopant employed, varying in the order of Yb, Gd, Zr, Mg and La shows the promotional effect and Ca, Y and Zn shows the inhibitory effect. Among the various dopants Yb-doped CeO_2 catalyst shows higher activity and the order of activity is Yb > Gd > Zr > Mg > La > CeO_2 (undoped) > Ca > Y > Zn.

Schuurman *et al.* [64] reported Pt/ceria/alumina catalysts for the WGSR. They reported that the catalysts are highly active compared to Pt/ceria. Petallidou *et al.* [65] investigated the effect of Ce/Ti ratio on the WGS activity of Pt/Ce_{x}-$Ti_{1-x}O_2$ catalysts. The effect of support chemical composition on the WGS activity of Pt/Ce-Ti catalysts as a function of reaction temperature is presented in Figure 3.10. The $Ce_{1-x}Ti_xO_{2-\delta}$ ($x=0.2$, 0.5 and 0.8)-supported Pt catalysts present higher activity in the 200-275 °C range compared to Pt supported on the single metal oxides of CeO_2 and TiO_2. 0.5 wt% $Pt/Ce_{0.8}Ti_{0.2}O_{2-\delta}$ solid presents the highest CO conversion in the entire temperature range of 200-350 °C.

Then they [66] investigated the effect of Ce/La ratio on the WGS activity of $Pt/Ce_{1-x}La_xO_{2-\delta}$. It was found that doping of ceria with 20 wt% La^{3+} has increased significantly the catalytic activity of 0.5 wt% $Pt/Ce_{0.8}La_{0.2}O_{2-\delta}$ solid in the 250-350 °C range, whereas addition of 50-80 at.% La^{3+} in ceria caused a negative effect on the CO conversion with respect to pure ceria.

Farias *et al.* [67] reported Pt/CeO_2-MgO catalysts. Addition of MgO to Pt/CeO_2 increased the activity and stability of the catalyst. It seems to favour the formate decomposition and lower the carbonate concentration on the catalyst surface during WGSR. The presence of magnesium also improved ceria reduction favouring the creation of OH groups, which are considered to be the active sites for the WGSR.

Hwang *et al.* [68] reported silica-modified Pt/CeO_2 catalysts for the WGSR. They evaluated catalysts at 270 °C in a reformer outlet composition. The order

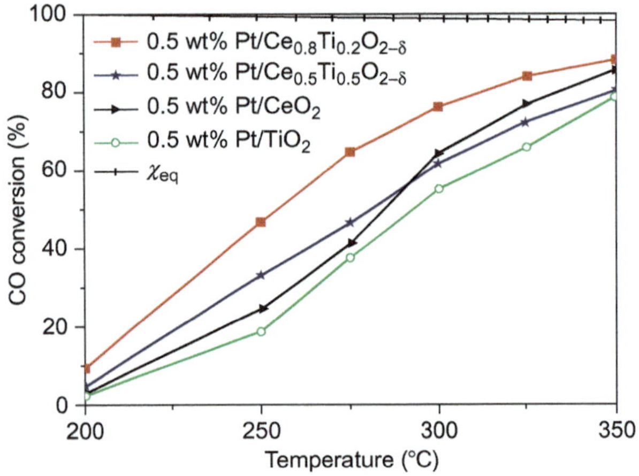

FIGURE 3.10 Effect of support chemical composition on the conversion of CO (X_{CO}, %) as a function of WGS reaction temperature over 0.5 wt% $Pt/Ce_{1-x}Ti_xO_{2-\delta}$ (x=0.0, 0.2, 0.5 and 1.0) catalysts; GHSV=40,000 h^{-1} (L/L_{cat}/h). *(Taken from Figure 2 of J. Phys. Chem. C 117 (2013) 25467-25477.)*

of activity is $Pt/Si_5Ce_{95} > Pt/Si_{10}Ce_{90} > Pt/Si_{20}Ce_{80} > Pt/Ce_{100} > Pt/Si_{100}$. Pt/Si_5Ce_{95} catalyst exhibited 2.5 times higher reaction rate compared to Pt/CeO_2 alone. Shinde and Madras [69] synthesized $Ce_{0.88}Si_{0.1}Pt_{0.02}O_2$ and $Ce_{0.88}Al_{0.1}Pt_{0.02}O_2$ using sonochemical method and evaluated for WGSR. Of the two catalysts Si-substituted catalyst exhibits better activity.

Buitrago *et al.* [70] dispersed various loading of ceria over activated carbon support and impregnated Pt over those supports. The catalytic activities are presented in Figure 3.11. The catalysts exhibit CO conversion more than 90% only at high temperatures. Among the various catalysts 40% loaded catalyst exhibited higher activity.

Bi *et al.* [71] reported Pt/CeO_2/SBA-15 catalysts for the low-temperature WGSR. Pt/CeO_2/SBA-15 catalyst prepared by deposition-precipitation exhibits better activity than Pt/CeO_2 catalyst. However, the catalyst synthesized by impregnation method exhibits similar activity to Pt/CeO_2. The catalytic activity of the Pt/CeO_2/SBA-15 catalyst for WGSR was greatly improved due to the CeO_2 particles nanosize and their effective interaction with Pt particles.

3.3.1.3 Pt/TiO_2

As explained above, reports on Pt/TiO_2 catalysts are available in the literature along with other noble metals. Iida *et al.* [72] investigated the effect of Pt precursor on the WGS activity of Pt/TiO_2 rutile catalysts. Pt/TiO_2 (R) catalysts prepared from H_2PtCl_6 and $Pt(C_5H_7O_2)_2$ showed relatively high activities. The catalytic activities decreased in the following order for precursors used: $H_2PtCl_6 > Pt(C_5H_7O_2)_2 >$ [Pt $(NH_3)_4]Cl_2 > [Pt(NH_3)_4](NO_3)_2 >$ *cis*-$[Pt(NO_2)_2(NH_3)_2]$. Furthermore, they found a relationship between catalytic activity and Pt dispersion.

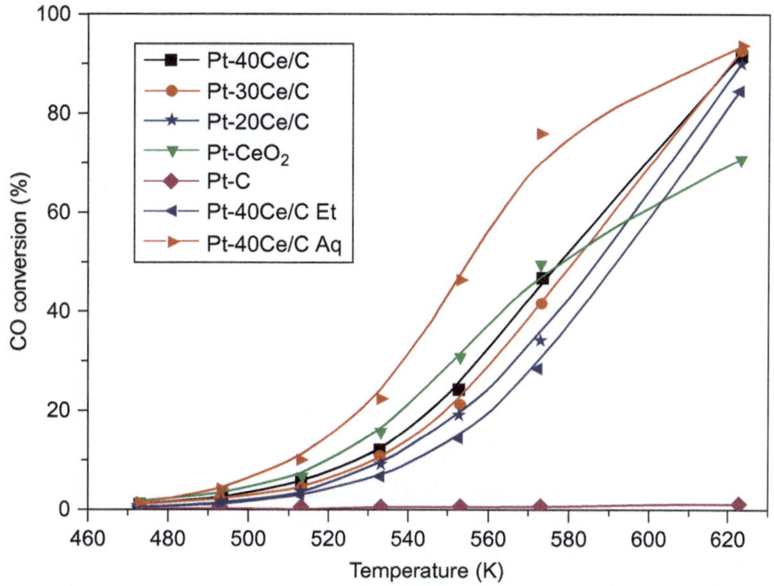

FIGURE 3.11 CO conversion versus reaction temperature. *(Taken from Figure 4 of Catal. Today 180 (2012) 19-24. The catalysts labelled Et and Aq were prepared using ethanol and water, respectively.)*

Panagiotopoulou *et al.* [73] reported the effect of titania particle size on the WGS activity of Pt/TiO_2 catalysts. The WGS activity increases significantly with decreasing primary crystallite size of the support. Then they [74] reported various MO_x/TiO_2 (M=Ce, Nd, Co, Y, Gd, Ho) in the temperature range of 150-500 °C using a feed composition consisting of 3% CO and 10% H_2O. Among the various catalysts $Pt/CeO_2/TiO_2$ catalyst exhibits higher activity compared to other mixed oxides.

Panagiotopoulou and Kondarides [75] investigated the effect of alkali metal addition on the WGS activity of Pt/TiO_2 catalysts. It is observed that, with the exception of the MgO-containing sample, addition of alkaline earths results in a shift of the CO conversion curve towards lower reaction temperatures, compared to that obtained for unpromoted Pt/TiO_2. This shift is less pronounced for BaO-containing catalyst and more important for CaO- and SrO-containing samples, which are able to reach equilibrium CO conversion at about 325 °C, i.e., at temperature significantly lower than that obtained for the unpromoted catalyst. It is observed that increasing CaO loading from 0 to 2 wt% results in a progressive shift of the CO conversion curve towards lower temperatures, whereas further increase of promoter content to 4% CaO has the opposite effect.

3.3.1.4 Pt/Other Supports

Very few literature reports are available on the evaluation of Pt catalysts supported on other carriers except TiO_2, ZrO_2 and CeO_2. Davis *et al.* [76] reported HfO_2 as a support for the Pt and compared with ZrO_2- and CeO_2-supported catalysts. However, Pt/HfO_2 catalysts exhibit lesser catalytic activity compared to Pt/ZrO_2 and Pt/CeO_2 catalysts. Stephanopoulos *et al.* reported silica-encapsulated platinum (Pt@SiO_2) catalysts for the low-temperature WGSR. They reported that Na-promoted Pt@SiO_2 catalysts exhibited better activity compared to Pt/CeO_2 and Pt/ZrO_2 catalysts. Then they reported Na ion exchanged carbon nanotubes as supports for Pt catalysts for the low-temperature WGSR. Remarkable improvement in the activity was observed with Na modification compared to Pt supported on pure carbon nanotubes.

3.3.1.5 Bimetallic Pt Catalysts

In 2002 Iida *et al.* [77] reported Pt-Re/TiO_2 (rutile) and Pt-Re/ZrO_2 catalysts for the low-temperature WGSR. Then they [78] discussed the mechanistic aspects. Irrespective of impregnation sequence the addition of Re to Pt/TiO_2 and Pt/ZrO_2 increases the WGS activity. The characterization measurements suggest that Re acts as the anchor for Pt particles due to the strong interaction between Pt and Re and increases the Pt dispersion. Then, they prepared Pt-Re/TiO_2 catalyst by using subcritical drying method. Interestingly, the subcritical drying catalyst prepared with $Pt(C_5H_7O_2)_2$ exhibited superior catalytic activity for low-temperature WGS at 250 °C compared with that of a commercial Cu/Zn catalyst.

Choung *et al.* [79] have reported that the addition of Re to Pt/CeO_2-ZrO_2 catalysts enhances activity for the low-temperature WGSR. Radhakrishnan *et al.* [45] also developed Pt-Re/CeO_2-ZrO_2 catalysts for low-temperature WGSR. It was found that rhenium nearly doubled the reaction rate of the supported platinum catalyst and that rhenium carbonyl was the optimal precursor for the deposition of the promoter on the catalyst surface. The optimal platinum to rhenium ratio was determined to be about 2:1 by weight with the nominal platinum loading at 2 wt%.

Lee [80] investigated Ce, Ni and Co bimetallic Pt/ZrO_2 catalysts for the low-temperature WGSR. Table 3.1 shows the catalytic activities of various bimetallic catalysts. Pt-Ce/ZrO_2 catalyst prepared by co-impregnation method is highly active at low temperature. The low reduction temperature of Pt with Ce in Pt-Ce/ZrO_2 seems to give rise to high activity in WGSR.

Kim *et al.* [81] then investigated the effect of support (Al_2O_3, SiO_2, TiO_2 (P-25), CeO_2, SiO_2-Al_2O_3, yttria-stabilized zirconia (YSZ) and ZrO_2) on the WGS activity of bimetallic Pt-Ce catalysts. Figure 3.12 shows the effect of support on the WGS activity of bimetallic catalysts. The WGS activity decreased in the following order: Pt-Ce/TiO_2 > Pt-Ce/YSZ > Pt-Ce/CeO_2 > Pt-Ce/ZrO_2 > Pt-Ce/Al_2O_3 > Pt-Ce/SiO_2 > Pt-Ce/SiO_2-Al_2O_3. Pt-Ce/CeO_2 showed the similar

WGS activity with Pt-Ce/Al_2O_3, Pt-Ce/ZrO_2 and Pt-Ce/SiO_2 below 550 K but its WGS activity increased further to be higher than those of the latter catalysts above 550 K. In the case of Pt-Ce catalyst supported on silica-alumina, no noticeable WGS activity was observed at all reaction temperatures below 673 K.

TABLE 3.1 Catalytic Activities of Various Bimetallic Catalysts (Reaction Conditions: Temperature 250 °C, GHSV: 3000 h^{-1}, Steam/CO Ratio 6, Gas Composition: 3% CO, 14% CO_2, 37% H_2, 23% H_2O, Ar Balance

Catalyst	CO conversion (%)
Pt/ZrO_2	11.4
Pt-Ce/ZrO_2	95
Pt-Ni/ZrO_2	10.1
Pt-Co/ZrO_2	8.7

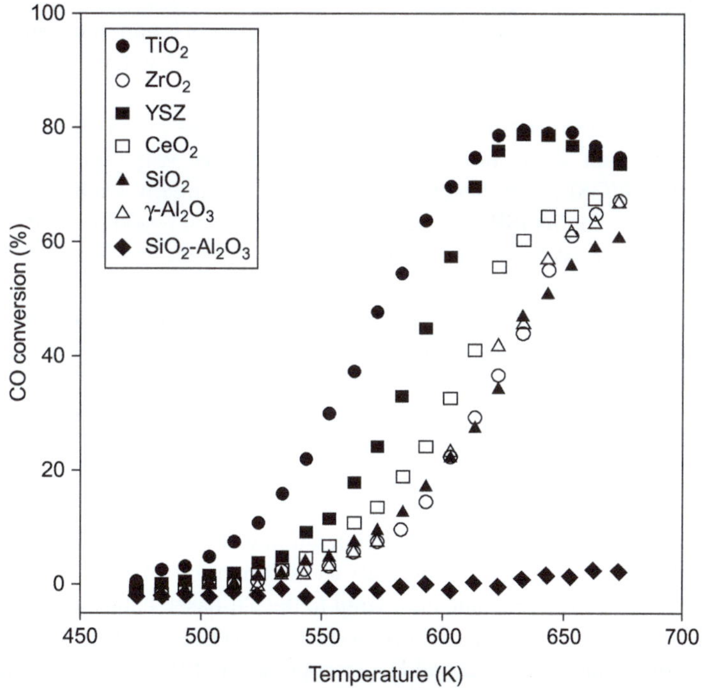

FIGURE 3.12 The CO conversion over 1 wt% Pt-Ce/support (Ce/Pt = 5) with increasing reaction temperature. The feed composition: 6.7 vol.% CO, 6.7 vol.% CO_2 and 33.2 vol.% H_2O in H_2. F/W – 1500 mL/min/g_{cat}. *(Taken from Figure 1 of Appl. Catal. B Environ. 90 (2009) 45-54.)*

This result implies that the kinds of support can play an important role for the WGSR. The WGS activity increases with increasing Ce/Pt ratio and the maximum activity was observed when the Ce/Pt was 5.

Caglayan and Aksoylu [82] synthesized Pt-Ni/Al_2O_3 catalysts. They reported that with increasing Ni loading the equilibrium conversions can be reached at lower temperatures. Wang *et al.* investigated the effect of support (Al_2O_3, SiO_2, TiO_2, CeO_2 and ZrO_2) on the WGS activity of Pt-Ni bimetallic catalysts. Based on catalyst weight, the Pt-Ni bimetallic catalysts follow the activity trend of CeO_2 > high surface area-ZrO_2 > TiO_2 ~ Al_2O_3 > SiO_2 > low surface area-ZrO_2.

Pazmiño *et al.* [83] investigated the effect of alkali metal addition on the WGS activity of Pt/Al_2O_3 and Pt/TiO_2 catalysts. Na promotion causes more increases in the CO conversion compared to other promoters. The addition of Na resulted in the highest enhancement of the WGS rate as compared to Li or K at the same alkali:Pt ratio for alumina-supported catalysts.

3.3.1.6 Deactivation of Pt Catalysts

One of the drawbacks of the Pt/TiO_2 compared to Pt/CeO_2 catalyst is it needs higher temperatures to partially reduce. Gononzalez *et al.* [84] reported that ceria-modified Pt/TiO_2 shows better thermal stability and low-temperature reducibility compared to titania and a higher WGS activity than titania or ceria supports.

Azzam *et al.* [85] conducted long-term catalytic runs over Pt/TiO_2 catalysts. They found that even though Pt/TiO_2 catalyst is very active for high-temperature WGSR, they deactivated with time on stream. Catalyst deactivation of Pt/CeO_2 for WGSR has been reported by several researchers. Ruettinger *et al.* [86] measured the Pt dispersion by CO adsorption and X-ray diffraction and showed that Pt deactivation is mainly due to the loss of noble metal surface area. The long-term stability of ceria-based catalysts for WGS operation in fuel cell applications was studied by Zalc *et al.* [87]. They prepared a variety of Pt-CeO_2 catalysts and tested for WGSR in the temperature region 250-400 ° C under real reforming conditions. Interestingly, all Pt catalysts exhibit identical first order deactivation. The deactivation rate was decreased when hydrogen was not presented in the gas mixture. They attempted to reactivate the catalyst by heating in steam and air which were unsuccessful. The catalyst deactivation is the major problem for the commercialization of Pt catalysts. Another reason for the deactivation of Pt-based catalysts is formation of side products, i.e., hydrocarbons such as methane compared to the Cu-based catalysts. Pt catalysts are well known for the higher Fisher-Tropsch activity for the synthesis of organic compounds from CO and H_2. Incorporation of Pt into partially reduced ceria lattice sites and preventing Pt in its ionic states is the better way to reduce the Fisher-Tropsch activity. The goal of several industries is to develop a Pt-based catalyst which can be stable up to 40,000 h. This is about 4.5 years of continuous operation. Recently, industries realize that addition of rare earth oxides such as La, Pr, etc. can reduce the sintering of ceria crystallites and deactivation of Pt-ceria-zirconium catalysts.

3.3.2 Ru-Based Catalysts

Basińska and Domka [88] reported first time Ru/Fe_2O_3 catalyst for the low-temperature WGSR. They found that there is no correlation between surface area of the catalysts and their WGS activity. Also the phase composition of the iron oxide changes during the WGSR. Then, they studied [89] the effect of alkali salts on the WGS activity of Au/Fe_2O_3 catalyst. Among the various dopants Na-doped catalyst shows the higher activity compared to K-, Rb-, Cs- and Li-doped catalysts. Then, they studied [90] the effect of support on the WGS activity of Ru catalysts. The order of activity of various supports is $Ru/La_2O_3 > Ru/Fe_2O_3 > Ru/Fe_2O_3 > Ru/MgO > Ru/C_{act}$. They also [91] found that doping of La and Sm into Fe_2O_3 improves the WGS activity of Ru/Fe_2O_3 catalyst.

3.3.3 Rh-Based Catalysts

In 1987 Dictor [92] performed kinetic studies over Rh/Al_2O_3 catalyst for WGSR for the first time. They proposed that the WGSR over Rh/Al_2O_3 catalysts only proceeds on surfaces previously exposed to oxygen. Overnight oxidation at 500 °C, which is thought to produce supported Rh_2O_3, yields a catalyst that has high shift activity but deactivates rapidly. On the other hand, extensive reduction of the catalyst, which presumably leads to Rh^0/Al_2O_3 or $Rh^{1+}/A1_2O_3$, produces a catalyst that is inactive for the WGSR. The catalyst with the greatest activity and stability is that which has first been reduced and then exposed to small quantities of O_2.

In 1993 Shido and Iwasawa [93] investigated the mechanism of the WGSR over Rh/CeO_2 catalysts. They made the following conclusions. The reaction occurs at Rh/CeO_2 interface and not on Rh metallic particles. OH groups form bidentate formates with ceria. Then, bidentate formate decomposes and forms H_2 and unidentate. Water molecules promote the desorption of formate into CO_2. Electron donor-acceptor interaction between Ce-O and H_2O is an important factor for the mechanism.

Galletti *et al.* [94] investigated the effect of synthesis method on the WGS activity of Rh/CeO_2 catalysts. They synthesized CeO_2 with solution combustion synthesis method and hard templating method. They deposited Rh by using wet impregnation method. With only H_2O and CO, the equilibrium curve was reached with both catalysts at about 330 °C, but CO conversion started at slightly lower temperature with solution combustion prepared catalyst. Moreover, higher CH_4 outlet concentration was obtained with the hard template prepared catalyst. Under realistic reforming conditions, different shapes of CO conversion curves were shown and with the complete mixture the equilibrium curve was exceeded for both catalysts. The catalytic activity results are presented in Figure 3.13. Hard template method-synthesized catalysts exhibit better activity than solution combustion-synthesized catalysts. However, hard template method-synthesized catalyst exhibits higher selectivity towards methane.

Recently, Karakaya *et al.* [95] investigated kinetics of the WGSR over Rh/Al_2O_3 catalysts. They proposed the reaction pathways for the WGSR.

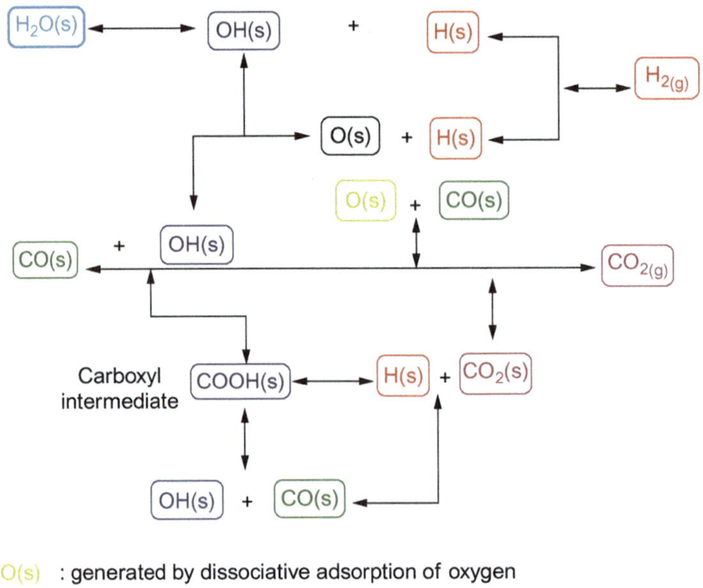

Taken from reference Appl. Catal. A Gen. 470 (2014) 31-44.

3.3.4 Pd Catalysts

In 2009 Han *et al.* [96] reported Pd on one-dimensional ceria for WGSR. 1D-ceria was synthesized using hydrothermal method in two stages: one is precipitation and another is ageing. Pd was deposited using deposition-precipitation method. The catalytic activity was increased with increasing temperature, becoming significant at 300 °C. There is a certain change in the amount of oxygen vacancies and cell dimensions after the WGSR. They proposed that ceria participates in the WGSR when the oxygen vacancies formed by CO reduction facilitate the breakdown of the H_2O to form H_2 and O^{2-} ions. They proposed that here the catalysts follow the redox mechanism. The hydrogen pretreatment temperature plays one of the key roles for the catalytic activity of pure 1D-ceria and Pd-loaded 1D-ceria. The special 1D feature

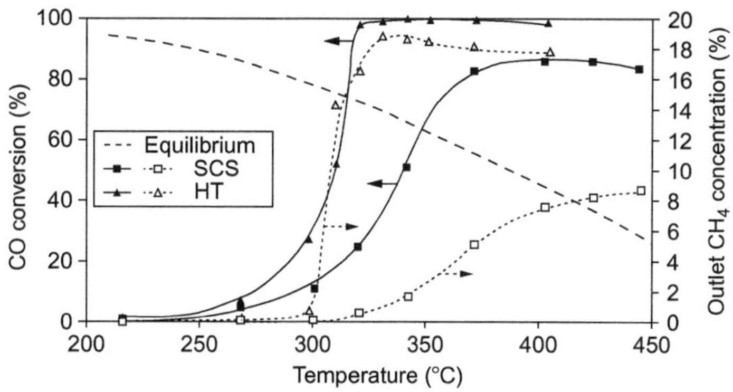

FIGURE 3.13 CO conversion and outlet CH_4 concentration versus temperature for Rh/CeO_2 catalysts prepared by SCS and HT method (inlet gas composition: 5% CO+20% H_2O+11% CO_2+40% H_2. N_2 balance; WSV=0.33 NL min^{-1} g^{-1}). *(Taken from Figure 5 of Catal. Today 176 (2011) 336-339.)*

could efficiently increase the catalytic activity of ceria because the effective surface area can be increased by reducing the aggregation problem of nanoparticles and also benefiting from the double surfaces of nanotubes, which decrease the reduction temperature of CeO_2. In addition, the increased activity can be correlated with the crystal faces present in the different nanocrystalline morphologies.

3.3.5 Au-Based Catalysts

Pt, Pd, Rh and Ru catalysts are very active in the temperature region 250-400 °C. In the past two decades supported gold catalysts received much attention since they have remarkably high activity for the WGSR. Au catalysts are very active in the temperature region 180-250 °C, where the Pt group metals are insufficiently active in this temperature region. After developing the gold catalysts by Haruta *et al.* [97] for CO oxidation it was soon realized that Au exhibits a very good WGS activity when the gold crystallite size lies between 1 and 5 nm. The main supports used for gold are Fe_2O_3, ZrO_2, TiO_2 and CeO_2.

3.3.5.1 Au/Fe_2O_3

In 1996 Andreeva *et al.* [98] reported first gold catalyst, i.e., Au/Fe_2O_3 catalyst for the low-temperature WGSR. They prepared Au/Fe_2O_3 catalyst by using CP method with the Au:Fe atomic ratio 1:22 and compared with the Au/Al_2O_3 catalyst. The following reaction conditions were used: 1 cm^3 catalyst bed; 4000 h^{-1} space velocity. The catalytic results suggest that the catalyst exhibited higher WGS activity at lower reaction temperatures (120 °C). The Au/Fe_2O_3 catalyst shows much higher activity than Au/Al_2O_3 catalyst. This activity is even higher than that of the commercial Cu-Zn-Al catalyst. The activity results

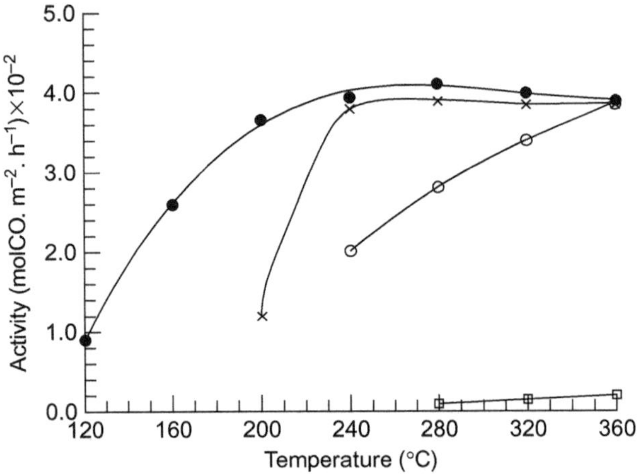

FIGURE 3.14 Temperature dependence of the catalytic activity of the samples studied: ●, Au/α-Fe_2O_3; ○, α-Fe_2O_3; □, Au/Al_2O_3; ×, CuO/ZnO/Al_2O_3. *(Figure 1 of J. Catal. 158 (1998) 354-355.)*

are presented in Figure 3.14. The activity measured by X-ray diffraction methods suggested that the iron oxide in Au/Fe_2O_3 presented as Fe_2O_3 and Fe_3O_4. The higher activity of Au catalyst is due to the specific interaction between Au and iron oxide.

Andreeva *et al.* investigated [99] the mechanism by evaluating Au/Fe_2O_3 catalyst with different characterization techniques, i.e., transmission electron microscopy (TEM), X-ray diffraction (XRD), TGA-DTA and Fourier transform infrared (FTIR) techniques. They proposed that initial precursor structure and morphology, fixed by the specific preparation method, influence both the Fe_2O_3 texture and properties as well as the dispersion of the gold particles. FTIR results suggest that there is an increase in the concentration of OH groups in Au/Fe_2O_3 samples compared to Fe_2O_3 alone. The presence of an increased concentration of active OH groups on Au/Fe_2O_3 can explain the enhanced catalytic activity of the gold-containing samples in the WGSR on the basis of the associative mechanism. Most likely, the low-temperature activity of the Au/Fe_2O_3 catalyst is related to an associative type of mechanism, connected with the formation of intermediate formate or carbonate species. The formation of an intermediate surface compound through the interaction between a CO molecule and an OH group is an important step of this associative mechanism. The presence of a sufficiently high concentration of surface active OH groups in the catalyst is an essential prerequisite for the appearance of high catalytic activity in the WGSR. Surface metal atoms of low coordination are more energy-rich and are unusually reactive. Upon oxidation or contact with water these atoms can coordinate more oxygen atoms or hydroxide species. These surface species

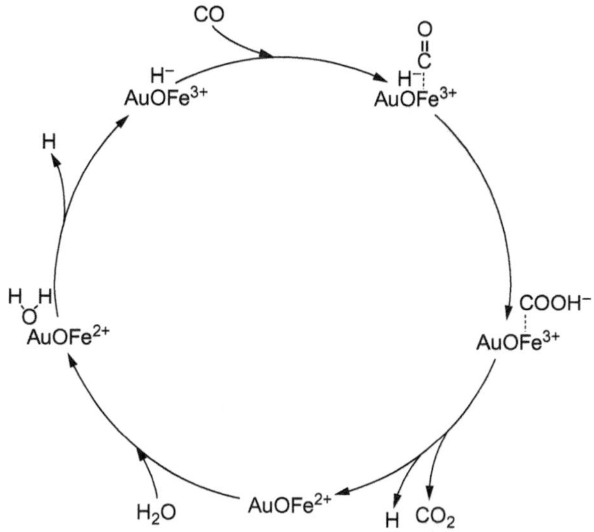

FIGURE 3.15 Mechanism of the low-temperature WGS reaction over Au/Fe_2O_3 catalysts. *(Figure taken from Appl. Catal. A Gen 134 (1996) 275-283.)*

participate actively in redox reactions on the surface. The mechanism is presented in Figure 3.15.

The essential aspects of mechanism are the dissociative adsorption of water on ultrafine gold particles, followed by the spillover of active OH groups onto adjacent sites of the ferric oxide. The formation and decomposition of intermediate species are accompanied by redox transfer of $Fe^{3+} \leftrightarrow Fe^{2+}$ in Fe_3O_4 and the reverse step during the dissociation of water molecule.

Andreeva *et al.* also prepared the Au/Fe_2O_3 [100] catalysts with modified deposition-precipitation method and compared with the co-precipitated catalysts. The catalyst prepared by the deposition-precipitation method exhibited higher activity compared to the co-precipitated one. The Fe 2p X-ray photoemission spectroscopy (XPS) spectra show the Fe_2O_3 phase in fresh catalysts and Fe_3O_4 phase in spent catalysts. Peak due to interaction between Au and Fe was also observed. Elemental mapping measurements show that gold was dispersed uniformly on the iron oxide surface in the case of catalyst synthesized by the deposition-precipitation method. The characterization results confirmed two points. (1) In the modified deposition-precipitation catalyst gold dispersion is very high and prevents cluster formation. (2) Major part of the gold remains on the surface, adsorbed on Fe_2O_3 thus being accessible to catalysis. On the whole in the deposition-precipitation method the interaction between Au and iron is weaker and likely more favourable for catalysis.

Andreeva *et al.* [101] investigated Au/Fe_2O_3 catalyst using *in situ* FTIR technique. They made the following conclusions. Hydrogen dissociates on gold atoms at room temperatures and reacts with absorbed oxygen atoms of the surface sites. CO is absorbed both on Au^0 and support sites. Water and OH groups are adsorbed on the supports, on gold sites and at the interface between them. They located the active sites for the hydrogen dissociation and for water-CO reactive interactions, on the basis of the experimental finding, at the surface of the metallic gold small particles. A redox, regenerative mechanism, similar to that discussed for copper-based catalysts, can explain all the experimental results.

Andreeva *et al.* [102] investigated the effect of addition of ZnO, ZrO_2 oxides to the Au/Fe_2O_3 and Au/ZrO_2 alone. Interestingly, Au/Fe_2O_3 catalyst exhibits much higher activity than Au/Fe_2O_3-ZnO, Au/Fe_2O_3-ZrO_2. Also, Au/Fe_2O_3 catalyst exhibits better activity than Au/ZrO_2, Au/ZnO catalyst. The catalytic activity results are presented in Figure 3.16. It could be concluded that the catalytic activity of the gold/metal oxide catalysts depends strongly not only on the dispersion of the gold particles but also on the state and structure of the supports.

The textural property measurements show that there is a reduction in pore volume of Au/ZnO, Au/ZrO_2, Au/Fe_2O_3-ZrO_2, Au/Fe_2O_3-ZnO after the WGSR which is responsible for the lower catalytic activity.

Scurrell and Venugopal [103] investigated the influence of reductive pretreatment on the WGS activity of Au/Fe_2O_3 catalyst. They prepared Au/Fe_2O_3 by using deposition-precipitation method and reduced in different organic/inorganic precursors before the WGSR. They found that the extent of surface reduction is dependent on the choice of the reducing agent. The

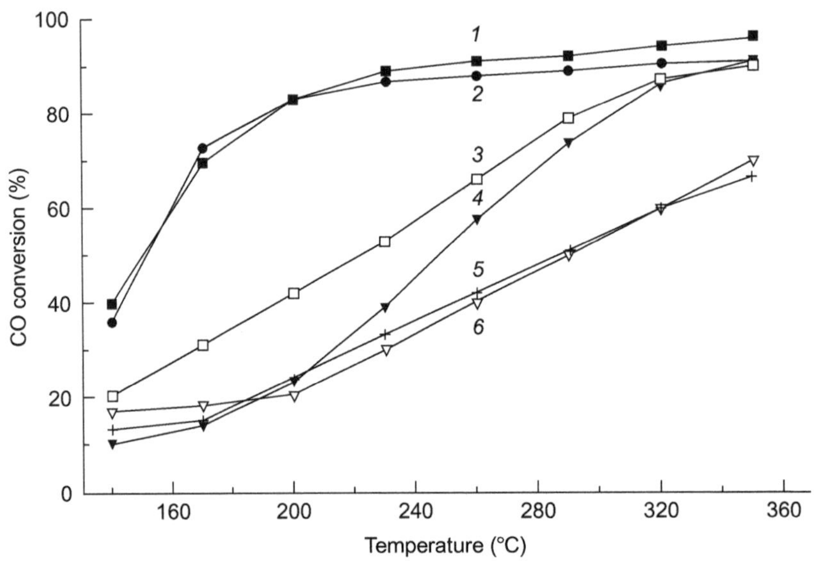

FIGURE 3.16 Temperature dependence of the catalytic activity (degree of conversion) of the samples: (1) Au/Fe_2O_3; (2) 1Au/ZrO_2; (3) Au/Fe_2O_3-ZnO; (4) Au/Fe_2O_3-ZrO_2; (5) 2Au/ZrO_2; and (6) Au/ZnO. *(Taken from Figure 4 of Appl. Catal. A Gen 202 (2000) 91.)*

support plays a major role in stabilizing the active component. Au is stabilized in the form of oxidic gold along with some metallic Au species present on Fe_2O_3 support. Ethanol appears to be the most efficient reducing agent.

Wei *et al.* [104] investigated the effect of calcination temperature on the WGS activity of Au/Fe_2O_3 catalysts. They prepared the catalysts by CP method and calcined at 423, 473, 523, 573, 623, 723 or 823 K. Figure 3.17 shows the effect of reaction temperature on the WGS activity of Au/Fe_2O_3 catalysts calcined at different temperatures. Interestingly with increasing calcination temperature the catalytic activity decreases drastically. The characterization measurements suggest that the higher calcination temperature causes the gradual growth of Au particles and the increase of the crystallinity of the support.

Scurrell *et al.* [105] investigated the effect of Ru addition on the WGS activity of Au/Fe_2O_3 catalyst. The activity results are presented in Table 3.2. They found that bimetallic catalyst exhibits higher WGS activity compared to Au/Fe_2O_3.

They investigated both the fresh and spent catalysts by using temperature programmed reduction technique. The bimetallic catalysts exhibit much lower Au reduction temperature compared to Au/Fe_2O_3. The X-ray diffraction measurements show that there is no evidence for mixed metal species (Au-Ru) either in fresh or spent catalysts. TPR analysis of the spent Au-Ru/iron oxide catalyst revealed the presence of the reduction signal due to Au_xO_y species at a T_{max} of 155 °C, which suggests the lowering in the reduction temperature which is found at 161 °C in the Au-Ru/iron oxide fresh catalyst.

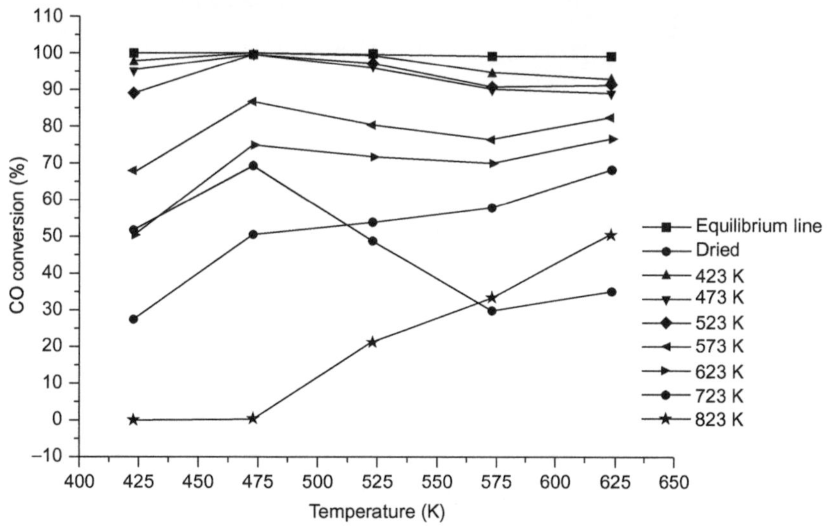

FIGURE 3.17 Convection of CO shift to CO_2 on 8 wt% Au/iron oxide dried at 393 K and calcined at different temperatures. Reactant: 10 vol.% CO in N_2; space velocity: 2000 h^{-1} mL/g_{cat}; vapour gas ratio: 1:1. *(Taken from Figure 1 of Appl. Catal. A Gen. 259 (2004) 121-130.)*

Characterization measurements suggest the occurrence of Au-Ru interactions in the bimetallic systems.

Zheng *et al.* [106] did extensive study on the effect of foreign oxide doping into the Fe_2O_3 lattice for the low-temperature WGSR on Au/Fe_2O_3 catalyst. They investigated Zr-, Mg-, Ca-, Ni-, La-, Cu-, Zn-, Al-, Ba-, Cr-, Co-, Ce-, Mo-, Bi-, Ti- and Mn-doped Au/Fe_2O_3 catalysts for low-temperature WGSR. The WGS activity results are presented in Figure 3.18. The WGS activity

TABLE 3.2 Comparison of Activities of Catalysts for the WGS Reaction at 120 and 180 °C

	CO conversion (%)	
Catalyst	*At 120 °C*	*At 180 °C*
Au-Ru/iron oxide	24.0	63.0
Ru/iron oxide	12.3	42.0
Au/iron oxide	8.2	16.0
Iron oxide	<0.5	1.0

Taken from Table 1 of Appl. Catal. A Gen. 245 (2003) 149-158.

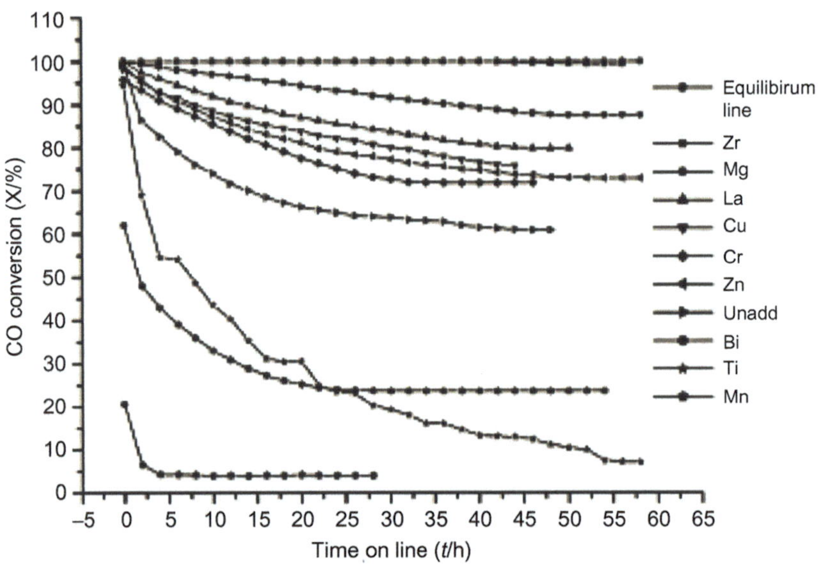

FIGURE 3.18 WGSR activity and stability of Au/Fe_2O_3 catalysts modified by various oxides. Reaction temperature, 473 K; vapour gas ratio, 1:1; space velocity, 10,000 h^{-1}. *(Taken from Figure 1 of Catal. Lett. 102 (2005) 99-108.)*

and stability of Au/Fe_2O_3 are improved significantly upon addition of ZrO_2 and to a lesser extent MgO, CaO, NiO, La_2O_3, Cr_2O_3 and CuO. In contrast, Bi, Ti and Mn oxides seriously decrease the catalytic activity, while addition of Zn, Al, Ba, Co, Ce and Mo oxides does not influence the catalytic activity and its stability. Based on the characterization of some of as-prepared and spent samples, they concluded that the catalytic activity of gold catalysts supported on composite oxide of Fe_2O_3-MO_x depends not only on the dispersion of the gold particles but also on the reduction property of composite oxide supports, regardless of the fluctuation of gold loading and some change of specific surface area and pore structure due to introduction of the modifying metal oxides. The improvement of catalytic stability is due to the comparative stabilization of high dispersion of gold particles and uneasily sintering of Fe_3O_4 crystallites during the catalytic operation.

Zheng *et al.* investigated the effect of Nb_2O_5 [107] addition on the best catalyst system Au/Fe_2O_3-ZrO_2 and concluded that 10 wt% Nb_2O_5 promoted higher CO conversion and reached equilibrium conversion at 200 °C (99.9%).

Zheng *et al.* prepared [108] Au/Fe_2O_3 catalyst by modified deposition-precipitation method and compared with co-precipitated, deposition-precipitation method samples. Under strictly controlled conditions, the Au/Fe_2O_3 catalyst prepared by modified deposition-precipitation displayed higher activity for the WGSR than that prepared by CP and deposition-precipitation. Characterization measurements suggest that in the catalyst prepared by modified deposition-precipitation, both amorphous and crystallized iron oxides coexist and has high surface area and small gold particles with uniform distribution and possesses a relatively strong interaction between gold particles and the support, leading to the highest catalytic activity for the WGSR.

Recently, Soria *et al.* [109] also investigated the effect of synthesis method on the WGS activity of Au/Fe_2O_3 catalysts. They prepared catalysts by deposition-precipitation method, liquid phase reductive depositions and double impregnation method. Among the various catalysts, the catalyst synthesized by deposition-precipitation method exhibits higher CO conversion.

3.3.5.2 Au/TiO_2

Along with Andreeva *et al.* [98] in 1996 Sakurai *et al.* [110] reported Au/TiO_2 catalyst for low-temperature WGSR. Au/TiO_2 catalyst shows much higher activity than traditional $Cu/ZnO/Al_2O_3$ catalyst at 373 K. The reaction rate at 373 K increased with increasing Au loading for a series of Au/TiO_2 catalysts prepared by the deposition-precipitation method, and Au/TiO_2 prepared by CP showed a lower rate despite the higher metal loading and the smallest gold particle sizes. This may be attributed to differences in exposure of gold particles on the surface and the crystalline nature of TiO_2 depending on the preparation method.

In 2004 Idakiev *et al.* [111] synthesized mesoporous titania using surfactant templating method. Gold was deposited from 1 to 5 wt% by using deposition-precipitation method. Interestingly, with increasing gold loading from 2.5 to 5 wt% the WGS activity decreases. As expected increase in reaction temperature increases the WGS activity. However, no serious change in CO conversion was observed with change in steam/CO ratio. Characterization results reveal that the WGS activity depends on the average Au particle size and gold dispersion.

Idakiev *et al.* prepared [112] ceria-modified mesoporous titania supports for the Au catalyst. The ceria addition strongly improves the catalytic performances due to: (i) the beneficial role of ceria additive in decreasing the degree of crystallinity of mesoporous titania and decreasing its particle size; (ii) the strong effect on the reducibility of the support owing to the interaction of ceria additive with the mesoporous titania; and (iii) a high degree of synergistic interaction between ceria and mesoporous titania. The high and stable WGS activity could be related to the high stability of the gold dispersion and to larger number of active sites located at the Au/ceria-modified mesoporous titania interface.

Idakiev *et al.* investigated [113] titania nanotubes for the low-temperature WGSR. They used laboratory made TiO_2 and commercial P25 titania to synthesize titania nanotubes. They deposited gold by using deposition-precipitation method. The catalytic results are presented in Figure 3.19. The laboratory made titania nanotube catalyst exhibits much higher activity compared to P25 made catalyst. They found that observed catalytic behaviour of gold-supported TNTs correlates with the structural peculiarity and the nature of the TNTs. They also suggest that catalytic performance of Au/TNTs may be improvable by increasing the diameters of the nanotubes or modifying the crystalline structure of the nanotubes to anatase phase.

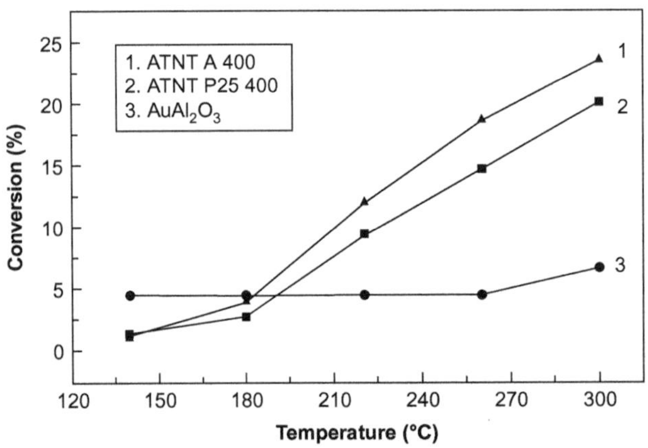

FIGURE 3.19 Temperature dependence of the catalytic activity of the samples. *(Taken from Figure 9 of Appl. Catal. A Gen. 281 (2005) 149-155.)*

Rebrov *et al.* [114] synthesized 1 wt% Au supported on mesoporous titania thin films for low-temperature WGSR. The catalyst is very stable for 72 h of operation.

Rodriguez *et al.* [115] synthesized Au(1 1 1) surface covered by 20-30% of CeO_2 and TiO_2 nanoparticles. Although clean Au(1 1 1) is not catalytically active, gold surfaces that are 20-30% covered by the oxide nanoparticles had activities per unit geometric area similar to or larger than those of Cu(1 1 1) or Cu(1 0 0) under similar conditions.

$$CO_{gas} \rightarrow CO_{ads} \quad (3.1)$$

$$H_2O_{gas} \rightarrow H_2O_{ads} \quad (3.2)$$

$$H_2O_{ads} \rightarrow OH_{ads} + H_{ads} \quad (3.3)$$

$$CO_{ads} + OH_{ads} \rightarrow HOCO_{ads} \quad (3.4)$$

$$HOCO_{ads} \rightarrow CO_{2,gas} + H_{ads} \quad (3.5)$$

$$2H_{ads} \rightarrow H_{2,gas} \quad (3.6)$$

The catalytic activity of the surface reaches a maximum, and then decreases and disappears once there is no Au exposed and multilayers of ceria or titania cover the metal substrate. High catalytic activity for low coverages of ceria and titania can be attributed to special chemical properties of the oxide NPs and cooperative effects at oxide-metal interfaces. They also proposed reaction pathways with the minimum-energy barriers which involve the following steps:

Sandoval *et al.* [116] investigated the influence of support on the WGS activity of Au catalysts. They synthesized Au catalysts supported on titania, ceria, alumina and silica by deposition-precipitation method. The catalytic activity results are presented in Figure 3.20. The silica-supported catalyst shows minimal activity. The order of activity is $TiO_2 > CeO_2 \gg Al_2O_3$ at reaction temperatures below 225 °C. However, at reaction temperature above 225 °C, the ceria-supported catalyst is more active than titania-supported catalyst.

Ma *et al.* [117] developed a series of Au/M_xO_y/TiO_2 (M=Al, Ca, Fe, Zn, Ga, Y, Zr, La, Ce, Pr, Nd, Sm, Eu, Gd, Dy, Ho, Er and Yb) catalysts for low-temperature WGSR. The specific reaction rates at reaction temperatures 200 °C and pretreated at 200 °C before the WGSR are presented in Figure 3.21. As can be seen from Figure 3.21, the specific rates on Au/M_xO_y/TiO_2 (M=Al, Ca, Fe, Zn and Ga) at 200 °C were lower than that on Au/TiO_2, whereas those on Au/M_xO_y/TiO_2 (M=Y, Zr, La, Ce, Pr, Nd, Sm, Eu, Gd, Dy, Ho, Er and Yb) were comparable or higher than that on Au/TiO_2.

At 500 °C pretreatment temperatures, many Au/M_xO_y/TiO_2 (M=Al, Ca, Y, Zr, La, Pr, Nd, Sm, Eu, Gd, Dy, Ho, Er and Yb) catalysts were more active than Au/TiO_2, although others (M=Ga, Fe, Zn and Ce) exhibited little or no promotional effect. They think that the maintenance of high activity on some Au/M_xO_y/TiO_2 catalysts upon high-temperature pretreatment is due to the stabilization of gold nanoparticles on M_xO_y/TiO_2 supports.

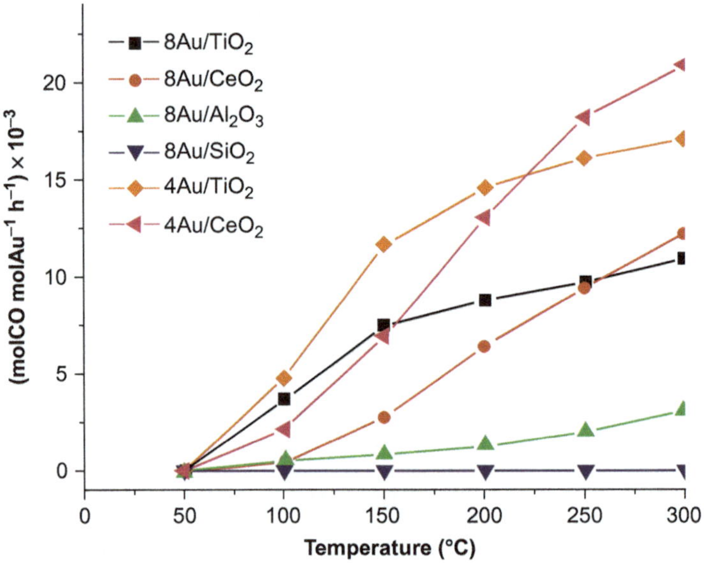

FIGURE 3.20 Catalytic activity in the WGS as a function of reaction temperature displayed by Au nanoparticles supported on TiO_2, CeO_2, Al_2O_3 and SiO_2. *(Taken from Figure 5 of J. Mol. Catal. A Chem. 278 (2007) 200-208.)*

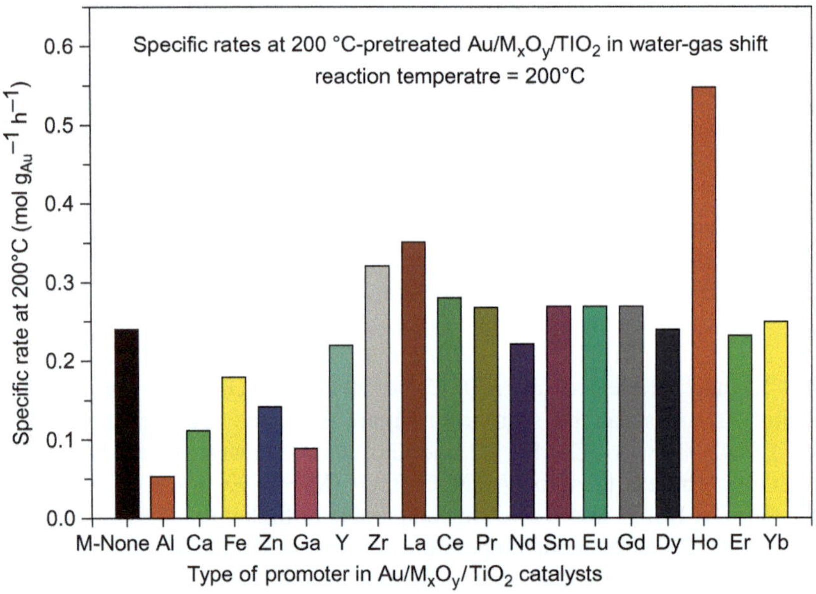

FIGURE 3.21 Specific rates calculated according to CO conversions at a reaction temperature of 200 °C. The CO conversions are read out from Figure 1. Au/TiO_2 and $Au/M_xO_y/TiO_2$ catalysts were pretreated at 200 °C prior to undergoing WGS reaction. *(Taken from Figure 2 of Catal. Lett. 136 (2010) 83-91.)*

Shekhar *et al.* [118] investigated the size and support effect of the Au catalysts and concluded that low coordinated corner Au sites are three and seven times more active than the perimeter Au sites for Au/Al_2O_3 and Au/TiO_2 catalysts, respectively, and that the corner sites are an order of magnitude more active for Au/TiO_2 versus Au/Al_2O_3. They also found that metallic Au species are active for the WGSR.

3.3.5.3 *Au/ZrO₂*

In 2006 Idakiev *et al.* [119] reported mesoporous ZrO_2-supported Au catalysts for WGSR. The mesoporous zirconia was synthesized by surfactant template method. They compared 2.5% and 5% Au/ZrO_2 with Au/TiO_2 catalysts. The catalytic activity results are presented in Figure 3.22.

The mesoporous zirconia-supported Au catalysts have higher activity than mesoporous titania-supported catalysts. They reported that the structure of mesoporous zirconia facilitates the formation of well-dispersed and stable gold particles on the surface upon calcination and reduction and thus strongly improves the catalytic performances. The catalyst deactivates with time; however, the deactivation is reversible. The initial activity can be fully restored after heating the deactivated catalysts in air at elevated temperatures.

In 2012 Idakiev *et al.* [120] investigated the Au supported on ceria-modified mesoporous zirconia catalysts for the WGSR. The WGS activity of various Au-supported mesoporous zirconia and ceria-modified mesoporous zirconia catalysts are presented in Figure 3.23.

Ceria-modified mesoporous zirconia catalysts show much higher activity than pure mesoporous zirconia catalysts especially at lower reaction temperatures. The characterization results reveal that ceria decreases the degree of zirconia crystallinity and particle size. Also ceria interacts with the mesoporous zirconia and causes the strong effect on the reducibility of the support. The high and stable WGS activity could be related to the high stability of the gold dispersion and to larger number of active sites located at the Au/ceria-modified mesoporous zirconia interface.

Then, Menegazzo *et al.* [121] investigated Au/ZrO_2 and Au/SO_4^{2-}-ZrO_2 catalysts. They show that both zirconia-supported and sulphated zirconia-supported catalysts show similar activity. Shen *et al.* [122] investigated the influence of zirconia phase on the WGS activity of Au/ZrO_2 catalysts. They synthesized Au supported on monoclinic zirconia and tetragonal zirconia by using deposition-precipitation method. Au/*m*-ZrO_2 shows better activity than Au/*t*-ZrO_2 catalyst. They found that higher CO adsorption capacity of the Au/*m*-ZrO_2 catalyst is responsible for the higher activity. The chemical state of gold that was strongly related to the pretreatment atmosphere also played an essential role in determining the catalytic activity for WGSR.

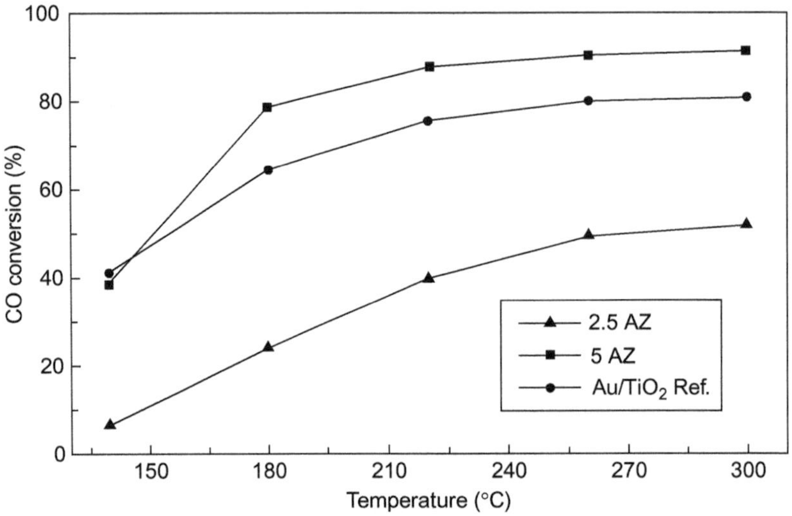

FIGURE 3.22 The WGS activity of the gold/mesoporous zirconia catalysts with different gold contents, compared to reference Au/TiO_2 type A from World Gold Council. *(Taken from Figure 7 of Appl. Catal. B Environ. 63 (2006) 178-186.)*

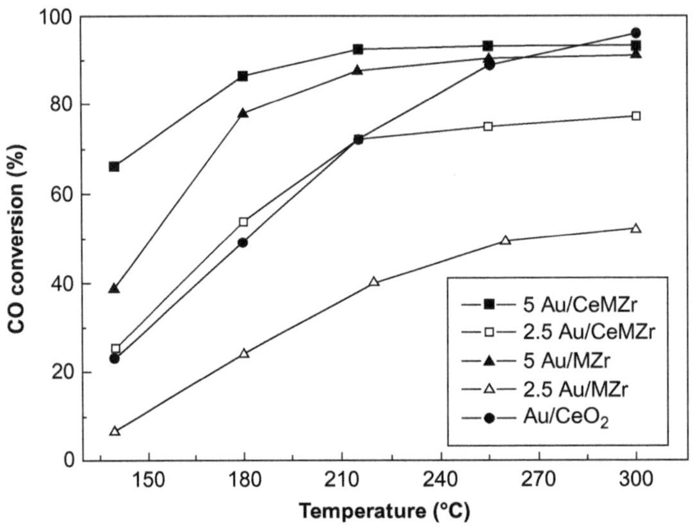

FIGURE 3.23 The WGS activity of the new gold-based catalysts supported on CeMZr with different gold contents (2.5 Au/CeMZr and 5 Au/CeMZr), compared with that of gold catalysts supported on simple oxides, CeO_2 (Au/CeO_2) and mesoporous ZrO_2 (5 Au/MZr and 2.5 Au/MZr). *(Taken from Figure 5 of J. Porous. Mater. 19 (2012) 15-20.)*

Then, Li *et al.* [123] investigated the influence of particle sizes on the WGS activity of Au/ZrO_2 catalyst. They proposed that the activity of Au/ZrO_2 can be improved by reducing the size of Au and ZrO_2 particles. Variations in the size of Au and ZrO_2 particle sizes modify the Au-ZrO_2 contact structure which is the most critical factor for the WGSR. Zhang *et al.* investigated the influence of ZrO_2 pre-calcination temperature on the WGS activity of Au catalysts. They found that by increasing the ZrO_2 pre-calcination temperature the catalytic activity of the resultant Au/ZrO_2 catalyst increases initially and then decreases and gives a maximum at the pre-calcination temperature of 350 °C. They proposed that zirconia pre-calcination temperature influences the Au dispersion, F-centre concentration and electron density of the Au cluster for the reduced catalysts.

3.3.5.4 Au/CeO_2

In 2001 Stephanopoulos *et al.* [124] reported for first time the Au/CeO_2 catalyst for WGSR. Ceria is a crucial component of emission control catalysts, mainly on account of its role in oxygen storage. The facile Ce^{4+}/Ce^{3+} redox reaction is believed to be the driving force leading to this behaviour. The oxygen produced via the redox process $2CeO_2 \leftrightarrow Ce_2O_3 + 1/2O_2$ can be utilized under reduction conditions. However, the role of ceria as a support for noble metal-based catalyst is not only related to its high oxygen storage capacity but also to the improved dispersion of the noble metal and the promotion of the WGSR. They prepared Au/CeO_2 catalyst by CP, deposition-precipitation and urea gelation/precipitation methods. They found that Au/CeO_2 catalyst is very active for WGSR and its activity can be correlated with its reducibility.

In 2002 Andreeva *et al.* [125] also reported the Au/CeO_2 catalyst for the low-temperature WGSR. They prepared 1, 3 and 5 wt% Au/CeO_2 catalysts and evaluated for low-temperature WGSR. Among the various samples, 5 wt % Au/CeO_2 sample exhibits higher WGS activity. The high activity of gold/ceria catalysts could arise from the high and stable gold dispersion. The results are presented in Figure 3.24.

Andreeva *et al.* did *in situ* FTIR studies over Au/CeO_2 catalysts in oxidative and reductive environments and made the following conclusions. In oxidative environment Au was strongly influenced by ceria and cause coordinatively more unsaturated sites on the surface. The reduction in hydrogen at 373 K results in negatively charged gold sites and oxygen vacancies on ceria surface. The WGSR proceeds at the boundary between small metallic gold particles and ceria, where CO adsorption on gold and H_2O dissociation on ceria defects take place. The interactions between small gold metallic particles and ceria appear to be of a crucial importance in explaining the remarkably high stability of Au/CeO_2 catalyst.

Andreeva *et al.* [126] prepared gold catalysts supported on CeO_2 and CeO_2-Al_2O_3 by CP by varying the ratio between Ce and Al. The catalytic activity

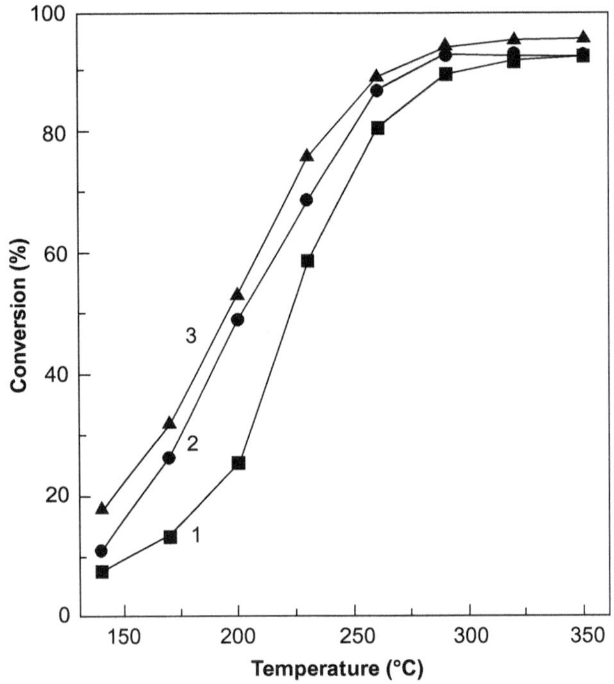

FIGURE 3.24 Effect of gold loading on the WGS activity: (1) $1Au/CeO_2$; (2) $3Au/CeO_2$; and (3) $5Au/CeO_2$. *(Taken from Figure 1 of Catal. Today 72 (2002) 51-57.)*

results of fresh and reduced catalysts are presented in Figure 3.25. The Au/CeO_2 catalyst exhibited better activity than ceria-alumina supported catalyst. The lower catalytic activity is due to the creation of deeper oxygen vacancies in ceria, oxidizing of water vapour being difficult under reaction conditions. After the first catalytic test they recovered the catalysts under the pretreatment of water vapour and air. It is clearly seen that the treatment under water vapour is not enough for the reactivation of gold-ceria-alumina catalysts, while the WGS activity of gold-ceria sample was fully recovered. The reactivation in air practically fully recovered the WGS activity of all the studied catalysts.

Stephanopoulos *et al.* [127] prepared various Au-supported ceria-lanthana catalysts and investigated for WGSR. They prepared various gold catalysts by varying the loading % from 1% to 8.3%. All the gold samples exhibit 50% CO conversion below 120 °C. Among the various catalysts $8Au\text{-}CeO_2\text{-}La_2O_3$ exhibited higher activity. Addition of Au improves the reducibility and oxygen storage capacity of the CeO_2. The amount of surface oxygen available for the reduction is controlled primarily by the crystallite size of ceria.

Stephanopoulos *et al.* then synthesized [128] Au/CeO_2 with low loading of Au (0.2-0.9 wt%). Interestingly, these materials exhibit similar activity to high loading Au catalysts in the reformate mixture. Then they evaluated these

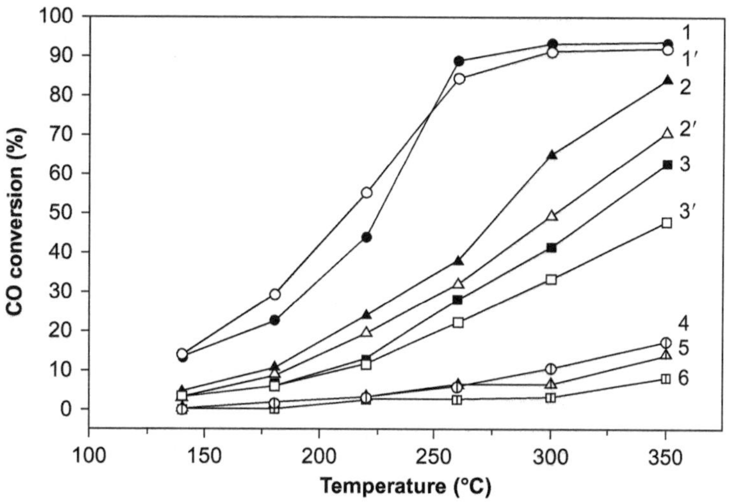

FIGURE 3.25 Catalytic activity of initial supports and gold-supported catalysts as prepared (1-3) and after reduction (1′-3′): (1, 1′) AuCe; (2, 2′) AuCeA110; (3, 3′) AuCeA120; (4) Ce; (5) CeA110; and (6) CeA120. *(Taken from Figure 1 of Appl. Catal. A Gen. 302 (2006) 127-132.)*

catalysts for long-term time-on-stream stability experiments. The results are presented in Figure 3.26. The conversion dropped 20% in the first 10 h and was then stabilized with very slow further decay. They found that the initial deactivation is not due to increase in crystallite size of ceria and it is due to the loss of surface of ceria which weakens the interaction between Au and ceria.

They also performed X-ray adsorption near edge structure and extended X-ray adsorption fine structure studies over Au/CeO_2 catalysts to investigate the structural changes during the WGSR at 100 and 200 °C. Au changes from cationic state to partially reduced state during the WGSR at 100 °C. Also the reduction of gold was controlled by the reactant gas composition.

Zheng *et al.* [129] also reported the effect of La doping on the WGS activity of Au/CeO_2 catalysts. They prepared the Au/CeO_2 catalysts by deposition-precipitation method with different La loadings and studied for WGSR. In the complete La doping range ceria retains cubic fluorite structure. La doping improves the activity of Au/CeO_2 by stabilizing ceria and modifying its morphology.

Wang *et al.* [130] also reported homogeneous mixed La/Ce oxides as active supports for the Au catalysts. They prepared supports using micro-emulsion template method. They studied the influence of pH at which the support was prepared on the WGS activity and found that 8.5 is the optimum pH. As expected, Au/CeO_2 shows much higher activity than Au/La_2O_3, which can be ascribed to the reducibility of the supporting oxides. For the mixed oxides, the activity of the catalysts initially increases with La doping up to ~25% La,

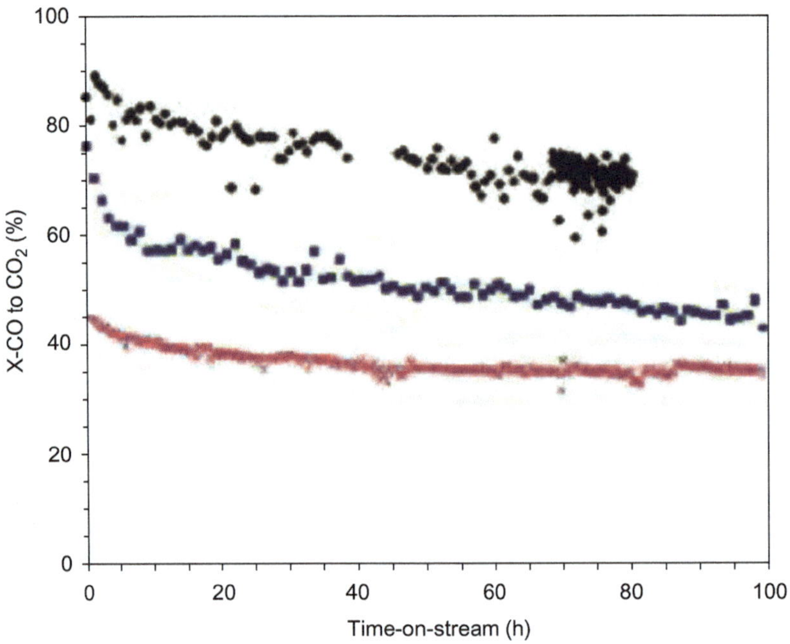

FIGURE 3.26 Stability of gold-ceria WGS catalysts (see Table 3.1 for properties). Space velocity: 15,000 h^{-1} (NTP); 5% CO-15% H_2O-35% H_2-He; reaction temperature: 250 °C. Diamonds: 8Au-CL(UGC); crosses: 4.7Au-CL(DP); squares: 0.44Au-CL(DP, NaCN); all catalysts calcined at 400 °C (see Table 3.1 for properties). *(Taken from Figure 7 of Appl. Catal. B Environ. 56 (2005) 57-68.)*

and then decreases again with further La addition. They found that the WGS activity is directly related to support reducibility.

Davis [131] group synthesized Au/CeO_2 catalyst by using chemical vapour deposition of an Au precursor. The partial reduction of ceria is necessary for generating bridging OH groups on the surface of ceria, which serve as the active sites. During this surface shell reduction process in hydrogen, bridging OH groups are formed, and two cerium atoms for each pair of Type II OH groups change from the Ce^{4+} to Ce^{3+} oxidation state. Addition of Au to ceria catalyzes the surface reduction process and reduces the peak reduction temperature.

Behm *et al.* [132] explained the mechanism of the WGSR over Au/CeO_2 by synthesizing several catalysts and by using XPS and DRIFT studies. The reaction mechanism involves the following: (i) direct CO adsorption on Au nanoparticles as well as adsorption on the support, followed by subsequent reverse spillover on the Au particles/clusters, (ii) the formation and decomposition of surface formates at the interface between Au particles and CeO_2 support ('adlineation sites') or on $Au^{\delta+}$ species, by reaction of CO_{ad} and OH_{ad}, and their subsequent decomposition to CO_2 and H_2, as well as (iii) the spillover and reversible spillover of the surface formates on the support and back, with the

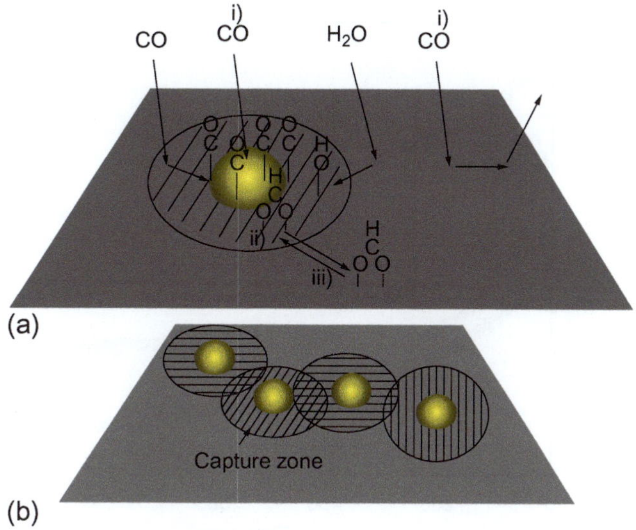

FIGURE 3.27 Schematic description of the reaction mechanism *(Taken from J. Catal. 244 (2006) 137-152.)*

formate adlayer on the ceria support acting as formate reservoir. The mechanism is presented in Figure 3.27.

Behm *et al.* [133] also investigated the mechanism of the WGSR by leaching technique. They used cyanide leaching technique to remove the non-ionic Au species and studied WGSR. The catalytic activity decreased after leaching. The catalytic activity decreased further after calcination of the leached sample. However, the catalytic activity can be restored by hydrogen treatment of the leached sample.

Behm *et al.* also [134] investigated the effect of support surface area on the WGS activity of Au/CeO_2 catalysts. The results of various Au/CeO_2 catalysts are presented in Figure 3.28. The surface area of catalysts varies from 24 to 284 m^2/g.

The characterization results reveal that the Au particle size is independent of the support surface area. However, the catalyst stability varies with catalyst surface area. The stability increases with surface area up to 78 m^2/g and then decreases again. The increase in the stability with increase in surface area is due to two factors: (i) an increasing formation of OH groups on the smaller and more defective (more Ce^{3+}) ceria particles, which will promote the decomposition of stable carbonates and (ii) an increasing stability of these carbonates with increasing surface area.

Behm *et al.* [135] also investigated the effect of pretreatment on the WGS activity of Au/CeO_2 at higher reaction temperature. The order of activity of different pretreatment conditions is $O_2 > H_2$ followed by O_2 $N_2 > CO > O_2$, followed by H_2. The long-term stability of the catalysts follows the reverse order. It is confirmed that the reductive pretreatments result in more stable catalysts.

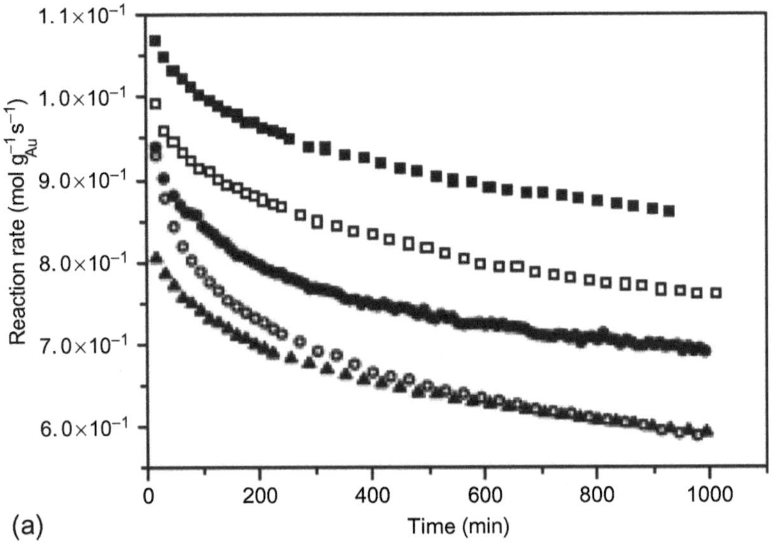

FIGURE 3.28 Activity of the Au/CeO_2 catalysts (2.7 wt% Au, surface area between 24 and 284 $m^2\ g^{-1}$) in dilute water gas (1 kPa CO, 2 kPa H_2O in N_2) at 180 °C. (a) Activity measurement of the Au/CeO_2 catalysts during 1000 min reaction: (■) 78 $m^2\ g^{-1}$, (□) 121 $m^2\ g^{-1}$, (●) 158 $m^2\ g^{-1}$, (▲) 188 $m^2\ g^{-1}$, (○) 284 $m^2\ g^{-1}$ and (b) reaction rates of the Au/CeO_2 catalysts after 1000 min reaction. *(Taken from Figure 5 of Top. Catal. 44 (2007) 183-198.)*

Thompsett *et al.* [136] prepared $Au/CeZrO_4$ catalyst by CP for low-temperature WGSR. $Au/CeZrO_4$ exhibits higher activity than Au/ZrO_2, Au/TiO_2, Au/CeO_2 and Au/FeO(OH). It also exhibits significant WGS activity at temperatures as low as 100 °C. They also found that the stability of $Au/CeZrO_4$ can significantly improve by changing the feed conditions. Pretreatment with WGS feed (2% CO, 2% CO_2, 8.1% H_2 and 7.5% H_2O) and removal of CO_2 from this feed make the catalyst to be very stable. The extent of deactivation can be altered by varying the feed conditions during a reaction. They also investigated the effect of synthesis conditions over WGS activity of $Au/CeZrO_4$. They prepared $Au/CeZrO_4$ catalyst by using CP, deposition-precipitation and UGC methods. Among the various methods, the catalyst prepared by deposition-precipitation method exhibits higher activity. They also found that drying procedure, calcination temperature, support Ce to Zr atomic ratio and choice of gold precursor also effect the WGS activity.

Idakiev *et al.* [137] reported mesoporous ceria and nanorods CeO_2 for the Au catalysts for WGSR. Both mesoporous- and nanorod-supported catalysts exhibit higher activity than regular Au/CeO_2. Of the mesoporous- and nanorod-supported catalysts, mesoporous-supported catalyst exhibited higher activity.

Idakiev *et al.* synthesized [138] CeO_2/TiO_2-ZrO_2 support for the Au particles. The support was synthesized through the surfactant templating technique combined with the use of mixed alkoxide solutions. The catalyst exhibits higher activity than Au/TiO_2, Au/ZrO_2 and Au/TiO_2-ZrO_2 catalysts. High catalytic efficiency can be explained by the high structural homogeneity, availability of more oxygen vacancies and an enlarged surface area.

Idakiev *et al.* prepared [139] Au/CeO_2-Fe_2O_3 catalyst for WGSR. They prepared supports using homogeneous CP method by varying the ratio between Ce and Fe. The catalytic activity results are presented in Figure 3.29. The order of activity is Au/CeO_2 > Au/$Ce_{50}Fe_{50}$ > Au/$Ce_{75}Fe_{25}$ > Au/$Ce_{25}Fe_{75}$ > Au/Fe_2O_3. The CO conversion over Au/CeO_2 was significantly higher than that observed over the other catalysts and it was quite close to the equilibrium curve in the high-temperature region. The activity of Au/Fe_2O_3 increased by increasing the temperature and achieved 88% CO conversion at 623 K. This effect can be attributed to significant difference in gold particles size and in the ability of the supports to assist oxygen vacancies formation.

Idakiev *et al.* also [140] investigated the effect of ZrO_2 addition to the CeO_2 support on the WGS activity of Au/CeO_2 catalysts. The WGS activity order is $AuCe_{50}Zr_{50}$ > $AuCe_{80}Zr_{20}$ > AuCe. They proposed that the ZrO_2 addition changes the Au dispersion. Idakiev *et al.* [101] also prepared Au catalysts supported on ceria-modified mesoporous zirconia and found that mesoporous

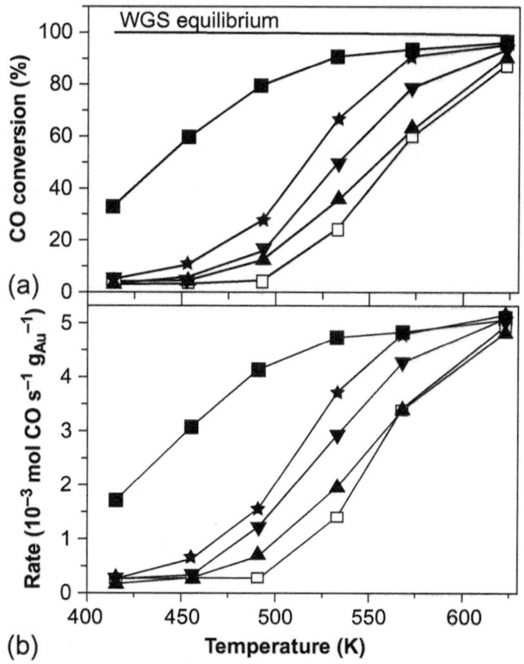

FIGURE 3.29 Temperature dependence of CO conversions (a) and specific rates (b) during WGSR over gold catalysts: (■) Au/CeO_2, (▼) Au/Ce75Fe25, (⋆) Au/Ce50Fe50, (▲) Au/Ce25Fe75 and (□) Au/Fe_2O_2. *(Taken from Figure 1 of Appl. Catal. B Environ. 101 (2011) 266-274.)*

zirconia-doped catalyst has better activity than Au/$CeZrO_4$ catalyst. They also synthesized Au catalysts supported on ceria-modified mesoporous titania and investigated for low-temperature WGSR.

Galletti *et al.* [141] investigated Au/CeO_2-Al_2O_3 and Au/CeO_2-ZrO_2 catalysts in the medium temperature range and found that Au/CeO_2-Al_2O_3 and Au/CeO_2-ZrO_2 catalysts reach equilibrium conversions at temperatures above 300 °C.

In 2010 Andreeva *et al.* [142] investigated Al and Sm dopants for the CeO_2 support. They also investigated the support preparation method on the WGS activity. They prepared supports by CP and mechano-chemical activation (MA) methods. The order of activity is AuCeSm-MA > AuCeAlMA > AuCeSmCP > AuCe ≫ AuCeAlCP. They proposed that the high activity of AuCeSmCP catalyst could be related with the high dispersion of Au nanoparticles leading to large number of active Au-support interface sites (Au^{1+}-Ce^{3+}), which results in high concentration of formate species. The higher concentration of Au^{1+}-Ce^{3+} sites on the surface of the AuCeSmCP catalyst facilitates the dissociation of water. Then they investigated Fe, Mn and Sn dopants for CeO_2 support. The dopants improve oxygen storage capacity of ceria and thus WGS activity. The order of WGS activities at 250 °C of the present catalysts is the following: AuCeFe-MA ≈ AuCeMnMA > AuCe > AuCeSnMA > AuCeFeCP > AuCeMnCP > Au-CeSnCP. Recently, Reina *et al.* [143] investigated Au/CeO_2-ZnO/Al_2O_3 catalysts for the WGSR.

3.3.5.5 Bimetallic Gold Catalysts

Researchers also investigated bimetallic gold catalysts supported on ceria for low-temperature WGSR. Aksoylu and Çağlayan [144] studied ceria-supported Au-Re catalysts for the low-temperature WGSR. They were prepared by bimetallic catalysts using three different methods, namely, co-impregnation, sequential impregnation and deposition-precipitation method. Catalytic performance results show that Re addition effects the WGS activity of Au/ceria strongly. Au addition by deposition-precipitation on impregnated Re/ceria led to higher dispersion, stronger interaction between Au and Re particles; therefore, high CO conversion levels were obtained.

Tsang *et al.* [145] reported Au-Pt and Au-Pd catalysts for the low-temperature WGSR. Among the screened catalysts Au-Pt/CeO_2 displayed the best catalytic activity. This study is the first study on bimetallic gold promotion on ceria for WGSR and they also report the high activity of gold-platinum on ceria for the low-temperature WGSR. Finally, they proposed that the role of the gold-containing promoter is to alter the local band structure of ceria facilitating its redox properties at low temperature.

Suo *et al.* [146] also investigated bimetallic Au catalysts for the low-temperature WGSR. Bimetallic Au-M (with M = Ni, Cu, Ag, Pt and Pd) catalysts supported on CeO_2 were prepared by step-by-step impregnation method. Among them Au-Pt/CeO_2 catalyst shows the highest activity. They also investigated the effect of impregnation procedure on WGS activity of Au-Pt/CeO_2 catalysts. Relatively, Au-Pt/CeO_2 catalyst exhibits higher activity than Pt-Au/CeO_2 or Au(Pt)/CeO_2 catalysts. Furthermore, the activity increases obviously with Pt loadings in Au-Pt/CeO_2. Reports are also available on Au-Co bimetallic catalysts.

Although Au/CeO_2 catalysts are very active catalysts for low-temperature WGSR the main problem associated with them is deactivation with time. Thompson and Kim [147] investigated the deactivation mechanism of Au/CeO_2 catalysts during low-temperature WGSR. They found that primary reason for deactivation is the formation of carbonates and formate species. These species appeared to be formed by CO and H_2 and their formation was facilitated by oxygen deficient sites on ceria. However, the catalytic activity was fully recovered by calcination of the deactivated materials in flowing air at elevated temperatures. Behm *et al.* [148] studies also support this observation. They investigated the deactivation mechanism by TEM, XRD, XPS and DRIFTS studies. Deactivation of Au/CeO_2-supported catalysts is related mainly to the formation of mono-dentate carbonates. These are adsorbed on the ceria support and possibly also on the active sites for formate formation and decomposition, and thus block the surface for the formation/decomposition of adsorbed formate species as reaction intermediates. Catalyst deactivation is also accompanied by a more or less pronounced decrease in Au^{n+} species and increase in Ce^{3+} species, depending on the reaction atmosphere. They also reported that sintering of Au or ceria particles can be ruled out as a significant contributor to catalyst deactivation.

Behm *et al.* [135] studied the effect of H_2, CO_2 and H_2O on the deactivation behaviour of oxidatively and reductively pretreated Au/CeO_2 catalyst. They found that pretreatment has huge influence on the WGS activity of gold catalysts. Reductive pretreatment leads to higher stability compared to oxidative pretreatment. Addition of CO_2 and H_2 decreases both the activity and stability of the catalysts. For oxidative pretreatment catalysts the addition of H_2 leads to more deactivation compared to addition of CO_2. However, for reductive pretreatment catalysts the trend is the opposite. Increase in the water concentration leads to better stability.

Thompset *et al.* [148] investigated the deactivation mechanism of the Au/$CeZrO_4$ catalyst by using both experimental and theoretical studies. The rate of catalyst deactivation as a function of water content is shown in Figure 3.30.

A linear increase in the rate of deactivation during long-term exposure to the WGS conditions was found as the water concentration was raised. They proposed that hydrolysis of the interface between the gold and the support reduces the metal-support interaction and causes the metal nanoparticle to detach from the surface. The intimate contact between the support and the gold is essential for high WGS activity, and the deactivation rate is directly proportional to the concentration of water in the reaction mixture. Over-reduction of the catalyst

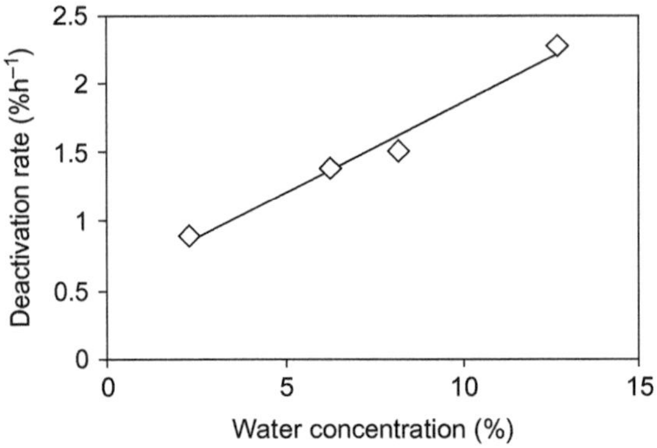

FIGURE 3.30 Rate of catalysts deactivation as a function of the water content of the WGS mix for a 2% Au/$CeZrO_4$ catalyst at 200 °C. *(Figure 3 of J. Catal. 250 (2007) 139-150.)*

and site blocking by formates and carbonates are not thought to be significant in the deactivation on-stream in the present case.

Then, they [148] performed *in situ* DRIFT studies to study the mechanism. Their DRIFTS-GC studies show that under varying feed conditions/reaction temperatures metallic Au was found to be the important species for the WGSR. $Au^{\delta+}$ species were shown to be less stable under the feed and were no longer observed at reaction temperatures above 200 °C when the catalyst was still active. They observed a reversible effect on the stability upon removal and re-introduction of CO_2 into the feed. The thermal deactivation was irreversible and significantly faster than the deactivation observed on 1 wt% Au/$CeZrO_4$ under varied feed conditions.

Scurrell and Venugopal [149] reported hydroxyapatite support for the Au catalyst. Stephanopolous *et al.* [150] reported La_2O_3 and $La_2O_2SO_4$ supports for the Au particles.

REFERENCES

[1] N. Ray, S. K. Roy, N. C. Ganguli, S. P. Sen, Deactivation of low temperature shift catalysts, J. Res. Inst. Catal. 21 (1973) 187–199.

[2] J. M. Moe, Low temperature CO conversion, Am. Chem. Soc. Div. Fuel Chem. 8 (1963) B29–B34.

[3] H. Uchida, N. Isogai, M. Oba, T. Hasegawa, The zinc oxide-copper catalyst for carbon monoxide-shift conversion. I. The dependency of the catalytic activity on the chemical composition of the catalyst, Bull. Chem. Soc. Jpn. 40 (1967) 1981–1986.

[4] T. M. Yureva, G. K. Boreskov, V. Sh. Gruver, Kinet. Katal. 10 (1969) 294.

[5] E. G. M. Kuijpers, R. B. Tjepkema, W. J. J. van der Wal, Structure-sensitivity of the water-gas shift reaction over highly active Cu/SiO_2 catalysts, Appl. Catal. 25 (1986) 139–147.

[6] E. Xue, M. O'Keeffe, J. R. H. Ross, Water-gas shift conversion using a feed with a low steam to carbon monoxide ratio and containing sulphur, Catal. Today 30 (1996) 107–118.

[7] M. J. L. Gines, N. Amadeo, M. Laborde, C. R. Apesteguia, Activity and structure-sensitivity of the water-gas shift reaction over Cu-Zn-Al mixed oxide catalysts, Appl. Catal. A Gen. 131 (1995) 283–296.

[8] G. C. Chinchen, M. S. Spencer, Sensitive and insensitive reactions on copper catalysts: the water-gas shift reaction and methanol synthesis from carbon dioxide, Catal. Today 10 (1991) 293–301.

[9] J. R. Mellor, N. J. Covflle, A. C. Sofianos, R. G. Copper, Raney copper catalysts for the water-gas shift reaction. I. Preparation, activity and stability, Appl. Catal. A Gen. 164 (1997) 171–183.

[10] A. A. G. Lima, M. Nele, E. L. Moreno, H. M. C. Andrade, Composition effects on the activity of Cu-ZnO-Al_2O_3 based catalysts for the water gas shift reaction: a statistical approach, Appl. Catal. A Gen. 171 (1998) 31–43.

[11] W. Fu, Z. Bao, W. Ding, K. Chou, Q. Li, The synergistic effect of the structural precursors of Cu/ZnO/Al_2O_3 catalysts for water-gas shift reaction, Catal. Commun. 12 (2011) 505–509.

[12] A. Budiman, M. Ridwan, S. M. Kim, J.-W. Choi, C. W. Yoon, J.-M. Haa, D. J. Suha, Y.-W. Suhe, Design and preparation of high-surface-area Cu/ZnO/Al_2O_3 catalysts using a modified co-precipitation method for the water-gas shift reaction, Appl. Catal. A Gen. 462-463 (2013) 220–222.

[13] H. Meland, T. Johannessen, B. Arstad, H. J. Venvik, M. Ronning, A. Holmen, Scientific bases for the preparation of heterogeneous catalysts, Stud. Surf. Sci. Catal. 162 (2006) 985.

[14] R. T. Figueiredo, M. S. Santos, H. M. C. Andrade, J. L. G. Fierro, Effect of alkali cations on the Cu/ZnO/Al_2O_3 low temperature water gas-shift catalyst, Catal. Today 172 (2011) 166–170.

[15] R. Knapp, S. A. Wyrzgol, A. Jentys, J. A. Lercher, Water-gas shift catalysts based on ionic liquid mediated supported Cu nanoparticles, J. Catal. 276 (2010) 280–291.

[16] Y. Tanaka, T. Utaka, R. Kikuchi, K. Sasaki, K. Eguchi, Water gas shift reaction over Cu-based mixed oxides for CO removal from the reformed fuels, Appl. Catal. A Gen. 242 (2003) 287–295.

[17] K. Zhi, Q.-S. Liu, Y.-G. Zhang, S. He, R. He, Effect of precipitator on the texture and activity of copper-manganese mixed oxide catalysts for the water gas shift reaction, J. Fuel Chem. Technol. 38 (2010) 445–451.

[18] T. Tabakova, V. Idakiev, G. Avgouropoulos, C. J. Papavasiliou, M. Manzoli, F. Boccuzzi, T. Ioannides, Highly active copper catalyst for low-temperature water-gas shift reaction prepared via a Cu-Mn spinel oxide precursor, Appl. Catal. A Gen. 451 (2013) 184–191.

[19] X. Du, Z. Yuan, L. Cao, C. Zhang, S. Wang, Water gas shift reaction over Cu-Mn mixed oxides catalysts: effects of the third metal, Fuel Process. Technol. 89 (2008) 131–138.

[20] R. He, Q. Liu, F. Wu, C. L. Zhou, K. Zhi, H. B. Yao, Effects of La-doping on the structure and performance of copper-manganese oxides for the water-gas shift reaction, Adv. Mater. Res. 512-515 (2012) 2131–2134.

[21] G. Aguila, A. Valenzuela, S. Guerrero, P. Araya, WGS activity of a novel Cu-ZrO_2 catalyst prepared by a reflux method. Comparison with a conventional impregnation method, Catal. Commun. 39 (2013) 82–85.

[22] K. Sagata, H. Yahiro, Redox properties of Fe-promoted Cu/Al_2O_3 catalysts active for water gas-shift reaction, Bull. Chem. Soc. Jpn. 85 (2012) 511–516.

[23] M. M. Hossain, S. Ahmed, Cu based mixed metal oxide catalysts for WGSR: reduction kinetics and catalytic activity, Can. J. Chem. Eng. 91 (2013) 1450–1458.

[24] N. Thouchprasitchai, A. Luengnaruemitchai, S. Pongstabodee, Water-gas shift reaction over Cu-Zn, Cu-Fe, and Cu-Zn-Fe composite-oxide catalysts prepared by urea-nitrate combustion, J. Ind. Eng. Chem. 19 (2013) 1483–1492.

[25] T. Shishido, M. Yamamoto, I. Atake, D. Li, Y. Tian, H. Morioka, M. Hondac, T. Sano, K. Takehira, Cu/Zn-based catalysts improved by adding magnesium for water-gas shift reaction, J. Mol. Catal. A Chem. 253 (2006) 270–278.

[26] D. C. Yeragi, N. C. Pradhan, A. K. Dalai, Low-temperature water-gas shift reaction over Mn-promoted Cu/Al_2O_3 catalysts, Catal. Lett. 112 (2006) 139–148.

[27] H. Yahiro, K. Murawaki, K. Saiki, T. Yamamoto, H. Yamaura, Study on the supported Cu-based catalysts for the low-temperature water-gas shift reaction, Catal. Today 126 (2007) 436–440.

[28] R. Kam, C. Selomulya, R. Amal, J. Scott, The influence of La-doping on the activity and stability of Cu/ZnO catalyst for the low-temperature water-gas shift reaction, J. Catal. 273 (2010) 73–81.

[29] Y. Li, Q. Fu, M. F. Stephanopoulos, Low-temperature water-gas shift reaction over Cu- and Ni-loaded cerium oxide catalysts, Appl. Catal. B Environ. 27 (2000) 179–191.

[30] A. L. M. da Silvaa, J. M. Assaf, Activity of Cu/CeO_2 and Cu/CeO_2-ZrO_2 for low temperature water-gas shift reaction, Stud. Surf. Sci. Catal. 167 (2007) 213–218.

[31] K. Zhi, Q. Liu, J. Li, R. He, F. Wu, The effect of the precipitator concentration on the activity of mesoporous Cu-Ce-La mixed oxide catalyst for water-gas shift catalyst, Appl. Mech. Mater. 291-294 (2013) 597–601.

[32] J.-K. Kim, Y. Kim, J.-W. Park, J.-S. Bae, D.-Y. Yoon, J.-G. Lee, J.-H. Kim, C. Han, Preparation of CuO-CeO_2-Al_2O_3 catalyst with mesopore structure for water gas shift reaction, Korean J. Chem. Eng. 26 (2009) 32–35.

[33] P. Djinovic, J. Batista, J. Levecb, A. Pintar, Comparison of water-gas shift reaction activity and long-term stability of nanostructured CuO-CeO_2 catalysts prepared by hard template and co-precipitation methods, Appl. Catal. A Gen. 364 (2009) 156–165.

[34] X. Lin, C. Chen, J. Ma, X. Fang, Y. Zhan, Q. Zheng, Promotion effect of Nb^{5+} for Cu/CeO_2 water gas shift reaction catalyst by generating mobile electronic carriers, Int. J. Hydrogen Energy 38 (2013) 11841–11852.

[35] C. G. Maciel, T. F. Silva, E. M. Assaf, J. M. Assaf, Hydrogen production and purification from the water-gas shift reaction on CuO/CeO_2-TiO_2 catalysts, Appl. Energy 112 (2013) 52–59.

[36] E. Poggio-Fraccari, F. Mariño, M. Laborde, G. Baronetti, Copper and nickel catalysts supported on praseodymium-doped ceria (PDC) for the water-gas shift reaction, Appl. Catal. A Gen. 460-461 (2013) 15–20.

[37] V. M. Shinde, G. Madras, Water gas shift reaction over multi-component ceria catalysts, Appl. Catal. B Environ. 123-124 (2012) 367–378.

[38] L. Jiang, C. Li, Z. Li, S. Zhang, Effect of additives on the activity of $CuO/Ce_{0.6}Zr_{0.4}O_2$ catalysts for the water-gas shift reaction, Chem. Eng. Technol. 36 (2013) 1891–1898.

[39] A. Andreev, V. Idakiev, K. Kostov, M. Gabrovska, Water-gas shift reaction over nickel hydroxides, Catal. Lett. 31 (1995) 245–252.

[40] A. Haryanto, S. D. Fernando, S. D. FilipTo, P. H. Steele, L. Pordesimo, S. Adhikari, Hydrogen production through the water-gas shift reaction: thermodynamic equilibrium versus experimental results over supported Ni catalysts, Energy Fuel 23 (2009) 3097–3102.

[41] S. D. Senanayake, J. Evans, S. Agnoli, L. Barrio, T.-L. Chen, J. Hrbek, J. A. Rodriguez, Water-gas shift and CO methanation reactions over Ni-CeO_2(1 1 1) catalysts, Top. Catal. 54 (2011) 34–41.

[42] V. M. Shinde, G. Madras, Nanostructured Pd modified Ni/CeO_2 catalyst for water gas shift and catalytic hydrogen combustion reaction, Appl. Catal. B Environ. 132-133 (2013) 28–38.

[43] J.-W. Park, S.-W. Lee, C.-B. Lee, J.-S. Park, D.-W. Lee, S.-H. Kim, S.-S. Kim, S.-K. Ryi, Single-stage temperature-controllable water gas shift reactor with catalytic nickel plates, J. Power Sources 247 (2014) 280–285.
[44] C. R. Prichard, C. N. Hinshelwood, The interaction of hydrogen and carbon dioxide on the surface of platinum, J. Chem. Soc. 127 (1925) 806–811.
[45] R. Radhakrishnan, R. R. Willigan, Z. Dardas, T. H. Vanderspurt, Water gas shift activity and kinetics of Pt/Re catalysts supported on ceria-zirconia oxides, Appl. Catal. B Environ. 66 (2006) 23–28.
[46] D. C. Greenoble, M. M. Estadt, D. F. Ollis, The chemistry and catalysis of the water gas shift reaction. 1. The kinetics over supported metal catalysts, J. Catal. 67 (1981) 90–102.
[47] R. J. Gorte, S. Zhao, Studies of the water-gas-shift reaction with ceria-supported precious metals, Catal. Today 104 (2005) 18–24.
[48] Q. Fi, H. Saltburg, M. F. Stephanopolous, Active nonmetallic Au and Pt species on ceria-based water-gas shift catalysts, Science 301 (2003) 935–938.
[49] A. Luengnaruemitchai, S. Osuwan, E. Gulari, Comparative studies of low-temperature water-gas shift reaction over Pt/CeO_2, Au/CeO_2, and Au/Fe_2O_3 catalysts, Catal. Commun. 4 (2003) 215–221.
[50] P. Panagiotopoulou, J. Papavasiliou, G. Avgouropoulos, T. Loannides, D. I. Kondarides, Water-gas shift activity of doped Pt/CeO_2 catalysts, Chem. Eng. J. 134 (2007) 16–22.
[51] O. Thinon, F. Deihl, P. Avenier, Y. Schuurman, Screening of bifunctional water-gas shift catalysts, Catal. Today 137 (2008) 29–35.
[52] M. Masuda, K. Migahrua, Isotopic study of water gas reaction catalyzed by platinum, Bull. Chem. Soc. Jpn. 47 (1974) 1058–1063.
[53] L. Mendelovici, M. Steinberg, Methanation and water-gas shift reactions over Pt/CeO_2, J. Catal. 96 (1985) 285–287.
[54] E. Xue, M. O. Keeffe, J. R. H. Ross, A study of Pt/ZrO_2 catalysts for water-gas shift reaction in the presence of H_2S, Stud. Surf. Sci. Catal. 130 (2000) 3813–3818.
[55] K.-R. Hwang, S.-Ki Ihm, S.-C. Park, J.-S. Park, Pt/ZrO_2 catalyst for a single-stage water-gas shift reaction: Ti addition effect, Int. J. Hydrogen Energy 38 (2013) 6044–6051.
[56] H. Xie, J. Lu, M. Shekhar, J. W. Elam, W. N. Delgass, F. H. Ribeiro, E. Weitz, K. R. Poeppelmeier, Synthesis of Na-stabilized nonporous t-ZrO_2 supports and Pt/t-ZrO_2 catalysts and application to water-gas-shift reaction, ACS Catal. 3 (2013) 61–73.
[57] P. Bera, S. Malwadkar, A. Gayena, C. V. V. Satyanarayana, B. S. Rao, M. S. Hegde, Low-temperature water gas shift reaction on combustion synthesized $Ce_{1-x}Pt_xO_{2-\delta}$ catalyst, Catal. Lett. 96 (2004) 213–219.
[58] I. Ivanov, P. Petrova, V. Georgiev, T. Batakliev, Y. Karakirova, V. Serga, L. Kulikova, A. Eliyas, S. Rakovsky, Comparative study of ceria supported nano-sized platinum catalysts synthesized by extractive-pyrolytic method for low-temperature WGS reaction, Catal. Lett. 143 (2013) 942–949.
[59] G. Jacobs, E. Chenu, P. M. Patterson, L. Williams, D. Sparks, G. Thomas, B. H. Davis, Water-gas shift: comparative screening of metal promoters for metal/ceria systems and role of the metal, Appl. Catal. A Gen. 258 (2004) 203–214.
[60] G. Jacobs, A. Crawford, L. Williams, P. M. Patterson, B. H. Davis, Low temperature water-gas shift: comparison of thoria and ceria catalysts, Appl. Catal. A Gen. 267 (2004) 27–33.
[61] S. Ricote, G. Jacobs, M. Milling, Y. Ji, P. M. Patterson, B. H. Davis, Low temperature water-gas shift: characterization and testing of binary mixed oxides of ceria and zirconia promoted with Pt, Appl. Catal. A Gen. 303 (2006) 35–47.

[62] A. Gayen, M. Boaro, C. Leitenburg, J. Llorca, A. Trovarelli, Activity, durability and microstructural characterization of ex-nitrate and ex-chloride $Pt/Ce_{0.56}Zr_{0.44}O_2$ catalysts for low temperature water gas shift reaction, J. Catal. 270 (2010) 285–298.

[63] D.-W. Jeong, H. S. Potdar, J.-O. Shim, W.-J. Jang, H.-S. Roh, H_2 production from a single stage water gas shift reaction over Pt/CeO_2, Pt/ZrO_2, and $Pt/Ce_{(1-x)}Zr_{(x)}O_2$ catalysts, Int. J. Hydrogen Energy 38 (2013) 4502–4507.

[64] A. S. Quiney, G. Germani, Y. Schuurman, Optimization of a water-gas shift reactor over a Pt/ceria/alumina monolith, J. Power Sources 160 (2006) 1163–1169.

[65] K. C. Petallidou, K. Polychronopoulou, S. Boghosian, S. Garcia-Rodriguez, A. M. Efstathiou, Water-gas shift reaction on $Pt/Ce_{1-x}Ti_xO_{2-\delta}$: the effect of Ce/Ti ratio, J. Phys. Chem. C 117 (2013) 25467–25477.

[66] K. C. Petallidou, A. M. Efstathiou, Low-temperature water-gas shift on $Pt/Ce_{1-x}La_xO_{2-\delta}$: effect of Ce/La ratio, Appl. Catal. B Environ. 140-141 (2013) 333–347.

[67] A. M. D. Farias, A. P. M. G. Barandas, R. F. Perez, M. A. Fraga, Water-gas shift reaction over magnesia-modified Pt/CeO_2 catalysts, J. Power Sources 165 (2007) 854–860.

[68] K.-R. Hwang, J.-S. Park, S.-K. Ihm, Si-modified Pt/CeO_2 catalyst for a single-stage water-gas shift reaction, Int. J. Hydrogen Energy 36 (2011) 9685–9693.

[69] V. M. Shinde, G. Madras, A single-stage water-gas shift reaction over highly active and stable Si- and Al-substituted Pt/CeO_2 catalysts, Chem. Cat. Chem. 4 (2012) 1968–1978.

[70] R. Buitrago, J. Ruiz-Martínez, J. Silvestre-Albero, A. Sepúlveda-Escribano, F. Rodríguez-Reinoso, Water gas shift reaction on carbon-supported Pt catalysts promoted by CeO_2, Catal. Today 180 (2012) 19–24.

[71] Y. D. Bi, W. Zhang, H. Y. Xu, W. Zhao Li, Nano crystalline CeO_2 in SBA-15: performance of Pt/CeO_2/SBA-15 catalyst for water-gas-shift reaction, Catal. Lett. 119 (2007) 126–133.

[72] H. Iida, K. Kondo, A. Igarashi, Effect of Pt precursors on catalytic activity of Pt/TiO_2 (rutile) for water gas shift reaction at low-temperature, Catal. Commun. 7 (2006) 240–244.

[73] P. Panagiotopoulou, A. Christodoulakis, D. I. Kondarides, S. Boghosian, Particle size effects on the reducibility of titanium dioxide and its relation to the water-gas shift activity of Pt/TiO_2 catalysts, J. Catal. 240 (2006) 114–125.

[74] P. Panagiotopoulou, D. I. Kondarides, A comparative study of the water-gas shift activity of Pt catalysts supported on single (MO_x) and composite (MO_x/Al_2O_3, MO_x/TiO_2) metal oxide carriers, Catal. Today 127 (2007) 319–329.

[75] P. Panagiotopoulou, D. I. Kondarides, Effects of promotion of TiO_2 with alkaline earth metals on the chemisorptive properties and water-gas shift activity of supported platinum catalysts, Appl. Catal. B Environ. 101 (2011) 738–746.

[76] M. C. Ribeiro, G. Jacobs, L. Linganiso, K. G. Azzam, U. M. Graham, B. H. Davis, Low temperature water gas shift: evaluation of Pt/HfO_2 and correlation between reaction mechanism and periodic trends in tetravalent (Ti, Zr, Hf, Ce, Th) metal oxides, ACS Catal. 1 (2011) 1375–1383.

[77] H. Iida, K. Tahara, H. Higashi, A. Igarashi, Adv. Technol. Mater. Mater. Proc. J. 4 (2002) 62.

[78] H. Iida, A. Igarashi, Structure characterization of $Pt\text{-}Re/TiO_2$ (rutile) and $Pt\text{-}Re/ZrO_2$ catalysts for water gas shift reaction at low-temperature, Appl. Catal. A Gen. 303 (2006) 192–198.

[79] S. Y. Choung, M. Ferrandon, T. Krause, Pt-Re bimetallic supported on CeO_2-ZrO_2 mixed oxides as water-gas shift catalysts, Catal. Today 99 (2005) 257–262.

[80] H. C. Lee, D. Lee, O. Y. Lim, S. Kim, Y. Tae, K. E.-Y. Ko, E. D. Park, ZrO_2-supported Pt catalysts for water gas shift reaction and their non-pyrophoric property, Stud. Sur. Sci. Catal. 167 (2007) 201–206.

[81] Y. T. Kim, E. Park, H. C. Lee, D. Lee, K. H. Lee, Water-gas shift reaction over supported Pt-CeOx catalysts, Appl. Catal. B Environ. 90 (2009) 45–54.

[82] B. S. Caglayan, A. E. Aksoylu, Water-gas shift reaction over bimetallic Pt-Ni/Al_2O_3 catalysts, Turk. J. Chem. 33 (2009) 249–256.

[83] J. H. Pazmiño, M. Shekhar, W. D. Williams, M. C. Akatay, J. T. Miller, W. N. Delgass, F. H. Ribeiro, Metallic Pt as active sites for the water-gas shift reaction on alkali-promoted supported catalysts, J. Catal. 286 (2012) 279–286.

[84] I. D. Gononzalez, R. M. Navarro, M. C. A. Galvan, F. Rosa, J. L. G. Fierro, Performance enhancement in the water gas shift reaction of platinum deposited over a cerium modified TiO_2 support, Catal. Commun. 9 (2008) 1759–1765.

[85] K. G. Azzam, I. V. Babich, K. Seshan, L. Lefferts, Single stage water gas shift conversion over Pt/TiO_2 – problem of catalyst deactivation, Appl. Catal. A Gen. 338 (2008) 66–71.

[86] W. Ruettinger, X. Liu, R. J. Farrauto, Mechanism of aging for a Pt-ceria-zirconia water gas shift catalyst, Appl. Catal. B Environ. 65 (2006) 135–141.

[87] J. M. Zalc, V. Sokolovski, D. G. Loffler, Are noble metal based, water gas shift catalysts practical for automotive fuel processing, J. Catal. 206 (2002) 169–171.

[88] A. N. Basińska, F. Domka, Changes in the activity of Ru/Fe_2O_3 catalysts modified with alkali metal salts in the water gas shift reaction, React. Kinet. Catal. Lett. 66 (1999) 3–11.

[89] A. N. Basińska, F. Domka, Influence of iron influence of iron oxide support preparation method on the properties of Ru/Fe_2O_3 catalysts for water gas shift reaction, React. Kinet. Catal. Lett. 67 (1999) 111–116.

[90] A. N. Basińska, L. KeÎpinÂski, F. Domka, The effect of support on WGSR activity of ruthenium catalysts, Appl. Catal. A Gen. 183 (1999) 143–153.

[91] A. N. Basińska, F. Domka, The effect of lanthanides on the Ru/Fe_2O_3 catalysts for water-gas shift reaction, Appl. Catal. A Gen. 179 (1999) 241–246.

[92] R. Dictor, A kinetic study of the water-gas shift reaction over Rh/$A1_2O_3$ catalysts, J. Catal. 106 (1987) 458–463.

[93] T. Shido, Y. Iwasawa, Reactant promoted reaction mechanism for water gas shift reaction on Rh, J. Catal. 141 (1993) 71–81.

[94] C. Galletti, P. Djinovi´ca, S. Specchiaa, J. Batista, J. Levec, A. Pintar, V. Specchia, Influence of the preparation method on the performance of Rh catalysts on CeO_2 for WGS reaction, Catal. Today 176 (2011) 336–339.

[95] C. Karakaya, R. Otterstatter, L. Maier, O. Deutschmann, Kinetics of the water-gas shift reaction over Rh/Al_2O_3 catalysts, Appl. Catal. A Gen. 470 (2014) 31–44.

[96] W.-Q. Han, W. Wen, J. C. Hanson, X. Teng, N. Marinkovic, J. A. Rodriguez, One-dimensional ceria as catalyst for the low-temperature water-gas shift reaction, J. Phys. Chem. C 113 (2009) 21949–21955.

[97] M. Haruta, N. Yamada, T. Kobayashi, S. Iijima, Gold catalysts prepared by coprecipitation for low-temperature oxidation of hydrogen and of carbon monoxide, J. Catal. 115 (1989) 301–309.

[98] D. Andreeva, V. Idakiev, T. Tabakova, A. Andreeva, Low temperature water gas shift reaction over Au-α-Fe_2O_3. Low-temperature water-gas shift reaction over Au/α-Fe_2O_3, J. Catal. 158 (1998) 354–355.

[99] D. Andreeva, V. Idakiev, T. Tabakova, A. Andreev, R. Giovanoli, Low-temperature water-gas shift reaction on Auα-Fe_2O_3 catalyst, Appl. Catal. A Gen. 134 (1996) 275–283.

[100] D. Andreeva, V. Idakiev, T. Tabakova, P. Christov, R. Giovanoli, Au/α-Fe_2O_3 catalyst for water-gas shift reaction prepared by deposition-precipitation, Appl. Catal. A Gen. 169 (1998) 9.

[101] F. Boccuzi, A. Chiorino, M. Manzoli, D. Andreeva, T. Tabakova, FTIR study of the low-temperature water-gas shift reaction on Au/Fe_2O_3 and Au/TiO_2 catalysts, J. Catal. 188 (1999) 176.

[102] T. Tabakova, D. Andreeva, V. Idakiev, I. Mitov, Influence of the microscopic properties of the support on the catalytic activity of Au/ZnO, Au/ZrO_2, Au/Fe_2O_3, Au/Fe_2O_3-ZnO, Au/Fe_2O_3-ZrO_2 catalysts for the WGS reaction, Appl. Catal. A Gen. 202 (2000) 91.

[103] A. Venugopal, M. S. Scurrell, Low temperature reductive pretreatment of Au/Fe_2O_3 catalysts, TPR/TPO studies and behaviour in the water-gas shift reaction, Appl. Catal. A Gen. 258 (2004) 241–249.

[104] J. Hua, K. Wei, Q. Zheng, X. Lin, Influence of calcination temperature on the structure and catalytic performance of Au/iron oxide catalysts for water-gas shift reaction, Appl. Catal. A Gen. 259 (2004) 121–130.

[105] A. Venugopal, J. Aluha, D. Mogano, M. S. Scurrell, The gold-ruthenium-iron oxide catalytic system for the low temperature water-gas-shift reaction: the examination of gold-ruthenium interactions, Appl. Catal. A Gen. 245 (2003) 149–158.

[106] J. Huaa, Q. Zheng, Y. Zheng, K. Wei, X. Lin, Influence of modifying additives on the catalytic activity and stability of Au/Fe_2O_3–MO_x catalysts for the WGS reaction, Catal. Lett. 102 (2005) 99–108.

[107] F. Zhang, Q. Zheng, K. Wei, X. Lin, H. Zhang, J. Li, Y. Cao, Improved performance of Au/Fe_2O_3 catalysts promoted with ZrO_2 andNb_2O_5 in the WGS reaction under hydrogen-rich conditions, Catal. Lett. 108 (2006) 131–136.

[108] J. Li, Y. Zhan, F. Zhang, X. Lin, Q. C. Zheng, Au/Fe_2O_3 water-gas shift catalyst prepared by modified deposition-precipitation method, Chin. J. Catal. 29 (2008) 346–350.

[109] M. A. Soria, P. Perez, S. A. C. Carabineiro, F. J. Maldonado-Hodar, A. Mendes, L. M. Madeira, Effect of the preparation method on the catalytic activity and stability of Au/Fe_2O_3 catalysts in the low-temperature water-gas shift reaction, Appl. Catal. A Gen. 470 (2014) 45–55.

[110] H. Sakurai, A. Ueda, T. Kobayashi, M. Haruta, Low-temperature water-gas shift reaction over gold deposited on TiO_2, Chem. Commun. (1997) 271–272.

[111] V. Idakiev, T. Tabakova, Z.-Y. Yuan, B.-L. Sub, Gold catalysts supported on mesoporous titania for low-temperature water-gas shift reaction, Appl. Catal. A Gen. 270 (2004) 135–141.

[112] V. Idakiev, T. Tabakova, K. Tenchev, Z.-Y. Yuan, T.-Z. Ren, B.-L. Su, Gold nanoparticles supported on ceria-modified mesoporous titania as highly active catalysts for low-temperature water-gas shift reaction, Catal. Today 128 (2007) 223–229.

[113] V. Idakieva, Z.-Y. Yuanb, T. Tabakovaa, B.-L. Su, Titanium oxide nanotubes as supports of nano-sized gold catalysts for low temperature water-gas shift reaction, Appl. Catal. A Gen. 281 (2005) 149–155.

[114] E. V. Rebrov, A. Berenguer-Murcia, B. F. G. Johnson, J. C. Schouten, Gold supported on mesoporous titania thin films for application in micro structured reactors in low-temperature water-gas shift reaction, Catal. Today 138 (2008) 210–215.

[115] J. A. Rodriguez, S. Ma, P. Liu, J. Hrbek, J. Evans, M. Pérez, Activity of CeOx and TiOx nanoparticles grown on Au(1 1 1) in the water gas shift reaction, Science 318 (2007) 1757–1760.

[116] A. Sandoval, A. Gomez-Cortes, R. Zanella, G. Dıaz, J. M. Saniger, Gold nanoparticles: support effects for the WGS reaction, J. Mol. Catal. A Chem. 278 (2007) 200–208.

[117] Z. Ma, H. Yin, S. Dai, Performance of $Au/M_xO_y/TiO_2$ catalysts in water-gas shift reaction, Catal. Lett. 136 (2010) 83–91.

[118] M. Shekhar, J. Wang, W.-S. Lee, W. D. Williams, S. M. Kim, E. A. Stach, J. T. Miller, W. N. Delgass, F. H. Ribeiro, Size and support effects for the water-gas shift catalysis over

gold nanoparticles supported on model Al_2O_3 and TiO_2, J. Am. Chem. Soc. 134 (2012) 4700–4708.

[119] V. Idakiev, T. Tabakova, A. Naydenov, Z.-Y. Yuan, B.-L. Su, Gold catalysts supported on mesoporous zirconia for low-temperature water-gas shift reaction, Appl. Catal. B Environ. 63 (2006) 178–186.

[120] V. Idakiev, T. Tabakova, K. Tenchev, Z.-Y. Yuan, T.-Z. Ren, B.-L. Su, Gold catalysts supported on ceria-modified mesoporous zirconia for low-temperature water-gas shift reaction, J. Porous. Mater. 19 (2012) 15–20.

[121] F. Menegazzo, F. Pinna, M. Signoretto, V. Trevisan, F. Boccuzzi, A. Chiorino, M. Manzoli, Highly dispersed gold on zirconia: characterization and activity in low-temperature water gas shift tests, ChemSusChem 1 (2008) 320–326.

[122] J. Li, J. Chen, W. Song, J. Liu, W. Shen, Influence of zirconia crystal phase on the catalytic performance of Au/ZrO_2 catalysts for low-temperature water gas shift reaction, Appl. Catal. A Gen. 334 (2008) 321–329.

[123] J. Li, T. Ta, W. Sang, E. Zhan, W. Shen, Au/ZrO_2 catalysts for low temperature WGS reaction: influence of particle sizes, Gold Bull. 1 (2009) 48–60.

[124] Q. Fu, A. Weber, M. F. Stephanopoulos, Nanostructured $Au–CeO_2$ catalysts for low-temperature water-gas shift, Catal. Lett. 77 (2001) 87–95.

[125] D. Andreeva, V. Idakiev, T. Tabakova, L. Ilievaa, P. Falaras, A. Bourlinos, A. Travlos, Low-temperature water-gas shift reaction over Au/CeO_2 catalysts, Catal. Today 72 (2002) 51–57.

[126] D. Andreeva, I. Ivanov, L. Ilieva, M. V. Abrashev, Gold catalysts supported on ceria and ceria-alumina for water-gas shift reaction, Appl. Catal. A Gen. 302 (2006) 127–132.

[127] Q. Fu, S. Kudriavtseva, H. Saltsburg, M. F. Stephanopoulos, Gold-ceria catalysts for low-temperature water-gas shift reaction, Chem. Eng. J. 93 (2003) 41–53.

[128] Q. Fu, W. Deng, H. Saltsburg, M. F. Stephanopoulos, Activity and stability of low-content gold-cerium oxide catalysts for the water-gas shift reaction, Appl. Catal. B Environ. 56 (2005) 57–68.

[129] Q. Zhang, Y. Zhan, X. Lin, Q. Zheng, The effect of La on Au-ceria catalyst for water gas shift reaction, Catal. Lett. 115 (2007) 143–147.

[130] Y. Wang, S. Lianga, A. Caoa, R. L. Thompson, G. Vesera, Au-mixed lanthanum/cerium oxide catalysts for water gas shift, Appl. Catal. B Environ. 99 (2010) 89–95.

[131] G. Jacobs, S. Ricote, P. M. Patterson, U. M. Graham, A. Dozier, S. Khalid, E. Rhodus, B. H. Davis, Low temperature water-gas shift: examining the efficiency of Au as a promoter for ceria-based catalysts prepared by CVD of an Au precursor, Appl. Catal. A: Gen. 292 (2005) 229–243.

[132] R. Leppelt, B. Schumacher, V. Plzak, M. Kinnea, R. J. Behm, Kinetics and mechanism of the low-temperature water-gas shift reaction on Au/CeO_2 catalysts in an idealized reaction atmosphere, J. Catal. 244 (2006) 137–152.

[133] A. Karpenko, R. Leppelt, V. Plzak, R. J. Behm, The role of cationic Au^{3+} and nonionic Au^0 species in the low-temperature water-gas shift reaction on Au/CeO_2 catalysts, J. Catal. 252 (2007) 231–242.

[134] A. Karpenkoa, R. Leppelta, V. Plzakb, J. Caia, A. Chuvilinc, B. Schumachera, U. Kaiserc, R. J. Behm, Influence of the catalyst surface area on the activity and stability of Au/CeO_2 catalysts for the low-temperature water gas shift reaction, Top. Catal. 44 (2007) 183–198.

[135] A. Abd El-Moemen, G. Kučerova, R. J. Behm, Influence of H_2, CO_2 and H_2O on the activity and deactivation behavior of Au/CeO_2 catalysts in the water gas shift reaction at 300 °C, Appl. Catal. B Environ. 95 (2010) 57–70.

[136] A. A. Fonsecaa, J. M. Fishera, D. Ozkayaa, M. D. Shannon, D. Thompsett, Ceria-zirconia supported Au as highly active low temperature water-gas shift catalysts, Top. Catal. 44 (2007) 223–235.

[137] Z.-Y. Yuan, V. Idakiev, A. Vantomme, T. Tabakova, T.-Z. Ren, B.-L. Su, Mesoporous and nanostructured CeO_2 as supports of nanosized gold catalysts for low-temperature water-gas shift reaction, Catal. Today 131 (2008) 203–210.

[138] V. Idakiev, T. Tabakova, K. Tenchev, Z. Y. Yuan, T. Z. Ren, A. Vantomme, B. L. Su, Gold nanoparticles supported on ceria-modified mesoporous-macroporous binary metal oxides as highly active catalysts for low-temperature water-gas shift reaction, J. Mater. Sci. 44 (2009) 6637–6643.

[139] T. Tabakovaa, M. Manzoli, D. Panevaa, F. Boccuzzib, V. Idakieva, I. Mitov, CO-free hydrogen production over Au/CeO_2-Fe_2O_3 catalysts. Part 2. Impact of the support composition on the performance in the water-gas shift reaction, Appl. Catal. B Environ. 101 (2011) 266–274.

[140] F. Vindigni, M. Manzoli, T. Tabakova, V. Idakiev, F. Boccuzzi, A. Chiorino, Gold catalysts for low temperature water-gas shift reaction: effect of ZrO_2 addition to CeO_2 support, Appl. Catal. B Environ. 125 (2012) 507–515.

[141] C. Galletti, S. Specchia, G. Saracco, V. Specchia, Gold-supported catalysts for medium temperature-water gas shift reaction, Top. Catal. 52 (2009) 688–692.

[142] D. Andreeva, M. Kantcheva, I. Ivanov, L. Ilieva, J. W. Sobczak, W. Lisowskic, Gold supported on ceria doped by Me^{3+} (Me = Al and Sm) for water gas shift reaction: influence of dopant and preparation method, Catal. Today 158 (2010) 69–77.

[143] T. R. Reina, S. Ivanova, J. J. Delgado, I. Ivanov, V. Idakiev, T. Tabakova, M. A. Centeno, J. A. Odriozola, Viability of Au/CeO_2–ZnO/Al_2O_3 catalysts for pure hydrogen production by the water-gas shift reaction, Chem. Catal. Chem. 6 (2014) 1401–1409.

[144] B. S. Çağlayan, A. E. Aksoylu, Water-gas shift activity of ceria supported Au-Re catalysts, Catal. Commun. 12 (2011) 1206–1211.

[145] M. A. H. Juan, C. M. Y. Yeung, S. C. Tsang, A study of co-precipitated bimetallic gold catalysts for water-gas shift reaction, Catal. Commun. 9 (2008) 1551–1557.

[146] Q. Yu, W. Chen, Y. Li, M. Jin, Z. Suo, The action of Pt in bimetallic Au–Pt/CeO_2 catalyst for water-gas shift reaction, Catal. Today 158 (2010) 324–328.

[147] C. H. Kim, L. T. Thompson, Deactivation of Au/CeOx water gas shift catalysts, J. Catal. 230 (2005) 66–74.

[148] A. Karpenko, R. Leppelt, J. Cai, V. Plzak, A. Chuvilin, U. Kaiser, R. J. Behm, Deactivation of an Au/CeO_2 catalyst during the low-temperature water-gas shift reaction and its reactivation: a combined TEM, XRD, XPS, DRIFTS, and activity study, J. Catal. 250 (2007) 139–150.

[149] A. Venugopal, M. S. Scurrell, Hydroxyapatite as a novel support for gold and ruthenium catalysts behaviour in the water gas shift reaction, Appl. Catal. A Gen. 245 (2003) 137–147.

[150] J. D. Lessard, I. Valsamakisz, M. F. Stephanopoulos, Novel Au/La_2O_3 and Au/$La_2O_2SO_4$ catalysts for the water-gas shift reaction prepared via an anion adsorption method, Chem. Commun. 48 (2012) 4857–4859.

Chapter 4

WGS Reaction over Co-Mo Sulphided Catalysts

4.1 INTRODUCTION

The utilization of hydrogen for various purposes such as liquid synthesis, fuel cell applications increased the demand of using the sulphur containing coal for gasification to produce syngas. Although Fe-Cr catalysts are stable to sulphur poisoning, large amount of sulphur deactivates the high temperature WGS catalysts. The low temperature WGS Cu catalysts cannot tolerate the low levels of sulphur. Hence, completely sulphur tolerant catalysts with high activity and selectivity for the WGS reaction would be desirable. Moreover, if such a catalyst was not only sulphur tolerant but if its activity was even enhanced by sulphur, then single-stage WGS reactors with only one subsequent step to remove CO_2 would be sufficient for pure hydrogen production. Then a family of group VI and VIII metals other than Fe and Cr are tested for water-gas shift reaction.

In 1954, German scientists Wustrow, Maedrich and Macura [1] pointed out that not only can the Mo-based catalysts be used as methanation, but also Fisher-Tropsch, hydro-desulphurization (HDS) catalysts have very good WGS activity. Then in 1964, British scientist Stern [2] filed a patent on alkali metal added Co-Mo sulphide catalysts for the water-gas shift reaction.

Then in 1970 Esso research and engineering filed [3] a series of patents over several years on the cobalt oxide/molybdenum oxide/caesium acetate supported alumina catalyst. This catalyst henceforth is called the Aldridge catalyst in tribute to its principal discoverer Clyde L. Aldridge at Esso. Actually these patents describe a family of catalysts. Aldridge believes that the catalyst should have the following components to be active at as low as temperatures below 140 °C.

(a) An alkali metal salt of an acid with a dissociation constant less than 1×10^{-3}.
(b) A hydrogenation/dehydrogenation component, which may be impregnated with the alkali salt.

The alkali salts include carbonates, bicarbonates, phosphates, sulphides, bisulphides, silicates, bisulphites, aluminates, hydroxides, acetates, tungstenates, etc. of sodium, potassium, rhodium and caesium. Suitable hydrogenation/dehydrogenation components can be elements of group V-B, VI-B or group VIII,

Water Gas Shift Reaction. http://dx.doi.org/10.1016/B978-0-12-420154-5.00004-8

either alone or in combination. V, Cr, Mo, W, Fe, Ni or Co may be used. However, mixtures of Fe, Ni or Co together with V, Cr, Mo are suitable.

According to Aldridge, the particular combinations of caesium carbonate, caesium acetate, potassium carbonate, potassium acetate together with Co-Mo are especially attractive. Some of the results of Aldridge experiments are presented in Table 4.1.

Aldridge made the following conclusions from the WGS activity measurements:

1. KoAc and Cs_2CO_3 are nearly as effective as CsoAc for the alkali component.
2. Fe-Mo, Co-V, Co-Cr and Ni-Mo combinations are not as effective as Co-Mo.
3. A catalyst pre-treated with H_2S and run with a reactant stream containing no H_2S is much more active than catalyst which was not pre-treated with H_2S and run with a reactant stream containing no H_2S.
4. A catalyst not pre-treated and run with a reactant stream containing H_2S also much more active than a catalyst which was not pre-treated and run with a reactant stream containing no H_2S.
5. Na_2CO_3 is as effective as caesium acetate for the alkali metal component.
6. Support has minimal effect on the WGS activity.

Then Overstreet performed [3] kinetic and theoretical studies over these catalysts and made following conclusions. The activity of the Aldridge catalysts is a

TABLE 4.1 WGS Activity of Alkali Metal Promoted Co-Mo Catalysts (Temperature: 525 °C, Pressure 550 psig, Space Velocity 27 h^{-1}, Gas Mixture 46% H_2, 53% CO, 1% H_2S)

Catalyst	CO conversion
Co-CsoAc	77.7
Mo-CsoAc	88.5
Co-Mo-CsoAc	99.5
Co-Mo-KoAc	99.5
Co-Mo-Cs_2CO_3	98.8
Fe-Mo-CsoAc	88.7
Ni-Mo-CsoAc	99
Co-V-CsoAc	54
Co-W-CsoAc	57.6

strong function of the amount of alkali metal component and an optimum amount of this component exists. A sudden change in activation energy occurs at relatively higher temperatures as the caesium to Co/Mo ratio decreases. Then Berispek [4] continued the work of Overstreet and made some important conclusions. The WGS activity of the catalysts is strongly dependent on the Cs to Mo ratio. The K-promoted catalysts are quite active compared to the Li and Na impregnated catalysts. The caesium-promoted Zn-Mo catalyst is not active as Co-Mo catalyst.

Then German Scientists Levison and Harris [5] filed patents separately in the same year (1974) on Ni- or Co-promoted Mo sulphided catalysts. In same year British Scientists Reita and Lorenz also filed a patent on Co, Mo and Ni sulphided catalysts for WGS reaction. Then in 1974 and 1975 Exxon Research & Engineering Co. filed two patents on alkali metal doped Co-Mo-Ni catalysts [6,7].

4.2 Mo CATALYSTS

In 1983 Hou *et al.* [8] published Mo/Al_2O_3 catalysts for WGS reaction in open literature. They deposited Mo over high surface area alumina by using impregnation method. Before the reaction the catalysts were sulphided using H_2/H_2S mixture at a constant temperature. During exposure of the catalyst to H_2S/H_2 gas mixtures a rapid uptake of sulphur was noted until the MO/S ratio approached approximately unity, assuming a uniform distribution of sulphur throughout the catalyst sample. The experimental results show that the CO conversion increases with increasing sulphur content during the treatment. As for CO_2 formation as a function of H_2O/CO inlet ratio at a constant H_2S level, the reaction rate was found to go through a maximum. Based on XPS measurements they also proposed mechanism.

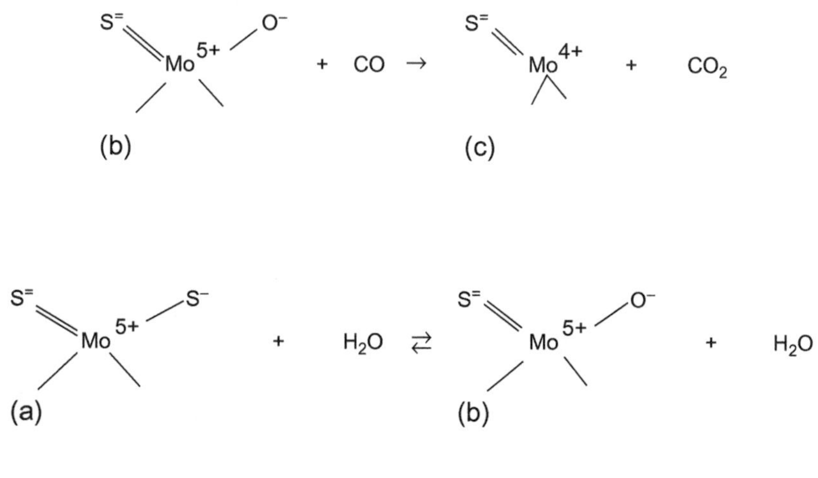

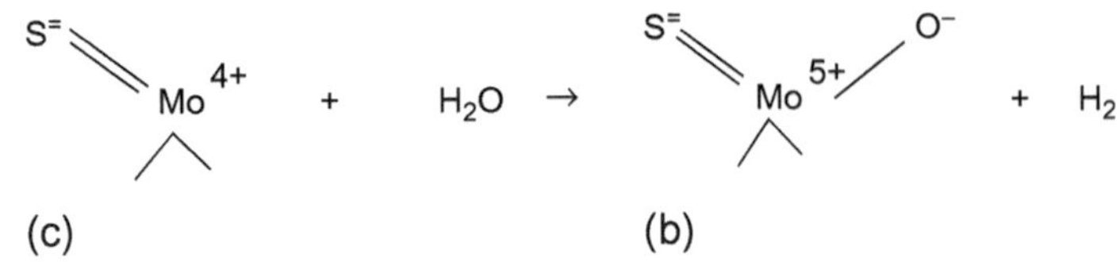

4.3 Co-Mo CATALYSTS

As explained earlier so many patents were filed on the Co-Mo catalysts for WGS reaction by so many researchers. After several years Hakkarainen and Salmi [9] performed transient kinetics over non-sulphided commercial Co-Mo/Al_2O_3 catalyst and published in open literature. The experiments show that the commercial catalyst acts in the WGS reaction mainly as an oxide catalyst, if sulphur is absent in the feed. The most abundant surface intermediates are adsorbed water and its cleavage products as well as CO_2 formed on the surface during the shift reaction. The formation rate of H_2 was always higher than the formation rate of CO_2 during the transient period of the kinetic experiments. The proposed reaction mechanism is as follows:

$$H_2O + {}^*O^* \leftrightarrow 2OH^*$$
$$CO + 2OH^* \leftrightarrow H_2 + CO(O^*)_2$$
$$CO(O^*)_2 \leftrightarrow CO_2 + {}^*O^*$$

where *O* denotes a metal oxide, OH* a surface hydroxyl (i.e., hydrogen on the metal oxide) and $CO(O^*)_2$ a surface carbonate.

Then Lund [10] from SUNY Buffalo University also explained the mechanism of the Co addition to the Mo/Al_2O_3 by using micro-kinetic model. He proposed that the promotional effect of Co is only observed when the steam partial pressure is high. According to the modelling, the most important effects of adding Co are an increase in the surface bond strength of hydroxyl groups relative to sulphydryl groups and a very small decrease in the bond strength of surface oxide. At higher steam partial pressures where the promotional effect is most pronounced, the Co-containing catalyst makes use of two oxidative pathways whereas the un-promoted catalyst uses pre-dominantly one pathway, and this difference can be used to explain the difference in rate between the two catalysts. At lower steam partial pressures the promoted catalyst actually exhibits a lower rate than the un-promoted catalyst.

In 2010 Chen *et al.* [11] performed density functional theory computation calculations to investigate the role of Co on the WGS activity of Mo/Al_2O_3 catalyst. They proposed that Co promoter does not operate directly and actively in H_2O dissociation and CO oxidation steps, and the Co promoter is not an active centre. From the comparison with those on pure MoS_2, the Co promoter changes the adsorption configuration from bridge to top and reduces the reaction barriers and facilitates the WGS reaction. Therefore, the Co promoter plays the role of a textual promoter in creating more active sites and accelerating the reaction rate.

Mo edge calculations reveal that CO, H_2O and H prefer to adsorb on the Co site rather than on the Mo site, while OH and O prefer to adsorb on the Mo site rather than on the Co site. In addition, CO_2 and HCOO (formate) and COOH (carboxy) prefer bridging adsorption on Co and Mo. H_2O prefers dissociative adsorption on the Mo edge with co-adsorbed OH and H. On the contrary, H_2O prefers desorption instead of dissociation on the S edge. Therefore, the Mo edge is the active surface for WGS reaction, and the S edge surface is therefore inactive. On the S edge, CO, H_2O, H, OH and O prefer to adsorb on the Mo site rather than on the Co site. However, CO_2 adsorbs on neither the Co site nor the Mo site. The redox mechanism and the COOH-mediated reaction path have been computed and compared. It is found that COOH is a kinetically favoured intermediate, while OH dissociation into surface O and H needs a higher barrier (0.62 vs. 1.34 eV). However, COOH dissociation needs a much higher barrier than CO oxidation into CO_2 (1.89 vs. 0.30 eV). On the basis of these barriers, the redox mechanism is the preferable WGS reaction path, and the rate-determining step is the second step dissociation of OH into surface O and H.

4.4 EFFECT OF SUPPORT AND OTHER METAL ADDITION

In 1997 Mellor *et al.* [12] reported Co-Mn and Co-Cr catalysts for WGS reaction. Their preliminary activity measurements suggest that Co-Mn (3:1) and Co-Cr (3:1) exhibit excellent WGS activity in the presence of sulphur. They prepared these catalysts with co-precipitation method using ammonia as a precipitating agent. Initially they evaluated Co-Cr catalyst in the absence of pre-sulphidation. As expected CO conversion increases from 50% to 90% with increasing reaction temperature from 400 to 550 °C. The Co-Mn catalyst also exhibited excellent WGS activity in the absence of sulphur at 400 °C for 100 h. Then they investigated pre-sulphided Co-Cr catalyst for WGS reaction. Pre-sulphidation was done using mixture of H_2S, H_2 and N_2. The activity results are presented in Figure 4.1. It can be seen from Figure 4.1 that the WGS activity increases with increasing sulphur concentration. Then they compared Co-Cr catalyst with industrial Co-Mo/Al_2O_3 catalyst. The industrial catalyst exhibits consistently higher CO conversions compared to Co-Cr catalyst. Both pre-sulphided Co-Cr and industrial Co-Mo catalysts exhibit decrease in the activity with increasing WHSV.

Wachowski *et al.* [13] studied Co-Mo sulphides supported on alumina-modified with La_2O_3 for the WGS reaction. Results reveal that only higher concentrations of La_2O_3 increase the WGS activity. After that they studied the effect of support (alumina, titania, zirconia) on the WGS activity of Mo and Mo-Ni catalysts. Mo-impregnated TiO_2 catalyst shows much better activity compared to alumina and zirconia supported catalysts. The addition of Ni to

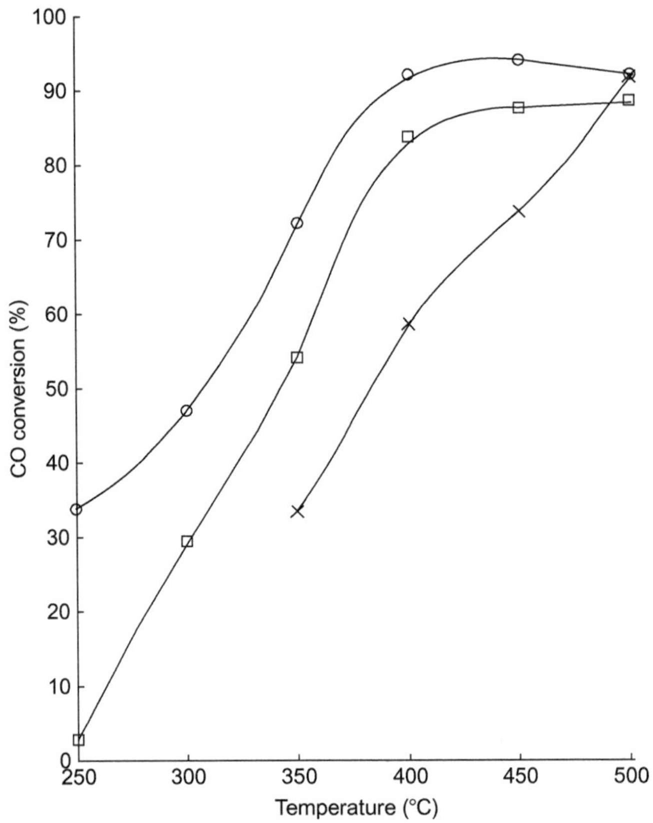

FIGURE 4.1 The effect of temperature and sulphur feed concentration on the WGS activity of a pre-sulphided cobalt-chromium catalyst (CO GHSV=500 h^{-1}; CO:H_2O=1:4.5; catalyst=2±0.1 ml). ○, total sulphur feed=0.743% (0.675% H_2S+0.068% COS): □, total sulphur feed=0.354% (0.321% H_2S+0.033% COS); ×, total sulphur feed=662 ppm COS. *(Taken from Figure 6 of J.R. Mellor, R.G. Copperthwaite, N.J. Coville, Appl. Catal. A Gen. 164 (1997) 69.)*

Mo increases the activity of almost twice in the case of alumina and titania but was almost four times higher for zirconia. Among the various catalysts, the highest activity was measured for titania supported with $Mo(CO)_6$ pre-cursor. The characterization measurements suggest that the high activity of TiO_2-supported catalysts can be related to better homogeneity in coordination of Mo species on titania than on alumina or zirconia. Deposition of nickel over studied supports resulted in the formation of different surface species. Independent of the support applied, the presence of nickel enhanced the reducibility of molybdenum and Ni-Mo-O species and in consequence the catalytic activity of the WGS reaction.

Then they investigated TiO_2-ZrO_2 [14] as a support for the Mo and Mo-Ni catalysts. They prepared TiO_2-ZrO_2 supports by co-precipitation method using ammonia as a precipitation agent. Then Ni and Mo were impregnated using wet impregnation method. In the case of Ni-Mo catalysts, molybdenum was deposited first, and next impregnation with Ni $(NO_3)_2$ was performed. The effects of Ti to Zr composition on the WGS activity of Mo, Ni and Mo-Ni catalysts are presented in Figure 4.2.

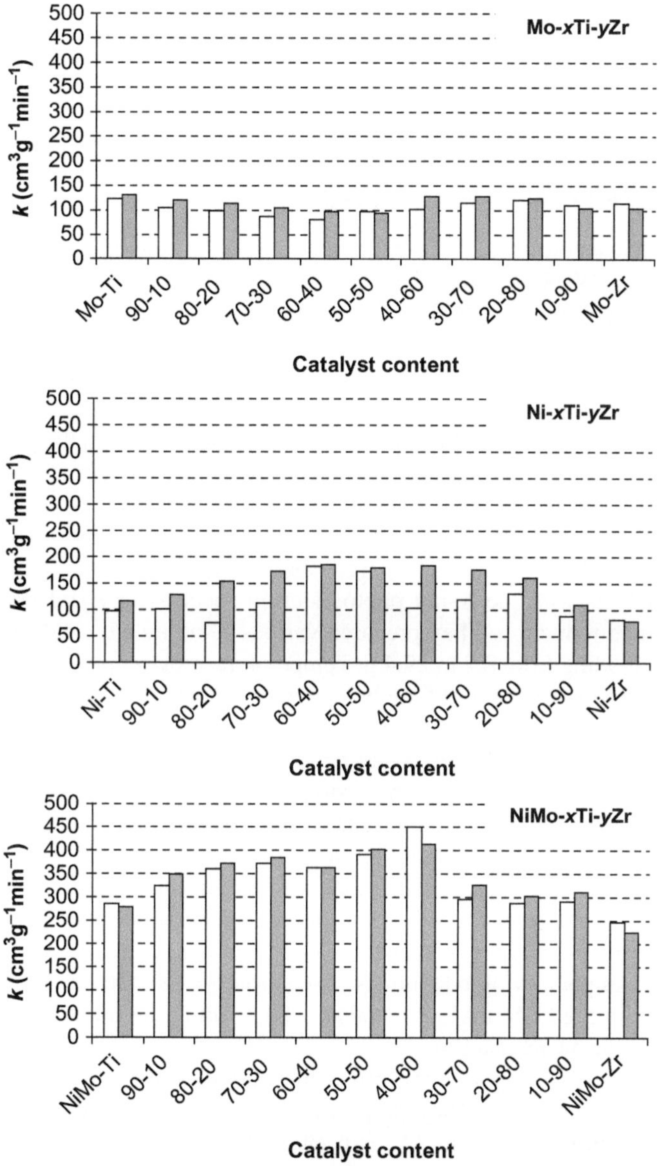

FIGURE 4.2 Catalytic activity of sulphided of Mo, Ni and Ni-Mo catalysts supported on TiO_2-ZrO_2 supports in the WGS reaction after 10 min (white columns) and 120 min (grey columns). *(Taken from Figure 5 of L. Wachowski, M. Laniecki, F. Domka, React. Kinet. Catal. Lett. 67 (1999) 123.)*

The best catalysts were supported on supports composed of 40 wt% of TiO_2 and 60 wt% ZrO_2 and containing 8 wt% of Mo and 3 wt% of Ni. They explained that the better activity of 40 wt% of TiO_2 and 60 wt% ZrO_2 catalyst is due to appropriate ratio of surface acid-base properties of the support. Their characterization measurements suggest that introduction of nickel ions onto the molybdenum catalysts causes the increase of dispersion of sulphided molybdenum species independently from support composition. The presence of Ni^{2+} on the surface of molybdenum catalysts decreases the temperature of reduction of Mo^{6+} ions and therefore facilitates the formation of MoS_2 and Ni-Mo-S species at lower temperature with better dispersion.

4.5 ROLE OF K AS PROMOTER

In 1984 Kantschewa *et al.* [15] reported potassium-promoted Ni-Mo/Al_2O_3 catalyst for the WGS reaction. The potassium-promoted Ni-Mo/Al_2O_3 catalyst exhibited much higher WGS activity compared to the un-promoted catalyst. The WGS activity increases with increasing potassium loading. The increase in the activity is due to two effects. First, the formation of formate species during adsorption of CO is strongly enhanced by K^+ ions. Second, the Mo^{5+} oxidation state seems to be stabilized in the presence of K^+ ions in the catalyst. The activity of the K-containing catalyst is higher by a factor of about 2 than that of Ni-Mo/Al_2O_3 in the sulphided state.

Then Xie *et al.* [16] also reported potassium-promoted Co-Mo/Al_2O_3 catalyst for the WGS reaction. They well characterized the fresh, sulphided and spent catalysts with the X-ray photo electron spectroscopy and temperature-programmed reduction. They prepared various K-promoted catalysts by varying the K loading from 8.2 to 29.4 g. Temperature-programmed reduction measurements the reducibility of molybdenum is enhanced by potassium in the CoMoK/Al_2O_3 catalyst. The sulphidation ability of molybdenum is enhanced by the presence of potassium. X-ray photo electron measurements suggest that after sulphidation the distribution of cobalt and molybdenum varied drastically. The Co distribution remained constant over the surface with increasing K loading. However, the Mo concentration increases over the surface with increasing K loading. The WGS activity increases with increasing K loading up to 14.1 g and further increase in the K loading decreases the WGS activity. The WGS activity results also suggest that the CO conversion decreases with time due to loss of sulphur. Molybdenum is susceptible to be sulphided after adding potassium in the CoMo/$A1_2O_3$ catalyst which should be one of the factors in promoting the WGS reaction.

Park *et al.* [17] reported the effect of K addition on the WGS activity of Co-Mo/Al_2O_3 catalysts. They prepared a series of $K_3Co_{10}M$ catalysts using co-impregnation on alumina. As expected addition of K increases the WGS activity of Co-Mo catalysts up to 10 wt% loading and further increase in the K loading decreases the activity. They also proved that K did not enhance the catalytic activity of either Mo or Co alone. Their TPR measurements suggest that K addition retarded the reducibility of Co and Mo. However, it results in good dispersion and higher surface area because of disappearance of polymolybdate cluster such as $Mo_7O_{24}^{6-}$ and the formation of small MoO_4^{2-} cluster which was confirmed by the Raman spectroscopy and electron diffraction.

In 1999 Andreev *et al.* [18] reported K-promoted Ni-Mo/Al_2O_3 catalysts for the WGS reaction in the presence of sulphur. They studied the influence of reaction medium. They conducted experiments in the absence of sulphur. The order of WGS activity in the absence of sulphur for non-promoted samples is NiMo > Mo > Ni. A decrease in the catalytic activity of the three samples was observed during the first 2 h. This decrease was not considerable even by the end of the test at high space velocities. Repeated sulphidation restored the initial activity of the samples within experimental error. K-promoted samples also behaved in similar manner. The experiment in the presence of water vapour reveals that water vapour induces hydrolysis of the sulphide phase to yield oxysulphide phase(s). The oxysulphide phase(s) manifest(s) also catalytic activity in the WGS reaction. They also found that during the WGS reaction nickel sulphides in the catalyst undergo hydrolysis to form nickel hydroxides. The nickel hydroxides show high catalytic activity in the WGS reaction. As said earlier, they pointed out that K promotes reducibility of Mo catalysts. Then they investigated effect of deposition and order of potassium introduction on the texture and activity of Mo/Al_2O_3 catalysts. They prepared MoK/Al_2O_3 and KMo/Al_2O_3. The order of activity of catalysts are MoK/Al_2O_3 > Mo/Al_2O_3 > KMo/Al_2O_3. The deposition of K and Mo on Al_2O_3 support leads, depending on the order of component introduction, to differences in position and distribution of the active phase in the porous texture of the support. The addition of K to Mo/Al_2O_3 leads to non-uniformity in Mo surface distribution due to migration of part of the MoO_x species to the external surface. The deposition of Mo to K contributes to the uniform distribution of molybdenum-oxygen species in the porous texture of the support. The uniform distribution in the oxidic pre-cursor is also kept constant to a high extent during the catalytic reaction.

Nikolova *et al.* investigated [19] mono-, bi- and tri-component (K)(Ni)(Mo)/g-Al_2O_3 systems after water-gas shift reaction tests in the presence of sulphur in the feed by physico-chemical characterization techniques. The catalytic activity results are presented in Figure 4.3. The order of activity in terms of CO conversions is KNi18Mo > Ni18Mo > Ni = 18Mo = K18Mo ≫ K: They calculated $Mo^{4+}/(Mo^{5+} + Mo^{6+})$ and Mo^{4+}/Mo^{5+} ratios from X-ray photo electron measurements and the decreases in the order: K18Mo > 18Mo > Ni18Mo > KNi18Mo: These results suggest that molybdenum reducibility is a necessary condition but it is not sufficient to determine the catalytic activity properties. The potassium addition stabilizes the Mo^{4+} oxidation state in the tested bi-component KMo sample, while the Mo^{4+} contribution is lowest in the tri-component KNiMo sample.

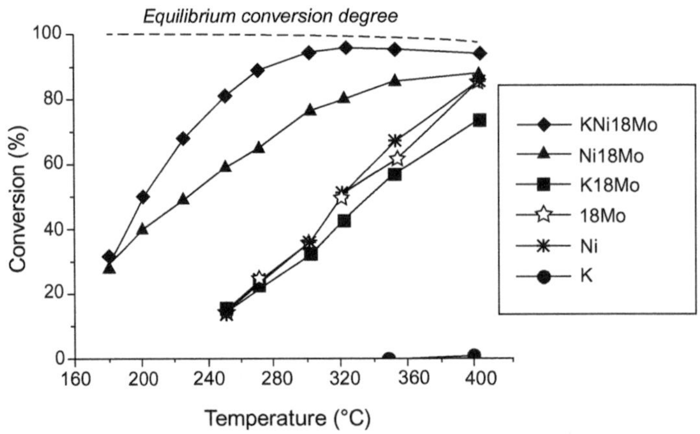

FIGURE 4.3 Catalytic activity of pre-sulphided (K)(Ni)(18Mo)/Al_2O_3 systems; steam/gas ratio 0.3; GHSV 4000 h^{-1}. *(Taken from Figure 1(a) of D. Nikolova, R. Edreva-Kardjieva, G. Gouliev, T. Grozeva, P. Tzvetkov, Appl. Catal. A Gen. 297 (2006) 135.)*

The different sulphur species such as sulphide sulphur, oxysulphides, polysulphides, paramagnetic sulphur and sulphate sulphur are present on the sample surface in the reaction redox environment (CO, H_2O, H_2S, H_2 and CO_2 agents). The sulphur species concentration depends on the sample type, e.g., the nature of the additives and the component number. They also calculated the sulphide/sulphate ratio and ratio increases in the order K18Mo < 18Mo < Ni18Mo < KNi18Mo: This order corresponds to the catalytic activity order. Obviously, the higher catalytic activity of the samples is accompanied by a higher sulphur lability of the surface. The tri-component KNi18Mo sample exhibits the highest catalytic activity. Its molybdenum reducibility is lowest and the sulphur lability is highest. On the contrary, the bi-component K18Mo sample manifests the lowest catalytic activity because the molybdenum reducibility is the highest and the sulphur lability is the lowest compared to other samples.

Then Nikolova *et al.* investigated [20] K-Ni-Mo/Al_2O_3 by varying Mo lading from 5% to18%. They also investigated the effect of reaction temperature from 180 to 400 °C, steam to gas ratio from 0.3 to 1. As expected increase in the Mo content increases the CO conversion. Among the various mono metallic Mo catalysts 18% Mo exhibits 85% CO conversion at 400 °C. The mono metallic Ni sample, in spite of low nickel amount, practically completely repeats the pattern of CO conversion over the 14Mo and 18Mo samples. Introduction of potassium as a second component decreases the activity of all bi-metallic KMo samples (atomic ratio K/Mo = 0.9-3.1) compared to the corresponding monometallic Mo samples. Evidently, potassium addition in the range of K/Mo = 0.9-3.1 does not cause any promoting effect on the activity of the Mo/Al_2O_3 system. As expected unlike K, addition of Ni to Mo-Al_2O_3 increases the CO conversion.

Depending on the K/(Ni+Mo) atomic ratio, introduction of potassium to the NiMo samples as a third component enables approaching the equilibrium conversion. The decrease in activity starts at 300 °C for KNi5Mo (K/(Ni+Mo)= 1.5) and KNi10Mo (K/(Ni+Mo)=1.1) and at 270 °C with KNi14Mo (K/(Ni +Mo)=0.8) and the KNi18Mo sample (K/(Ni+Mo)=0.7).

The increase in catalytic activity of the monometallic molybdenum samples with the increase in molybdenum content of the oxidic precursor may be related to a change in the coordination of the Mo^{6+} ions by analogy with pioneering studies. The observed positive effect of K present on the NiMo catalyst, depending on the Mo content, i.e., the K/(Ni+Mo) ratio, cannot be assigned to the Mo^{6+}(Oh)→Mo6+(Td) partial transition of the polymeric species into the monomeric ones in the oxidic pre-cursor since the KMo samples possess lower activity than the Mo ones. This effect cannot be related to the domination of the Ni2+(Td) species over the Ni2+(Oh) ones in the oxidic pre-cursor. The K promotion is mainly due to the extent of interaction between the components during the calcination step that causes formation of K-Ni-Mo-O species on the support surface of the oxidic pre-cursors.

They also investigated the effect of steam to CO ratio on the WGS activity of KMoNi catalysts and the results are presented in Figure 4.4. The catalytic activity of all the samples at 400 °C is independent of the steam/gas ratio values.

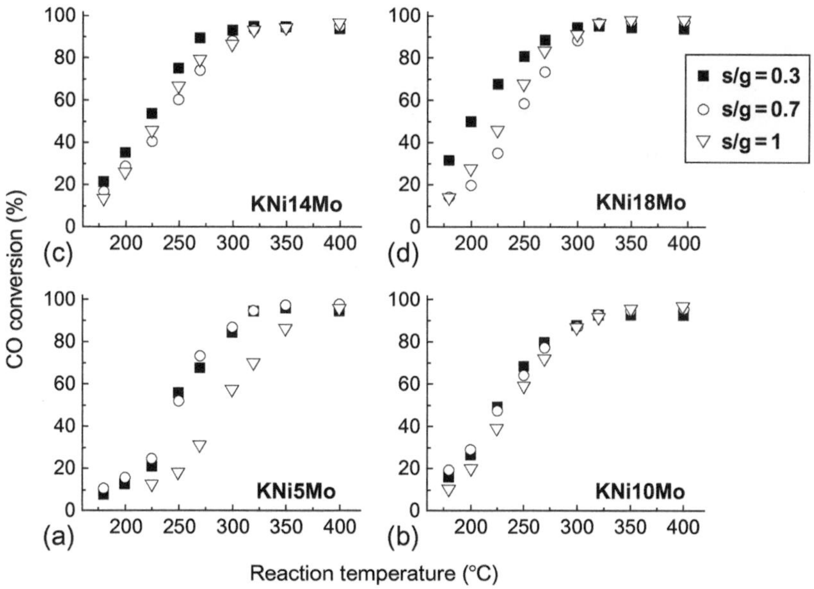

FIGURE 4.4 Influence of the steam to gas ratio (s/g=0.3, 0.7, 1) on the catalytic activity of the KNiMo samples at various reaction temperatures, GHSV=4000 h^{-1}: (a) KNi5Mo; (b) KNi10Mo; (c) KNi14Mo; (d) KNi18Mo. *(Taken from Figure 5 of D. Nikolova, R. Edreva-Kardjieva, T. Grozeva, React. Kinet. Mech. Catal. 103 (2011) 71.)*

They concluded that composition of the KNi14Mo sample can be accepted as optimal because of its comparable catalytic activity with KNi18Mo in the range 270-500 °C.

In 2008 Wang *et al.* [21] reported the effect of MgO addition to the Al_2O_3 support on the WGS activity of Co-Mo-K/Al_2O_3 catalysts. They prepared MgO-Al_2O_3 by mechanical mixing method. Then Co, Mo and K were impregnated using impregnation method. The catalysts were pre-sulphided using a H_2/CO/CS_2 mixture. The order of activity of K-free catalysts is Co-Mo/MgO-Al_2O_3 > Co-Mo/$MgAl_2O_4$ > Co-Mo/Al_2O_3. As expected increasing K loading increases the CO conversion. The optimum potassium loadings expressed as the amounts of K_2O were 4%, 6% and 6% on Co-Mo-K/MgO-Al_2O_3, Co-Mo-K/Al_2O_3 and Co-Mo-K/$MgAl_2O_4$ catalysts, respectively. Then they investigated the effect of reaction temperature on these catalysts. The results are presented in Figure 4.5. Interesting behaviour was observed with the temperature variation. Co-Mo-K/Al_2O_3 exhibits a very high CO conversion at lower reaction temperature and the CO conversion decreases with increasing temperature. On the other hand, Co-Mo/MgO-Al_2O_3 exhibits very low CO conversion at lower reaction temperature and CO conversion increases with increasing temperature. However, Co-Mo-K/MgO-Al_2O_3 catalysts exhibit reasonably higher CO conversions at all reaction temperatures. Under the conditions of temperature of 400 °C, total pressure of 2.0 MPa, GHSV of 15,000 h^{-1} and steam to gas ratio of 0.3, a steady loss of catalytic activity with the reaction time was observed over three catalysts. The rate of deactivation of Co-Mo-K_6/Al_2O_3 catalyst is much quicker than that of MgO-Al_2O_3-supported catalysts. Their ESR characterization shows that the Mo^{5+} species are not connected with the WGSR activity.

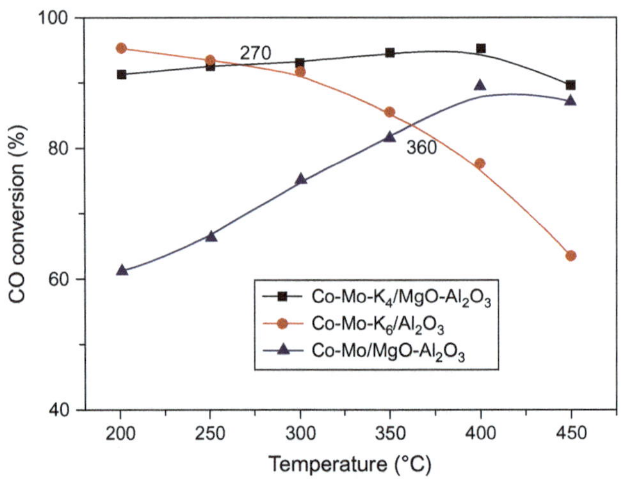

FIGURE 4.5 Effect of temperature on the CO conversion over the Co-Mo-based catalysts. *Reaction conditions*: total pressure = 2.0 MPa, GHSV = 3000 h^{-1} and steam to gas ratio = 0.5. *(Taken from Figure 3 of H. Wang, Y. Lian, Q. Zhang, Q. Li, W. Fang, Y. Yang, Catal. Lett. 126 (2008) 100.)*

Lian *et al.* [22] investigated the effect of support calcination temperature on the WGS activity of Co-Mo/MgO-Al_2O_3 catalysts. The activity testing results showed that the catalysts with the mixed oxide supports calcined at 600-800 °C exhibited high activity for the reaction. The physico-chemical characterization results indicated that MgO-Al_2O_3 mixed oxide supports calcined at 600-800 °C existed as amorphous compound $MgAl_2O_4$-xMgO-$y Al_2O_3$, on which more middle strength acid-base sites were formed. Then they investigated the effect of Mg/Al ratio on the WGS activity of Co-Mo/MgO-Al_2O_3 catalysts. The conversion of CO increases rapidly with increasing Mg/Al atom ratio at 475 K. When the Mg/Al atom ratio was raised to 0.452-0.525 at 475 K, the conversion of CO reached an optimum value, 98.2%. Hereafter, the conversion of CO decreased with the increase of Mg/Al atomic ratio. On the other hand, at the reaction temperatures of 575 and 625 K, the Mg/Al ratio has no effect on the CO conversion, the curves are rather smooth. Their characterization results suggest that Co-Mo-S species has direct effect on catalytic activity. The addition of MgO is propitious to the formation of the active Co-Mo-S phase of the catalyst. Their TPR and XPS results show that the reducibility of molybdenum, e.g., $[Mo^{4+}+Mo^{5+}]/[Mo^{6+}]$ ratio increases with the increase of Mg/Al atom ratio. At the same time, the ratio of peak areas of $[S^{2-}+S_2^{2-}]/[S^{6+}]$ increase in the same order. This suggests that higher reducibility of molybdenum and higher lability of sulphur are essential for higher catalytic activity of the catalysts. On the other hand, as described earlier Andreev *et al.* suggested that the Mo reducibility is a necessary condition but it is not sufficient to determine the catalytic activity properties.

Then Lian *et al.* pre-sulphided Co-Mo/MgO-Al_2O_3 catalyst with *ex situ* pre-sulphidation technique and investigated for WGS reaction [23]. The *ex situ* activation involves synthesis of catalyst using $(NH_4)_2MoS_4$ pre-cursor in ammonium sulphide solution and activation of catalyst in the presence of only H_2. Then they compared the WGS activities of *in situ* and *ex situ* activated catalysts. The *in situ* activation involves synthesis of catalyst with non-sulphide pre-cursor and activated the catalyst in the presence of H_2/H_2S mixture. Both the catalysts exhibit increase in the CO conversion with increasing steam to CO ratio at a fixed temperature. And both catalysts exhibit similar CO conversion. Also, both the catalysts exhibit a decrease in CO conversion with increasing reaction temperature. However, *ex situ* catalyst exhibited 87.2% CO conversion at 450 °C and *in situ* catalyst exhibited 82% only.

A steady loss of catalytic activity with the reaction time was observed over two catalysts and the rate of deactivation of *in situ* activated catalyst is much quicker than *ex situ* activated catalyst. In ammonium sulphide solution the total amount of sulphur is in excess in comparison with their stoichiometric value for MoS_2 plus CoS. After impregnation in ammonium sulphide solution, both the Co_9S_8 and thiomolybdate complexes will be formed which is conformed from X-ray diffraction measurements. However, after pre-reduction in the presence of H_2 these phases were disappeared. It can be understood that the well-dispersed CoMoS phase was formed as a result of the redispersion of Co_9S_8 particles over the edges of newly formed MoS_2 crystallites, thus it suggests that the dispersion of Co-Mo sulphides was optimized in the case of *ex situ* activated catalyst. Therefore, the pre-sulphided Co-Mo/MgO-Al_2O_3 catalyst exhibited a higher activity and stability for WGS reaction.

Then Wang *et al.* investigated [24] the influence of W addition to the WGS activity of Co-Mo/Al_2O_3 catalysts. The WGS activities of the W-promoted Co-Mo-K/Al_2O_3 catalysts increased with increasing atomic ratio of W/Mo and reached a maximum at 0.05 of atomic ratio W/Mo. At higher atomic ratios of W/Mo, the CO conversions decreased with increasing content of tungsten. When the amount of W on the CoMoK-$W_{0.05}$ catalyst was replaced by the same amount of Mo, the obtained CoMoK-$Mo_{0.05}$ and CoMo$_{1.05}$K catalysts (CO conversions are 89.3% and 89.9%, respectively) exhibited lower activities than that of the CoMoK-W0.05 catalyst (CO conversion, 95.7%) but higher than that of the base catalyst (CO conversion, 86.4%). The higher activities of the catalysts with higher Mo content can be attributed to a simple increase of total active sites. Their H_2-TPR and Raman characterization showed that the addition of W decreased the Mo atoms at octahedral sites and increased the Mo atoms at tetrahedral sites.

Zhang *et al.* [25] investigated the influence of Ce on the WGS activity of Co-Mo/Al_2O_3 and Co-Mo-K/Al_2O_3. CeO_2 influences the activity of the Ce-Co-Mo/Al_2O_3 catalyst compared with the Co-Mo/Al_2O_3 catalyst, especially at low temperatures. The WGS activity of Ce-promoted Co-Mo/Al_2O_3 catalysts increased with increasing CeO_2 loading, reached a maximum at 3.0% CeO_2, and then decreased with additional increases in the metal loading. They proposed that higher catalytic activities of catalysts with high Ce content can be attributed to an increased number of total active sites. Further addition of CeO_2 is detrimental to the CO shift reaction. As expected, the addition of Ce to the Co-Mo-K/Al_2O_3 increases the CO conversion and is close to the equilibrium conversion (95.7%) at 300 °C. However, the increased activity (even at lower reaction temperatures) is independent of the CeO_2 and K_2O content (total loadings of 7-10%). Introduction of ceria as an additional promoter increases the activity of bi-promoter Co-Mo-Ce-K samples (K_2O content 6-8) compared to the corresponding mono-promoter Co-Mo-K sample; this difference is more pronounced at lower K_2O content. As expected, the WGS activity is sensitive to steam/CO ratio, an increase in the steam/CO ratio results in the WGS activity. They also investigated the effect of CH_4 on the WGS activity of Co-Mo-K-Ce/Al_2O_3. They suggested that the fluctuating composition of CH_4 in the feed gas did not significantly influence the catalyst activity. Notably, the formation of CH_4 was not detected in the water-gas shift reaction when the CH_4 in the feed was replaced with N_2. This showed that the reformer reaction (the CO methanation reaction) does not occur in the water-gas shift reaction process over the Co-Mo-Ce-K/Al_2O_3 catalyst.

Reports are also available on investigation of metal sulphides other than Co-Mo for the WGS reaction. Valsamakis and Stephanopoulos [26] investigated La and Pr sulphided for high temperature WGS reaction. They prepared catalysts using hydrolysis and surfactant-assistant method. They investigated La and Pr catalysts from ambient to 800 °C, for both samples, the reaction started at approximately 400 °C and reached equilibrium at 800 °C under the conditions examined. The investigated Pr catalyst in the presence of different amounts of H_2S in a simulated fuel gas composition of 11%CO-7%CO_2-26%H_2O-26%H_2-He was examined in the test sequence shown in Figure 4.6. Introduction of 12 and 120 ppm of H_2S in the feed gas, panels 2 and 4 of Figure 4.6, did not affect the conversion over an extended period of time. The conversion was maintained at approximately 12% for about 2 h upon 12 ppm H_2S introduction and was around 10% after 5 h with 120 ppm H_2S in the feed. As shown in panel 5, when the water supply was cut, the catalyst was active and stable in the reverse water-gas shift reaction even in the presence of 120 ppm H_2S. Then they conducted additional sulphur tolerance test at 600 °C. The addition of 700 ppm of H_2S at 600 °C resulted in CO conversion drop from 45% in the sulphur-free gas to 20% in the sulphur-laden gas mixture for $La_2O_2SO_4$ catalyst. However, no permanent poisoning of the surface took place, and upon removal of the H_2S, the CO conversion was fully recovered. Since the catalyst surface is already in the oxysulphided phase, the observed loss in activity may be caused by inhibition due to competitive adsorption. They proposed that the catalyst operates in a clear redox mode derived from the high oxygen storage capacity of the oxysulphide/oxysulphate interconversion. No loss of sulphur from the catalyst takes place with time-on-stream, indicative of a stable catalyst composition. They concluded that compared to the CoMo sulphides, the lanthanide oxysulphides

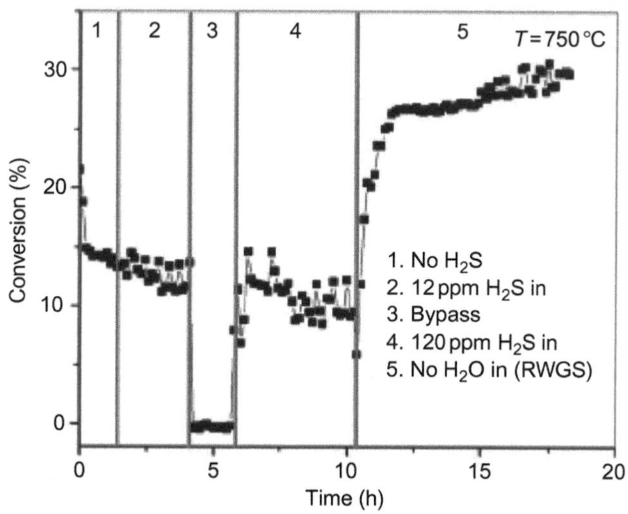

FIGURE 4.6 Sulphur tolerance test of $Pr_2O_2SO_4$ catalyst in the presence of various H_2S amounts. Gas composition: 11%CO-7%CO_2-26%H_2O-26%H_2-He; contact time: 0.04 g s ml^{-1}. *(Taken from Figure 9 of I. Valsamakis, M.F. Stephanopoulos, Appl. Catal. B Environ. 106 (2011) 255.)*

are superior catalytic materials since the former are not stable (lose of sulphur due to sulphide hydrolysis) in sulphur-free fuel gas streams.

4.6 Co-Mo CATALYSTS IN INDUSTRIAL CONDITIONS

Hla *et al.* [27] investigated Co-Mo catalysts using simulated coal-derived syngas. They used commercial Co-Mo catalysts with the composition CoO ~ 2%, MoO ~ 8%, MgO ~ 24%, Al_2O_3 ~ 50%, balance promoter wt%. They sulphided the catalyst in the following way. The samples were first heated under N_2 at a rate of 2.5 °C min^{-1} up to 110 °C and maintained at this temperature for 1 h to dry the sample. The N_2 stream was then switched to the sulphiding gas mixture consisting of H_2S (4%) and H_2 (96%) with a gas flow at atmospheric pressure of 5 L h^{-1} g^{-1} of catalyst. The reactor temperature was then increased from 110 to 450 °C with a heating rate of 2.5 °C min^{-1}. The catalyst was then sulphided at 450 °C for 4 h. They evaluated Co-Mo catalysts with four feed compositions namely dry-feed syngas, slurry-feed syngas, backend of membrane reactor, backend of WGS reactor. The compositions of four syngases are presented in Table 4.2. The Co-Mo catalyst exhibits higher CO conversion in dry-feed syngas composition where CO inlet content is highest and CO_2 content if low. It exhibits lowest CO conversion in backend of WGS reactor gas composition where CO content is lowest and H_2 content is highest.

A series of experiments were carried out using the dry-feed coal-derived syngas containing H_2S concentrations ranging from 330 to 2670 ppm on a wet basis (1000-8000 ppm on a dry basis) at 450 °C. The results are presented in Figure 4.7. From Figure 4.7, it can be seen that the performance of SSC1 depends significantly on the concentration of H_2S in the syngas. At higher concentrations of H_2S in the feed gas (at 1330 and 2670 ppm), the catalyst activity increases 2×10^3 times, respectively, compared to its activity at the lowest concentration of H_2S tested (340 ppm). Then they concluded that using sour shift catalysts for enhancing the WGS reaction could be a better option only if the minimum sulphur content in the syngas were kept above 1000 ppmv in the WGS reactor to ensure that a consistently high CO conversion rate is achieved.

They also measured kinetics using dry-feed coal-derived syngas and downstream of WGS reactor. The kinetic expression would be

$$R = 0.008 \pm 0.0004 \exp\left(\frac{-60.3 \pm 1.3}{R\prime T}\right) P_{CO}^{0.75\pm0.12} P_{H_2O}^{0.31\pm0.08} P_{CO_2}^{-0.07\pm0.02} P_{H_2}^{-0.09\pm0.02} (1-\beta)$$

for $T = 450$ °C (723 K), where $R' = 8.3144 \times 10^{-3}$ kJ mol^{-1} K^{-1}, β is defined as $\beta = 1/K \cdot P_{CO_2} P_{H_2} / P_{CO} P_{H_2O}$, where K is the equilibrium constant for the WGS reaction.

de la Osa *et al.* [28] also investigated commercial Co-Mo catalysts using industrial coal derives syngas feed. Catalyst pre-treatment and reaction

TABLE 4.2 Experimental Inlet Gas Composition (Dry Basis) Tested with a Constant Level of H_2S and Constant Wet Gas Velocity

Conditions	CO (% d.b.)	H_2 (% d.b.)	CO_2 (% d.b.)	N_2 (% d.b.)
Dry-feed coal-derived syngas	65	30	2	3
Slurry-feed coal-derived syngas	44	37	16	3
Backend of catalytic membrane reactor (retentate side)	7	12	78	3
Backend of WGS reactor	4	55	38	3
Weight of SSC1 samples = 1.2 g				
Constant wet gas velocity = 79.7 cm s^{-1} at 450 °C and 1 atm				
H_2O:C ratio (constant for all conditions) = 3				
Inlet H_2S concentration (w.b.) = 1000 ppm				

Taken from Table 2 of S.S. Hla, G.J. Duffy, L.D. Morpeth, A. Cousins, D.G. Roberts, J.H. Edwards, Int. J. Hydrogen Energy 36 (2011) 6638.

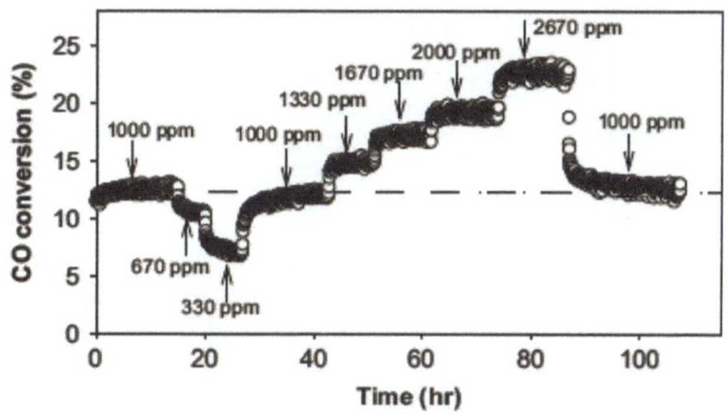

FIGURE 4.7 Performance of SSC1 at various concentrations of H_2S (330-2670 ppm w.b.) in the simulated dry-feed coal-derived syngas at 450 °C and at approximately 1 atm. *(Taken from Figure 4 of S.S. Hla, G.J. Duffy, L.D. Morpeth, A. Cousins, D.G. Roberts, J.H. Edwards, Int. J. Hydrogen Energy 36 (2011) 6638.)*

conditions, i.e., reaction conditions, steam to CO ratio, space velocity were studied in a bench scale setup. The effect of reduction with H_2 or sulphide activation with H_2S or carbonyl sulphide (COS) was also studied. They performed WGS reaction using ELCOGASIGCC plane dry gas feed composition. The reaction was performed under 19 bar operating pressure of the industrial gasification. The bench scale reactor setup was shown in Figure 4.8. As expected, increasing reaction temperature increases the CO conversion. On the other hand, steam/CO ratio does not have much effect on the CO conversion. However, methane formation was observed at lower steam/CO ratios at temperatures above 400 °C for catalyst not activated by sulphur. For the catalyst activated by 63 ppm H_2S showed no methane formation for entire temperature and steam to CO ratio ranges. At reaction temperature 350 °C, the space velocity has no influence on the CO conversion for both sulphided and non-sulphided catalysts. However, after 350 °C, increase in space velocity decreases the CO conversion for both catalysts. The CO conversion remains constant with increasing sulphur concentration during pre-activation until 50 ppm and further increase in the sulphur concentration leads to an increase in CO conversion. The sulphur source H_2S or COS has no or minimal activity on the WGS activity of Co-Mo catalysts. They also performed WGS reaction with non-sulphided Co-Mo catalyst and observed similar conclusions.

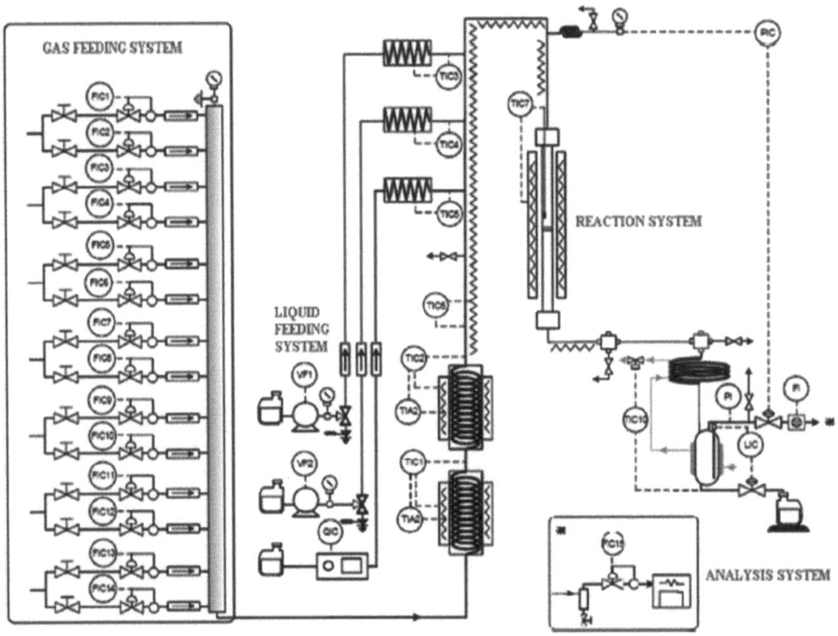

FIGURE 4.8 Simplified process diagram of the bench scale setup for WGS reaction. *(Taken from Figure 1 of A.R. de la Osa, A. De Lucas, J.L. Valverde, A. Romero, I. Monteagudo, P. Sanchez, Int. J. Hydrogen Energy 36 (2011) 44.)*

Liu *et al.* [29] tested Co-Mo catalysts in more than 30 coal to chemical plants with very broad range of steam to carbon ratios and various H_2S contents. Recently gasification technologies give rise to the gasified products with low steam to carbon ratio, in most cases less than 2, which reduces the number of the recycles needed. These new gasification technologies introduce challenges to the sour water-gas shift catalyst to operate in the lean steam to carbon ratio feeding gas. Zong [30] developed a series of Co-Mo catalysts with various supports for use with lean steam to carbon ratio gases in sour water-gas shift reaction. It is found that modification of the alumina support with TiO_2 can significantly improve the catalyst shift activity and give high activity even with low H_2S content. The catalyst can start up the reaction at much lower temperature than the benchmark catalyst and is resistant to carbon deposition. They also studied the effect of TiO_2 addition on the sour water-gas shift Co-Mo catalyst in parallel industrial reactor systems in the same plant. The industrial operation of sour shift catalysis is shown in Figure 4.9.

For the Co-Mo/Al_2O_3, it is shown that CO conversion is significantly lower compared to the Co-Mo/Al_2O_3-TiO_2 when H_2S is lower than 0.06 vol%. However, the catalyst activity increases faster with the rise of H_2S in the feed gas, and even when the H_2S reaches 0.35 vol%, the CO conversion is still lower than that over Co-Mo/Al_2O_3-TiO_2. Because this catalyst showed lower activity,

FIGURE 4.9 Industrial operation of the sour water-gas shift catalyst. *(Taken from Figure 1 of B. Liu, Q. Zong, P.P. Edwards, F. Zou, X. Du, Z. Jiang, T. Xiao, H. AlMegren, Ind. Eng. Chem. Res. 51 (2012) 11674.)*

especially with the low H_2S content in the feed gas, the feed gas has H_2S added when H_2S is lower than 0.3 vol%. Industrial operation results also showed that for Co-Mo/Al_2O_3 catalyst, CO conversion drops more than 10% when H_2S is below 0.06%. However, for the TiO_2-modified shift catalyst, when the catalyst bed temperature is above 630 °C and H_2O/gas changes from 1.0 to 1.2, and H_2S varies from 0.01% to 0.3%, in the 2-year operation. In this case, CO conversion is around 90% with very small changes. This is the reason that several companies replace the original Co-Mo/Al_2O_3 with Co-Mo/Al_2O_3-TiO_2. Increasing reaction temperatures increases the catalytic activity. The modification of the catalyst with TiO_2 makes the active components easy to sulphurize and have higher H_2S capacity. The TiO_2 promoting effect may be due to its higher resistance to the hydration than Al_2O_3. The TiO_2-promoted catalyst has higher shift activity, stable performance and fewer changes in physical properties after 2 years of industrial operation. There is aggregation of MoO_3 and carbon deposition over the shift reaction during the industrial operation, and the presence of TiO_2 in the catalyst support can reduce carbon deposition and stabilize the Mo active phase.

The commercial Co-Mo catalysts operate in the temperature range 250-350 °C and at pressures from atm to 40 bar. The typical process conditions for a Texaco partial oxidation process that generate syngas from heavy oil which use sour Co-Mo WGS catalyst are shown in Table 4.3. Three Co-Mo catalysts beds are used. The syngas from partial oxidation reactor contains 0.25% of H_2S. The inlet CO concentration of 46% is reduced to 1% at the exit of third bed. However, the Co-Mo catalyst converts H_2S and CO into COS. Hence, COS hydrolysis has to be performed after the water-gas shift reaction. However, if we

TABLE 4.3 Texaco Partial Oxidation Process Syngas Compositions

	Bed 1	Bed 2	Bed 3
Inlet feed composition			
CO	46	16	3.1
CO_2	6.9	26	34.2
H_2	47	57.9	62.6
CH_4	0.1	0.1	0.1
Sulphur	0.25		
Inlet steam/gas ratio	0.96	0.7	0.61
Pressure (bar)	35	34	33
Inlet temperature (°C)	266	288	278
Outlet temperature (°C)	411	367	292
Outlet CO (mol%)	16	3.1	1

use higher operating pressures and higher steam to syngas ratios the resulting COS concentrations are well below 0.1 ppmv.

4.7 SUPPORTED ZEOLITE CATALYSTS FOR THE WGS REACTION

Reports are also available in the literature on the Mo catalysts supported on the zeolites. Laniecki *et al.* performed most of the research on these catalysts. They prepared ammonium exchanged NaY zeolite catalysts [31]. They prepared Mo catalysts by impregnation method either with ammonium hepta molybdate or $Mo(CO)_6$. Pre-sulphidation was done at 675 K. The reaction was done with 54.5% CO, 2% H_2S and 43.5% H_2. The ratio of H_2O:CO was 1. Catalysts prepared with $Mo(CO)_6$ pre-cursor exhibit higher activity compared to catalysts prepared with ammonium hepta molybdate. Also the proton-exchanged zeolite catalysts exhibited much higher activity compared to pure zeolite. The activity for Nay-supported samples is almost three times lower than for the protonated supports. The increased activity of catalysts based on protonated supports suggests that OH groups are involved in the water-gas shift reaction. However, catalysts originating from HNaY and $Mo(CO)_6$ exhibit a decrease of activity with time. They proposed that the decrease in activity is due to coke formation. They also proposed that the dispersion of MoS_2 on zeolite support has direct correlation with the WGS activity. They also proposed redox mechanism for the WGS reaction over Mo catalysts supported on zeolites.

They also reported Ni^{2+}-exchanged NaY zeolite catalysts for WGS reaction [32]. Application of nickel-exchanged zeolite as the supports for Mo sulphided species significantly enhances catalytic activity. The catalysts pre-treated at 475 K show much higher activity compared to catalysts pre-treated at 675 K. They proposed that dispersion of Mo sulphided species, location of Ni^{2+} ions, ability of Ni ions toward nickel sulphide formation determine the WGS activity. They also proposed that support acidity influences the WGS activity.

4.8 MOLYBDENUM CARBIDE CATALYSTS

Reports are also available in the literature, on the Mo carbide catalysts. In 2000 Patt *et al.* [33] first time reported Mo carbide catalysts for WGS reaction. The synthesis carbide catalysts are as follows. Ammonium para-molybdate was first dissolved in warm distilled water. The liquid was slowly evaporated and the remaining solid was calcined in dry air for 3 h at 500 °C and sieved to retain material with a mesh size of 60/230. Then the samples was carburized using an equimolar mixture of CH_4 and H_2 flowing at 150 $cm^3\ min^{-1}$. They measured the WGS reaction rates using a synthetic methane steam reformer exhaust stream. The stream contains 62.5% H_2, 31.8% deionized H_2O and 5.7% CO. The Mo_2C catalyst exhibits better WGS activity compared to Cu-Zn-Al_2O_3 catalyst. The areal reaction rate for the Mo_2C catalyst was 0.22 $\mu mol\ m^{-2}\ s^{-1}$ at 295 °C. There was no decrease in the activity for the Mo_2C catalyst during

48 h on-stream. They also showed that there was no apparent modification of the bulk phase structure. Then Moon *et al.* investigated effect of carburization temperature on the WGS activity. The activities over molybdenum carbides catalysts follow the order Mo_2C (640 °C) > Mo_2C (650 °C) > Mo_2C (660 °C) > Mo_2C (630 °C) > Mo_2C (615 °C) > Mo_2C (670 °C). At carburization temperatures lower than 640 °C, carburization of molybdenum oxide will not be completed, whereas at higher temperatures, in excess of 650 °C, excess carbon deposits on the carbide systems, thereby resulting in low activity. They found that Mo_2C catalyst showed higher stability than the commercial LTS catalyst during the thermal cycling test. Their XPS measurements suggest that deactivation of the Mo_2C catalyst may be caused by the transition of MoO_xC_y, $Mo^{IV}(MoO_2)$ and Mo_2C on the surface of the catalyst to $Mo^{VI}(MoO_3)$ with the reaction of H_2O in reactant.

In 2006 Nagai *et al.* [34] reported Co-Mo carbide catalysts for WGS reaction. They found that the CO conversion for the Co-Mo catalysts increased with increasing Co content up to 50%, and then decreased at 75% cobalt content. The $Co_{0.5}Mo_{0.5}$ catalyst was the most active among the carbide catalysts with various Co contents. They also concluded that the $Co_{0.25}Mo_{0.75}$ catalyst carburized at 873 K was more active than the catalysts carburized at the other temperatures. They proposed that the incomplete carburization of the Co-Mo oxide formed the amorphous Co-Mo oxycarbide, leading to a high activity for the WGS reaction. Water reacted with the adsorbed CO or carbidic carbons to produce hydrogen and carbon dioxide in the gas phase.

Then they proposed Ni-Mo carbide catalyst [35] for the low temperature WGS reaction. Among the various catalysts $Ni_{0.25}Mo_{0.75}$ catalyst carburized at 873 K was more active than the other catalysts. Usually, Mo carbide catalysts deactivate with time-on-stream. However, $Ni_{0.25}Mo_{0.75}$ catalyst carburized at 923 K exhibits stable activity for 300 min of time-on-stream. The Ni contents of 15% and 25% increased the catalytic activity but further Ni content led to a drop in activity. They proposed that the promotion of the WGS reaction activity was due to the formation of Ni-Mo oxycarbide.

Then they investigated the influence of K and Zr on the WGS activity of Co-Mo carbide catalysts [36]. Interestingly, Zr-promoted Al_2O_3/Co-Mo carbide catalysts exhibit lesser activity compared to Al_2O_3/Co-Mo carbide catalyst. However, K-promoted Al_2O_3/Co-Mo carbide catalysts exhibit higher activity compared to Al_2O_3/Co-Mo carbide catalyst. Also Co-Mo-Zr-K-Al_2O_3 exhibits much higher activity compared to all other catalysts. The addition of potassium facilitates the CO dissociation on the Mo carbide. Their characterization measurements show that zirconia prevents the oxidation of Mo_2C and the transformation of Co_6Mo_6C carbide to Co_3Mo. The Co^0 and Co^{2+} percentages for the AlKZr-promoted CoMo catalyst were higher than those for either the AlK- or the AlZr-promoted catalyst in which the Co^{3+} ion was pre-dominant. Consequently, the carburized AlKZr-CoMo catalyst contained the active Mo_2C, Co_6Mo_6C1 carbide and small particles of Co metal covering the zirconia, potassium and alumina. The oxidation of the Mo carbides was resisted by zirconia and the high dissociation of water was promoted by potassium.

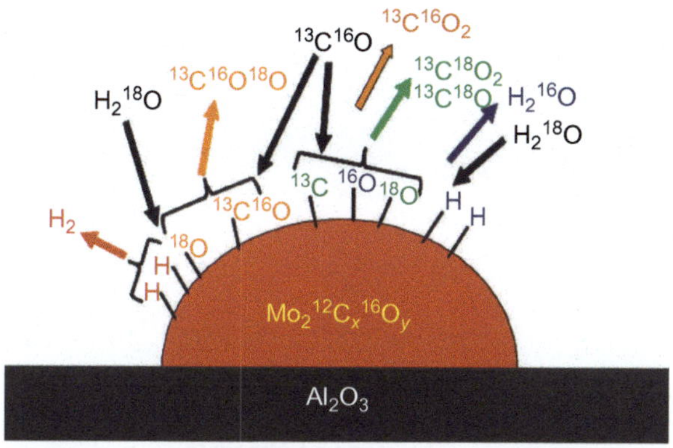

FIGURE 4.10 Dissociation of CO and H_2O during water-gas shift reaction on carburized Mo/Al_2O_3. *(Taken from Figure 7 of T. Namiki, S. Yamashita, H. Tominaga, M. Nagai, Appl. Catal. A Gen. 398 (2011) 155)*

Namiki *et al.* [37] proposed mechanism of WGS reaction over Mo_2C catalyst using XPS, *in situ* mass and FT-IR spectroscopic techniques using 13C- and 18O-labelled CO and H_2O. The typical adsorption of various isotopic labelled species is presented in Figure 4.10.

DRIFTS experiment show the formation of formate and carbonate species for the 973 K-carburized 4.8 wt% Mo/Al_2O_3 catalyst and alumina in the stream of CO. The injection of $H_2{}^{18}O$ into a stream of $^{13}C^{16}O$ on the 8.5 wt% Mo_2C/Al_2O_3 catalyst showed the formation of $^{13}C^{18}O$, $^{13}C^{18}O^{16}O$, $^{13}C^{18}O_2$ and $H_2{}^{18}O$. These results show that $^{13}C^{16}O$ and $H_2{}^{18}O$ were dissociated on the 973 K-carburized 4.8 and 8.5 wt% Mo/Al_2O_3 catalysts. Regarding the ^{16}O-labelled atom, $H_2{}^{16}O$, $^{12}C^{16}O$, $^{13}C^{16}O_2$ and $^{13}C^{16}O^{18}O$ were formed, but $^{12}C^{16}O_2$ and $^{12}C^{16}O^{18}O$ were not observed. Slightly higher intensities of $^{12}C^{18}O$ and $^{12}C^{16}O$ were observed with no peaks of $^{12}C^{16}O_2$, $^{12}C^{16}O^{18}O$ and $^{12}C^{18}O_2$ with respect to the formation of the ^{12}C-labelled reaction products. They proposed that the WGSR on the carburized 4.8 and 8.5 wt% Mo/Al_2O_3 catalysts preceded the redox route based on the dissociation of CO and H_2O as well as the dissociation-association mechanism.

Then Semina *et al.* [38] reported the effect of synthesis method on Mo_2C catalysts. According to them the highest catalyst activity in the WGS reaction was found for the Mo_2C/Mo system synthesized via the simultaneous electroreduction of carbonate and molybdate ions. The activity of this catalyst was three orders of magnitude higher than the activity of the commercial Cu/ZnO/Al_2O_3 catalyst. The stability tests performed for the catalysts showed that this system remains active for at least 5000 h. In addition, no methanation side reaction is found.

Schweitzer *et al.* [39] investigated Pt/Mo_2C catalyst for the low temperature WGS reaction. The Pt/Mo_2C catalyst exhibits much higher activity compared to Pt supported on CeO_2, TiO_2 and Al_2O_3 catalysts. The higher activity of Pt/Mo_2C

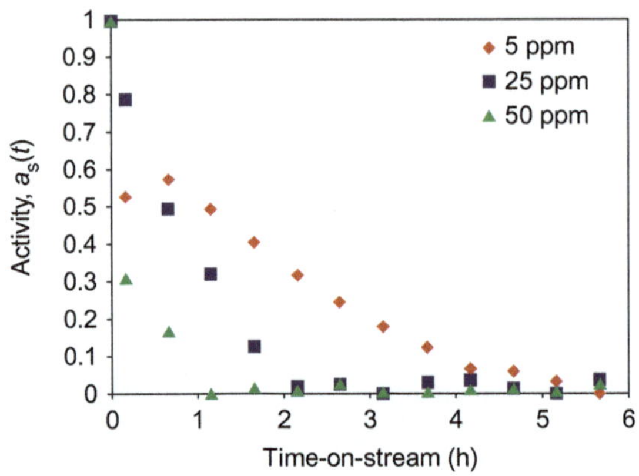

FIGURE 4.11 Activity $a_s(t)$ for the Pt/Mo_2C catalyst as a function of time-on-stream after the introduction of 5, 25 and 50 ppm H_2S to the reformate. *(Taken from Figure 6 of J.A. Schaidle, A.C. Lausche, L.T. Thompson, J. Catal. 272 (2010) 235.)*

is due to high density of active sites at the perimeter of the Pt particles (i.e. at the interface between the Pt and the Mo_2C support). The Pt particles appeared to be raft-like due to strong interactions with the Mo_2C surface which is advantageous due to its high surface area to volume ratio. Then they investigated alumina-supported Pt/Mo_2C catalysts. Rates for the Pt-Mo_2C/Al_2O_3 catalysts ranged from 52 to 69 μmol CO $(g\ s)^{-1}$ at 240 °C with the 4 wt% Pt-Mo_2C/Al_2O_3 catalyst demonstrating the highest rate.

Then they investigated effect of sulphur on the WGS activity of Pt-Mo_2C catalysts [40]. The WGS activity of Pt-Mo_2C at various sulphur concentrations is presented in Figure 4.11. As expected increase in the sulphur concentration leads to decrease in activity. After the introduction of 5 ppm H_2S, the hydrogen production rate for the Mo_2C catalyst decreased by 90% within 10 min. After 32 h on-stream, the catalyst regained some of its activity, reaching a H_2 production rate that was 25% of its sulphur-free reactant steady-state rate. When H_2S was removed from the reactant, the rate for the Mo_2C catalyst quickly decreased to zero. Upon treating the spent Mo_2C catalyst with 15% CH_4/H_2 at 590 °C for 4 h, 25-30% of its initial rate was recovered. Characterization of the spent catalysts suggested that deactivation of the Mo_2C catalyst was primarily due to the adsorption of sulphur on Mo_2C sites and the formation of surface MoS_2. The MoS_2 sites were active in the presence of sulphur. Deactivation of the Pt/Mo_2C catalyst appeared to be primarily due to the irreversible sulphidation of Pt nanoparticles.

REFERENCES

[1] W. Wustrow, O. Maedrich, H. Macura, German Patent 1,085,287, 1954.
[2] E. Stern, British Patent 961,860, 1964.
[3] A. D. Overstreet, A screening study of a new water-gas shift catalyst, (M.S. Thesis)Virginia Polytechnic Institute and State University, Blacksburg, VA, 1974.

[4] V. Berispek, Studies of an Alkali Impregnated Cobalt-Molybdate Catalyst for the Water Gas Shift and the Methanation Reactions, (M.S. Thesis), Virginia Polytechnic Institute and State University, Blacksburg, VA, 1976.

[5] G.S. Levison, L. Harris, German Patent 1,162,505, 1974.

[6] K.L. Riley, C.L. Aldridge, German Patent 2,453,784, Exxon Research & Engineering Co., 1975.

[7] M.A. Segura, C.L. Aldridge, U.S. Patent 4,054,644, Exxon Research & Engineering Co., 1977.

[8] P. Hou, D. Meeker, H. Wise, Kinetic studies with a sulfur-tolerant water gas shift catalyst, J. Catal. 80 (1983) 280–285.

[9] R. Hakkarainen, T. Salmi, Water-gas shift reaction on a cobalt-molybdenum oxide catalyst, Appl. Catal. A Gen. 99 (1993) 195–215.

[10] C. R. F. Lund, Effect of adding Co to MoS_2/Al_2O_3 upon the kinetics of the water-gas shift, Ind. Eng. Chem. Res. 35 (1996) 3067–3073.

[11] Y. Y. Chen, M. Dong, J. Wang, H. Jiao, On the role of a cobalt promoter in a water-gas-shift reaction on Co-MoS_2, J. Phys. Chem. C 114 (2010) 16669–16676.

[12] J. R. Mellor, R. G. Copperthwaite, N. J. Coville, The selective influence of sulfur on the performance of novel cobalt-based water-gas shift catalysts, Appl. Catal. A Gen. 164 (1997) 69–79.

[13] L. Wachowski, M. Laniecki, F. Domka, Activity of Co-Mo catalysts supported on alumina modified with La_2O_3, React. Kinet. Catal. Lett. 67 (1999) 123–127.

[14] M. Łaniecki, M. M. Grycz, F. Domka, Water-gas shift reaction over sulfided molybdenum catalysts I. Alumina, titania and zirconia-supported catalysts, Appl. Catal. A Gen. 196 (2000) 293–303.

[15] M. Kantschewa, F. Delannay, H. Jeziorowski, E. Delgado, S. Eder, G. Ertl, H. Knozinger, Nature and properties of a potassium-promoted NiMo/$A1_2O_3$ water gas shift catalyst, J. Catal. 87 (1984) 482–496.

[16] X. Xie, H. Yin, B. Dou, J. Huo, Characterization of a potassium-promoted cobalt molybdenum/alumina water-gas shift catalyst, Appl. Catal. 77 (1991) 187–198.

[17] J. N. Park, J. H. Kim, H.-I. Lee, A study on the sulfur-resistant catalysts for water gas shift reaction IV. Modification of CoMo/g-Al_2O_3 catalyst with K, Bull. Korean Chem. Soc. 21 (2000) 1239–1244.

[18] A. A. Andreev, A. J. Kafedjiysky, R. M. Edreva-Kardjiev, Active forms for water-gas shift reaction on NiMo-sulfide catalysts, Appl. Catal. A Gen. 179 (1999) 223–228.

[19] D. Nikolova, R. Edreva-Kardjieva, G. Gouliev, T. Grozeva, P. Tzvetkov, The state of (K)(Ni) Mo/g-Al_2O_3 catalysts after water-gas shift reaction in the presence of sulfur in the feed: XPS and EPR study, Appl. Catal. A Gen. 297 (2006) 135–144.

[20] D. Nikolova, R. Edreva-Kardjieva, T. Grozeva, Water-gas shift activity of K-promoted (Ni) Mo/c-Al_2O_3 systems in sulfur-containing feed, React. Kinet. Mech. Catal. 103 (2011) 71–86.

[21] H. Wang, Y. Lian, Q. Zhang, Q. Li, W. Fang, Y. Yang, MgO-Al_2O_3 mixed oxides-supported Co-Mo-based catalysts for high-temperature water-gas shift reaction, Catal. Lett. 126 (2008) 100–105.

[22] Y. Lian, H. Wang, W. Fang, Y. Yang, Effect of calcination temperature of support on the performance of Co-Mo/MgO-Al_2O_3 water-gas shift catalyst, Chin. J. Catal. 30 (2009) 549–554.

[23] Y. Lian, H. Wang, W. Fang, Y. Yang, Effect of Mg/Al atom ratio of support on catalytic performance of Co-Mo/MgO-Al_2O_3 catalyst for water gas shift reaction, J. Nat. Gas Chem. 19 (2010) 161–166.

[24] H. Wang, Y. Lian, Q. Zhang, Q. Li, W. Fang, Y. Yang, W-promoted Co-Mo-K/c-Al_2O_3 catalysts for water-gas shift reaction, Catal. Commun. 10 (2009) 1864–1867.

[25] Y. Zhang, G. Zhang, Y. Zhao, X. Li, Y. Sun, Y. Xu, Ce-K-promoted Co-Mo/Al_2O_3 catalysts for the water gas shift reaction, Int. J. Hydrogen Energy 37 (2012) 6363–6371.

[26] I. Valsamakis, M. F. Stephanopoulos, Sulfur-tolerant lanthanide oxysulfide catalysts for the high-temperature water-gas shift reaction, Appl. Catal. B Environ. 106 (2011) 255–263.

[27] S. S. Hla, G. J. Duffy, L. D. Morpeth, A. Cousins, D. G. Roberts, J. H. Edwards, Investigation into the performance of a Co-Mo based sour shift catalyst using simulated coal-derived syngases, Int. J. Hydrogen Energy 36 (2011) 6638–6645.

[28] A. R. de la Osa, A. De Lucas, J. L. Valverde, A. Romero, I. Monteagudo, P. Sanchez, Performance of a sulfur-resistant commercial WGS catalyst employing industrial coal-derived syngas feed, Int. J. Hydrogen Energy 36 (2011) 44–51.

[29] B. Liu, Q. Zong, P. P. Edwards, F. Zou, X. Du, Z. Jiang, T. Xiao, H. AlMegren, Effect of titania addition on the performance of CoMo/Al_2O_3 sour water gas shift catalysts under lean steam to gas ratio conditions, Ind. Eng. Chem. Res. 51 (2012) 11674–11680.

[30] Q. Zong, Patent application CN1999-112351, 2001.

[31] M. Laniecki, W. Zmierczak, Deactivation of sulfur tolerant WGS catalysts based on Ni-Y-Mo zeolites, Stud. Surf. Sci. Catal. 31 (1991) 799–802.

[32] M. Laniecki, Ni-Y-Mo zeolite as catalysts for the WGS reaction, Stud. Surf. Sci. Catal. 33 (1994) 363–370.

[33] J. Patt, D. J. Moon, C. Phillips, L. Thompson, Molybdenum carbide catalysts for water-gas shift, Catal. Lett. 65 (2000) 193–195.

[34] M. Nagai, A. M. Zahidul, K. Matsuda, Nano-structured nickel-molybdenum carbide catalyst for low-temperature water-gas shift reaction, Appl. Catal. A Gen. 313 (2006) 137–145.

[35] M. Nagai, K. Matsuda, Low-temperature water-gas shift reaction over cobalt-molybdenum carbide catalyst, J. Catal. 238 (2006) 489–496.

[36] M. Nagai, A. M. Zahidul, Y. Kunisaki, Y. Aoki, Water-gas shift reactions on potassium- and zirconium-promoted cobalt molybdenum carbide catalysts, Appl. Catal. A Gen. 383 (2010) 58–65.

[37] T. Namiki, S. Yamashita, H. Tominaga, M. Nagai, Dissociation of CO and H_2O during water-gas shift reaction on carburized Mo/Al_2O_3 catalyst, Appl. Catal. A Gen. 398 (2011) 155–160.

[38] G. L. Semina, A. R. Dubrovskiic, P. V. Snytnikova, S. A. Kuznetsovc, V. A. Sobyanin, Using catalysts based on molybdenum and tungsten carbides in the water-gas shift reaction, Catal. Ind. 4 (2012) 59–66.

[39] N. M. Schweitzer, J. A. Schaidle, O. K. Ezekoye, X. Pan, S. Linic, L. T. Thompson, High activity carbide supported catalysts for water gas shift, J. Am. Chem. Soc. 133 (2011) 2378–2381.

[40] J. A. Schaidle, A. C. Lausche, L. T. Thompson, Effects of sulfur on Mo_2C and Pt/Mo_2C catalysts: water gas shift reaction, J. Catal. 272 (2010) 235–245.

Chapter 5

Ultra High Temperature WGS Reaction

5.1 INTRODUCTION

Recently production of H_2 using membrane reactors has received much importance in recent years because hydrogen can selectively permeate through membrane, thus making complete conversion possible. Several theoretical and simulation studies show that the operation of membrane reactor at higher temperatures and higher pressure has several advantages compared to lower temperature because of faster kinetics and higher hydrogen permeance. These membranes typically operate at 450-550 °C temperatures and 1-20 bar pressures. We cannot use the commercial Fe-Cr-Cu catalyst for these applications because of the pyrophoric nature of Cu. Development of catalysts for membrane reactors operating at these high pressures and temperatures has to meet stringent requirements. We have been developing modified ferrite catalysts for these applications for the past few years.

5.2 INFLUENCE OF PROMOTERS

Fe is the active metal for high-temperature WGS reaction. Hence, we introduced a variety of metal dopants (M = Cr, Mn, Co, Ni, Cu, Zn and Ce) for iron oxide (spinel lattice) and screened their effectiveness for high-temperature WGS reaction [1]. The idea was to examine if ferrite formation can occur with dopants and promote the $Fe^{3+} \leftrightarrow Fe^{2+}$ redox couple. The substitution of 'Fe' sites in the ferrite structure with other transition/non-transition/inner transition metal atoms leads to the crystallization of an inverse (or mixed) spinel. The stoichiometry of an inverse spinel can be represented as $A_{(1-\delta)}B_{\delta}[A_{\delta}B_{(2-\delta)}]O_4$, where δ is the degree of inversion, while A and B represent typical divalent and trivalent cations, respectively. The catalysts were synthesized by co-precipitation method using nitrates as precursors. The synthesized catalysts were evaluated for ultra high temperature WGS reaction in the temperature region 400-550 °C and GHSV 60,000 h^{-1}.

Before performing WGS reaction the prepared catalysts which are in hematitic form have to be activated to magnetite form by controlled reduction in a stream of process gas such as CO, CO_2, H_2 and H_2O vapour. The ratios of water

Water Gas Shift Reaction. http://dx.doi.org/10.1016/B978-0-12-420154-5.00005-X

vapour to hydrogen and CO_2 to CO for the activation step determine the equilibrium of the Fe^{2+} and Fe^{3+} ions in octahedral sites. With commonly used process gases the catalyst is more reduced than the reduction to reach the state of equilibrium. In this way more Fe_3O_4 is created, which forms the stable state. It is important to avoid over reduction of the magnetite active material during the process to lower oxides, carbides or metallic iron species. The metallic iron species especially are active catalysts for the Fischer-Tropsch process thus leading to methanation in the WGS reaction. Hence, in the present study all the catalysts were activated by carrying out controlled reduction of hematite to magnetite in the presence of a process gas (CO, CO_2, H_2 and steam). XRD powder diffraction patterns of all the activated catalysts show only peaks due to the magnetite phase. The representative reactions are shown as follows:

$$3Fe_2O_3 + H_2 \rightarrow 2Fe_3O_4 + H_2O, \quad \Delta H = -16.3\,\text{kJ/mol},$$
$$3Fe_2O_3 + CO \rightarrow 2Fe_3O_4 + CO_2, \quad \Delta H = +24.8\,\text{kJ/mol}.$$

The ratio of oxidant to reductant (also called 'reduction factor,' R) in the process gas mixture has to be just optimal, in order to facilitate the reduction of hematitic to magnetitic phase and should not catalyze further reduction. In the present study $R = 1.4$ was utilized, where $R = [(CO) + (H_2)]/[(CO_2) + (H_2O)]$.

The WGS activity of various Fe-M catalysts (M = Cr, Mn, Co, Ni, Cu, Zn and Ce) is presented in Figure 5.1. As expected CO conversion increases with increasing reaction temperature. Also the WGS activity decreases with decreasing steam/CO ratio (not shown). The lower WGS activity observed at 350 °C can be related to the difficulty in sustaining the active magnetite phase, as observed from temperature programmed reduction (TPR) experiments. It is also suggested that at lower temperatures the activity of magnetite is limited by the

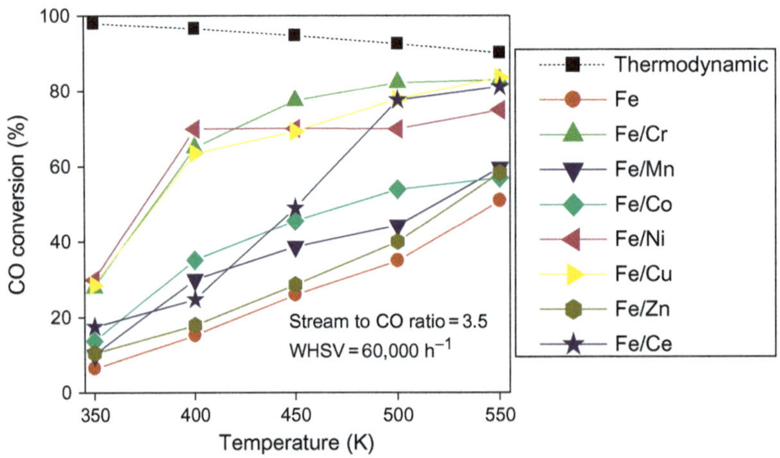

FIGURE 5.1 WGS activity of various Fe/M catalysts.

dissociation of steam. XRD data analyses of activated catalysts have established the existence of $Fe_{2.73}M_{0.27}O_4$ (M = Cr, Mn, Co, Ni, Cu, Zn and Ce) type spinels. The synergistic effect between iron and the other metal substituent leads to the formation of either inverse or mixed spinel of composition $A_{(1-\delta)}B_\delta[A_\delta B_{(2-\delta)}]O_4$. Incorporation of metal cations into the hematite (α-Fe_2O_3) crystal structure modifies the magnetic field and also influences the reducibility of hematite particles, as was observed in Mössbauer spectroscopy and TPR studies, respectively. Among the various catalysts Fe/Ce and Fe/Cr catalysts exhibited the higher activity.

Then we aimed to find the optimum amount of Cr and Ce dopant that can be incorporated into the hematite lattice to promote maximum WGS activity [2]. For this purpose, various Fe/Cr and Fe/Ce catalysts were prepared by ammonia-assisted co-precipitation method with variable Fe/M (M = Cr and Ce) atomic ratios. The chosen atomic ratios for the Fe/Cr catalysts are 10:0.2, 0:0.5, 10:1 and 10:2 and for the Fe/Ce catalysts are 10:0.5, 10:1, 10:2 and 10:2.5. The WGS activity results are presented in Figure 5.2. The WGS activity observed was found to be determined by the nature and amount of dopant used in a given case. The WGS activity increased with the increasing amount of chromium up to a Fe/Cr atomic ratio of 10:0.5, while further increase in the amount of chromium resulted in a slight decrease in the WGS activity. On the other hand, the order of WGS activity for the Fe/Ce catalysts is $Fe_2O_3 <$ Fe/Ce (10:0.5) < Fe/Ce (10:1) < Fe/Ce (10:2) > Fe/Ce (10:2.5). Interestingly, both iron and ceria possess highly facile $Fe^{III} \leftrightarrow Fe^{II}$ and $Ce^{IV} \leftrightarrow Ce^{III}$ redox couples, respectively; the interaction between these two couples is responsible for the improved WGS activity. Additionally, the rapid increase in WGS activity with increase in reaction temperature from 400 to 500 °C is due to the improvement in the OSC of ceria at higher reaction temperatures. XRD and cell parameter estimates reveal that cerium can be incorporated into hematite up to a Fe/Ce atomic ratio of 10:2. A further increase in the amount of cerium leads to demixing, i.e., a separate CeO_2 phase forms in the bulk. TPR results also reveal that chromium acts as a structural stabilizer for the Fe_2O_3, whereas cerium promotes the reduction of hematite to magnetite. Mossbauer spectroscopy results show that the addition of both Cr and Ce replaces Fe^{2+} and Fe^{3+} ions at octahedral sites in equal amounts.

5.3 ROLE OF Cu

Then we incorporated Cu into the Fe/M catalysts (M = Cr, Mn, Co, Ni, Cu, Zn and Ce) since Cu is a promoter for the Fe/Cr catalysts for the high-temperature WGS reaction in the industries [3–5]. Interestingly, Cu acts as a promoter for all the modified ferrite catalysts except Fe/Ce catalyst. It acts as an inhibitor for the Fe/Ce catalyst. These results show that all of the copper co-doped ferrites behave like Fe/Cr/Cu except Fe/Ce/Cu, which behaves differently for high-temperature WGS reaction. We explained this interesting behaviour of copper

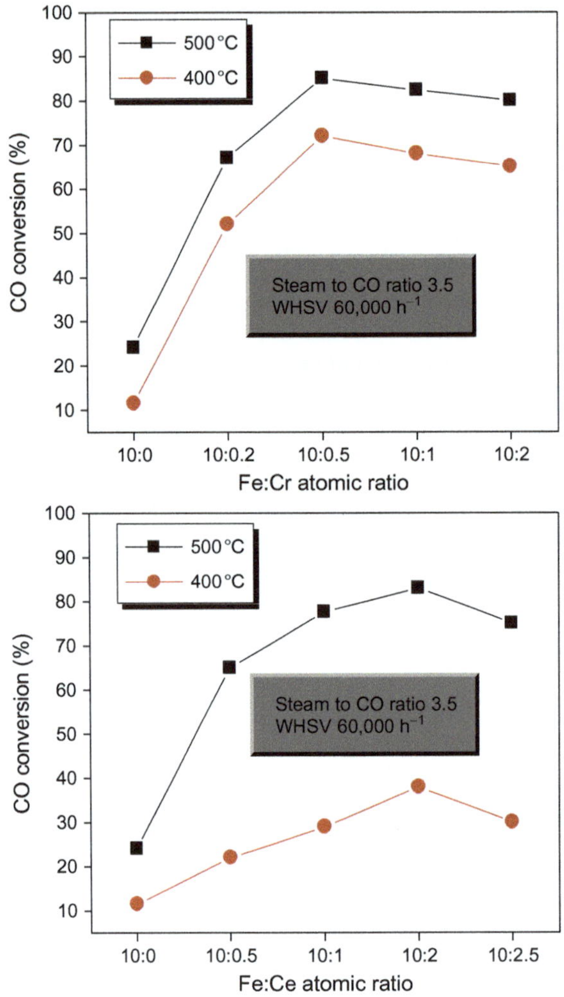

FIGURE 5.2 WGS activity of Fe/Ce and Fe/Cr catalysts.

based on the various characterization techniques like Mossbauer, TPR and XRD. Among the various catalysts, Fe/Ce, Fe/Cr/Cu and Fe/Ni/Cu reached equilibrium conversions at the highest reaction temperature (550 °C) investigated. When the reaction temperature increased from 400 to 500 °C, the activity of Fe/Ce increased from 30% to 82%. The higher activity of Fe/Ce at higher reaction temperature is because both iron and ceria undergo a facile charge transfer reaction between $Fe^{III} \leftrightarrow Fe^{II}$ and $Ce^{IV} \leftrightarrow Ce^{III}$ redox couples, respectively; the synergism between the two couples could be responsible for the improved WGS activity. Additionally, at higher temperatures, the rapid transformation of oxygen exchange between Ce^{3+}/Ce^{4+} redox couple, as well as the improvement in the oxygen storage capacity of ceria, will help the iron to keep its shift activity high.

We conducted X-ray diffraction measurements over activated Fe/M catalysts (M = Cr, Mn, Co, Ni, Cu, Zn and Ce) after pre-reduction in process gas. The patterns are presented in Figure 5.3 [6]. All of the activated catalysts except Fe/Ce/Cu catalyst exhibit intense diffraction lines at 2θ values 30°, 35°, 43°, 57°, 62° and 73°. These peaks could be indexed as that of cubic system with face-centred lattice. For M-modified ferrites and Cu co-doped modified ferrite samples, the strongest Bragg peak occurs at $2\theta = 35.9°$ corresponding to the (311) reflections. The above indexed peak is consistent with the Joint Committee for Powder Diffraction (JCPD) data. On the other hand, the Fe/Ce/Cu catalyst exhibits a few additional diffraction lines at $2\theta = 36°$, 41° and 60°. These peaks primarily belong to the wustite (FeO) phase. These results indicate that simultaneous precipitation of Fe along with Ce and Cu leads to the formation of FeO during the activation. We also measured Mossbauer spectra over the activated Fe/M catalysts (M = Cr, Mn, Co, Ni, Cu, Zn and Ce). The Mossbauer spectra are presented in Figure 5.4. All the samples exhibit peaks due to magnetite expect Fe/Ce/Cu. The Fe/Ce/Cu catalyst exhibits peaks due to magnetite and wustite (FeO). The typical spectra consist of two well-developed six-line magnetic hyperfine structure typical of bulk materials. The high field component is due to Fe^{3+} ions at tetrahedral sites, while the low field one arises from Fe^{2+} and Fe^{3+} ions present at octahedral sites. There is not much change in the δ and H values for Fe at tetrahedral sites after Cu co-doping. On the other hand,

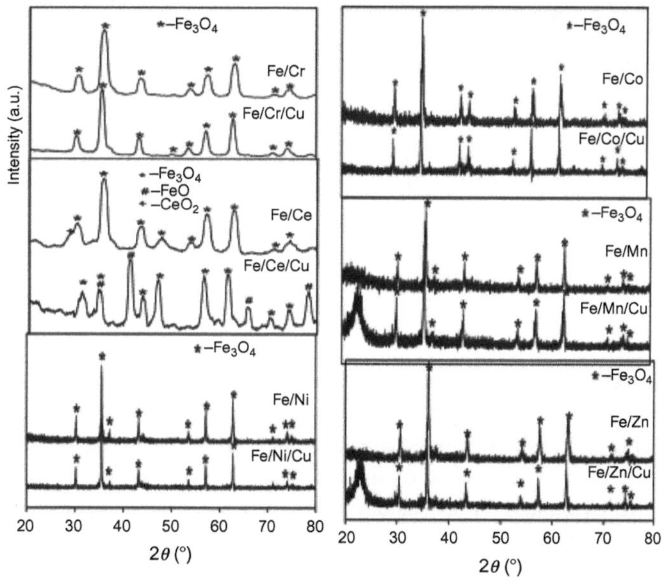

FIGURE 5.3 X-ray diffraction patterns of the various Fe/M/Cu catalysts.

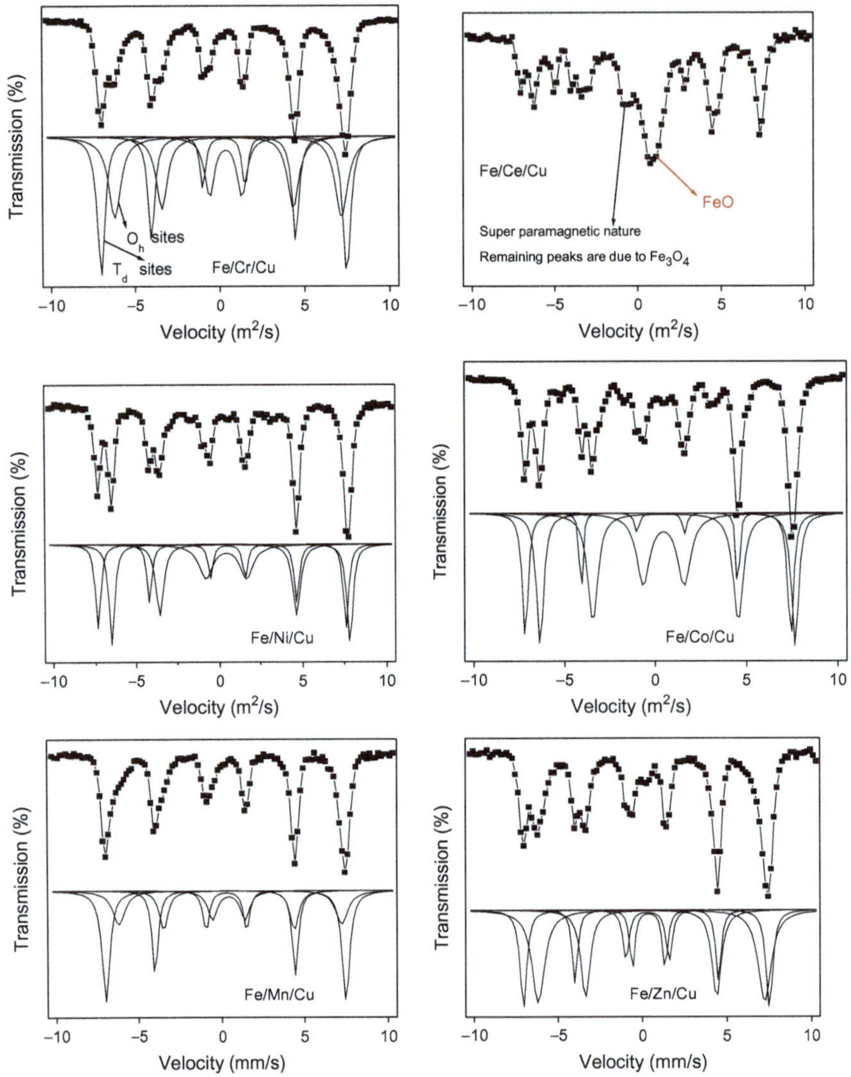

FIGURE 5.4 Mossbauer spectra of various Fe/M/Cu catalysts.

δ and *H* values for octahedral sites in the Cu co-doped-modified ferrite catalysts are noticeably higher than those for M-modified ferrite catalysts. Our XRD and Mossbauer effect results suggest that Cu enters at M-modified octahedral sites in magnetite during activation of the catalysts.

Then we evaluated Fe/Ce catalyst for long-term time on stream stability experiments. We conducted experiment at a steam to CO ratio of 3.5 and a temperature of 500 °C. The long-term time on stream stability WGS activity results of Fe/Ce catalyst are presented in Figure 5.5 [7]. Fe/Ce catalyst exhibits

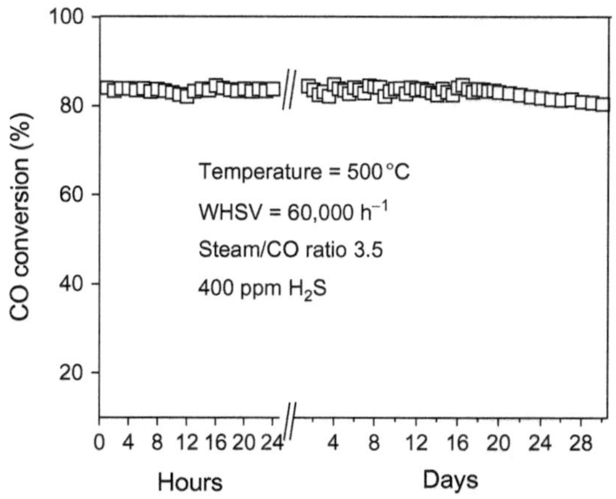

FIGURE 5.5 Long-term time on stream stability profile of the Fe/Ce at a steam to CO ratio of 3.5.

excellent stability for 30 days of time on stream at a steam to CO ratio of 3.5. The CO conversion is decreased only from 84% to about 80% during the first 30 days of reaction. The excellent stability of Fe/Ce is because both iron and ceria undergo a facile charge transfer reaction between $Fe^{3+} \leftrightarrow Fe^{2+}$ and $Ce^{4+} \leftrightarrow Ce^{3+}$ redox couples, respectively. The experiments were also carried out in the presence of 400 ppm of sulphur to check the resistance of Fe/Ce towards sulphur. Interestingly, Fe/Ce catalyst exhibited excellent stability towards sulphur for 30 days. The Fe/Ce catalysts in the presence of sulphur exhibited same activity as Fe/Ce catalyst in the absence of sulphur. Here also, the CO conversion is decreased only from 83% to about 79% during 30 days of reaction. These results reveal that sulphur has no effect on the activity of Fe/Ce catalyst for the WGS reaction. Then we evaluated Fe/Ce catalyst at a steam to CO ratio of 1.5.

Figure 5.6 represents the long-term time on stream WGS stability profiles of Fe/Ce at a steam to CO ratio of 1.5. The catalyst started to deactivate from the 4th day onwards. The conversion decreased from 70% to 20% from 4th day to the 17th day. From 17th day onwards it got stabilized and no further decrease in the conversion observed up to the 30th day. Our characterization measurements suggest that rapid sintering of magnetite particles, high rate of carbon formation and local structural rearrangement of iron ions are responsible for the early deactivation of the Fe/Ce catalyst at a steam to CO ratio of 1.5. At a steam to CO ratio of 3.5 there are sufficient number of water molecules; hence, $Ce^{4+} \leftrightarrow Ce^{3+}$ redox couple also participate in the WGS reaction and stabilize the Fe^{3+}/Fe^{2+} redox couple during the reaction loop. At a steam to CO ratio of 1.5 the number of water molecules is very low; CO dissociation on the surface leads to the formation of carbon. Due to this process the catalytic cycle does not complete which leads to destabilization of Fe^{3+}/Fe^{2+} and Ce^{4+}/Ce^{3+} redox couple and finally local structural rearrangement of iron ions.

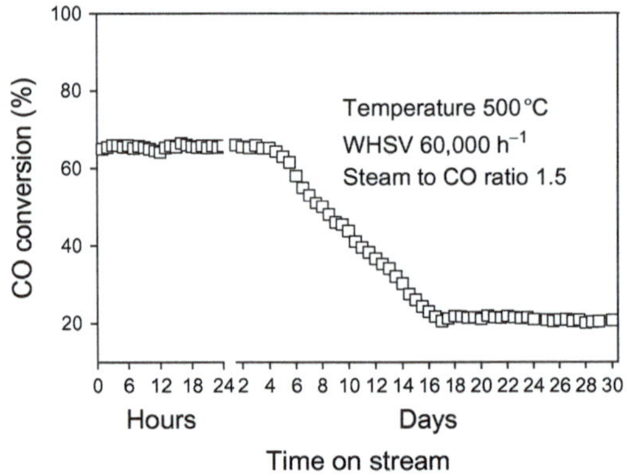

FIGURE 5.6 Long-term time on stream stability profile of the Fe/Ce at a steam to CO ratio of 1.5.

5.4 ROLE OF CO-PROMOTER

To develop a suitable catalyst at a steam to CO ratio of 1.5, we incorporated several metals like Cr, Co, Mo, Zr and Hf into the Fe/Ce lattice and investigated for the high-temperature WGS reaction [8]. The WGS activity results are presented in Figure 5.7. In general with increasing temperature, the WGS activity is found to increase. No pressure drop across the catalyst bed is observed during the experiments nor was CH_4 detected in the effluent stream. Interestingly, all the doped Fe/Ce catalysts reached equilibrium conversions at the highest temperature investigated (550 °C). Among the various dopants, both Cr- and Co-doped Fe/Ce catalysts exhibited high activity at lower reaction temperatures (400-500 °C). The remaining dopants, Zr, Hf and Mo in Fe/Ce catalysts had little or no effect on the WGS activity at steam/CO ratio = 3.5. They exhibited the same activity as Fe/Ce. Between the two dopants, Cr and Co, Cr-doped

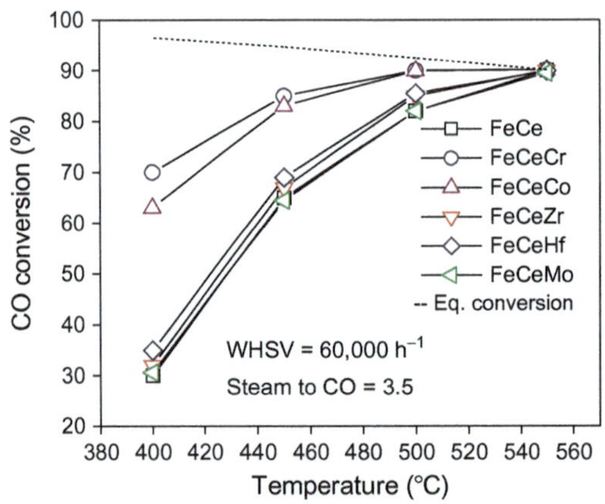

FIGURE 5.7 WGS activity profiles of the Fe/Ce/M catalysts.

Fe/Ce catalyst exhibited the highest WGS activity. TPR measurements suggest that the addition of Zr, Hf to the Fe/Ce lowers the reduction temperature of surface ceria. X-ray photoelectron (XPS) spectra show that both Cr and Co inhibit the carbonate formation in the magnetite during the activation as well as the WGS reaction, whereas Zr, Hf and Mo can inhibit the formation of carbonate in the magnetite during the activation but not during the WGS reaction. Mössbauer spectroscopy studies show local structural distortions in the magnetite phase as the lattice contracts or expands. These distortions are observed from changes in the internal magnetic field and isomer shift at octahedral Fe sites upon Cr and Co doping in Fe/Ce catalysts. Mössbauer spectroscopy also shows that both Cr and Co are able to retain their super-paramagnetic fraction after the WGS reaction. Our results show both Cr and Co substitutes at octahedral sites in magnetite during the activation and prevent the magnetite phase from sintering and promoting the WGS activity of Fe/Ce. Other dopants such as Zr, Hf and Mo have little or no effect on either the crystallographic structure or catalytic activity of Fe/Ce.

5.5 LONG-TERM TIME ON STREAM STABILITY TEST

Then we evaluated Fe/Ce/Cr catalyst for long-term time on stream stability WGS experiments [7]. Figure 5.8 shows the long-term stability high temperature WGS profile of Fe/Ce/Cr catalyst at a steam to CO ratio of 1.5. Remarkably, Fe/Ce/Cr catalyst exhibited excellent stability for 30 days. No decrease in the CO conversion is observed within the permissible error during the 30 days of reaction. To investigate whether the stability of Fe/Ce/Cr is due to both Ce and Cr or due to only Cr we perform the long-term stability experiment over Fe/Cr catalyst. Interestingly, Fe/Cr exhibits early deactivation. Hence, the excellent stability of Fe/Ce/Cr is because of promotional effect of ceria and stabilization effect of chromium. Cr stabilizes the magnetite during the WGS reaction and

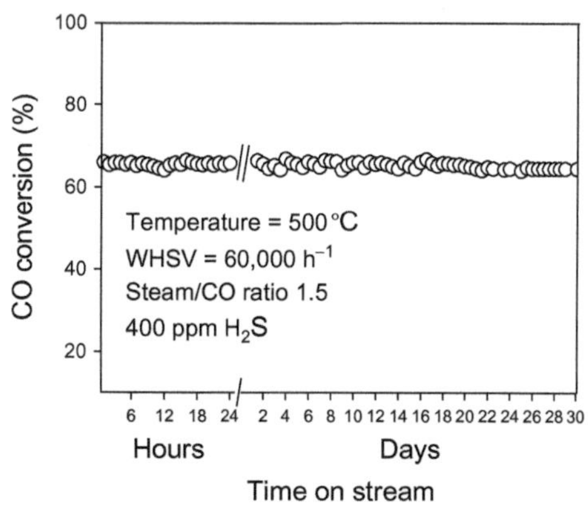

FIGURE 5.8 Long-term time on stream stability profile of the Fe/Ce/Cr at a steam to CO ratio of 1.5.

ceria undergoes a facile charge transfer reaction between $Ce^{4+} \leftrightarrow Ce^{3+}$ redox couple and promotes the WGS activity of magnetite. Hence, incorporation of Ce and Cr into iron oxide and controlled reduction promote and stabilize the magnetite during the WGS reaction both at high and low steam to CO ratios. Interestingly, no methane formation was observed for 30 days of reaction unlike the Fe/Ce catalyst.

REFERENCES

[1] A. Khan, P. Chen, P. Boolchand, P. Smirniotis, Modified nano-crystalline ferrites for high-temperature WGS membrane reactor applications, J. Catal. 253 (2008) 91–104.

[2] A. Khan, P. Smirniotis, Relationship between temperature-programmed reduction profile and activity of modified ferrite-based catalysts for WGS reaction, J. Mol. Catal. A Chem. 280 (2008) 43–51.

[3] G. K. Reddy, K. Gunasekara, P. Boolchand, P. Smirniotis, Cr- and Ce-doped ferrite catalysts for the high temperature water-gas shift reaction: TPR and Mossbauer spectroscopic study, J. Phys. Chem. C 115 (2011) 920–930.

[4] G. K. Reddy, K. Gunasekara, J. Dong, P. Boolchand, P. Smirniotis, High temperature water gas shift reaction over nanocrystalline copper codoped-modified ferrites, J. Phys. Chem. C 115 (2011) 7586–7595.

[5] G. K. Reddy, P. Smirniotis, Effect of copper as a Dopant on the water gas shift activity of Fe/Ce and Fe/Cr modified ferrites, Catal. Lett. 141 (2011) 27–32.

[6] G. K. Reddy, P. Boolchand, P. Smirniotis, Unexpected behavior of copper in modified ferrites during high temperature WGS reaction – aspects of $Fe3^{+} \leftrightarrow Fe^{2+}$ redox chemistry from Mössbauer and XPS studies, J. Phys. Chem. C 116 (2012) 11019–11031.

[7] G. K. Reddy, S. J. Kim, J. Dong, P. G. Smirniotis, J. B. Jasinski, Long-term WGS stability of Fe/Ce and Fe/Ce/Cr catalysts at high and low steam to CO ratios – XPS and Mössbauer spectroscopic study, Appl. Catal. A Gen. 415–416 (2012) 101–110.

[8] G. K. Reddy, P. Boolchand, P. Smirniotis, Sulfur tolerant metal doped Fe/Ce catalysts for high temperature WGS reaction at low steam to CO ratios – XPS and Mössbauer spectroscopic study, J. Catal. 282 (2011) 258–269.

Chapter 6

WGS Reaction in Membrane Reactors

6.1 INTRODUCTION

Membrane reactor (MR) is a catalytic reactor that additionally contains cylinder of some porous material within it, the tube within the shell of a shell-and-tube heat exchanger. This porous inner cylinder is the membrane that gives the membrane reactor its name.

A simple example of catalytic ceramic membrane reactor is shown in Figure 6.1. The membrane is a barrier that allows only certain components to pass through it. The selectivity of the membrane is controlled by its pore diameter, which can be on the order of Angstroms, for micro-porous layers, or on the order of microns for macro-porous layers. Membrane reactors combine reaction with separation to increase conversion. One of the products of a given reaction is removed from the reactor through the membrane, thus forcing the equilibrium of the reaction 'to the right' (according to Le Chatelier's principle), so that more of that product is generated. A catalytic membrane reactor has a membrane that has either been coated with or is made of a material that contains catalyst, which means that the membrane itself participates in the reaction. Some of the reaction products (those that are small enough) pass through the membrane and exit the reactor on the permeate side.

Based on the membrane properties, water-gas shift (WGS) membrane reactors are classified into two categories, namely, CO_2 selective membrane reactors and H_2 selective membrane reactors. In the CO_2 selective membrane reactors, CO_2 was removed from the catalytic membrane reactor and the reaction mixture becomes H_2-rich steam. This may cause over reduction of Fe- or Cu-based catalysts. However, in the H_2-selective membrane, CO_2 will be present at a higher concentration in the reaction medium, affecting the reaction rate.

The membrane reactor offers many potential advantages: reduced capital and downstream separation costs, as well as enhanced yields and selectivity. From the viewpoint of the WGS process in an membrane reactor, a reaction product moves to the permeate side, enabling the WGS reaction to proceed towards completion and so making it possible to achieve the following: (1)

Water Gas Shift Reaction. http://dx.doi.org/10.1016/B978-0-12-420154-5.00006-1

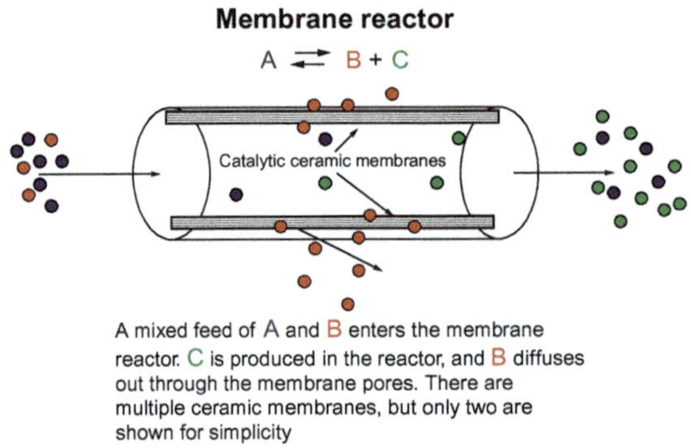

FIGURE 6.1 The schematic diagram of the membrane reactor.

higher conversion than a TR working under the same operating conditions or (2) the same conversion as a TR but working under milder operative conditions.

Kikuchi *et al.* [1] is the first one who reported the WGS reaction in membrane reactors using thin film of palladium. They used Fe-Ce catalysts inside the membrane. They used steam to CO ratio 1 and argon as a sweep gas. Interestingly, decreasing space velocity in the membrane reactor results in the conversion level exceeding equilibrium CO conversion and increased with increasing time factor. However, the molar fraction of H_2 in un-permeated gas decreases with decreasing space velocity. The equilibrium CO conversion is a function of the molar fraction of the hydrogen in the reaction system. Figure 6.2 shows the effect of pressure on the CO conversion in the membrane reactor. As expected, increasing reaction pressure increases the CO conversion. The CO conversion reached 100% at 5 atm. The effectiveness of the membrane reactor is influenced by the relative rate of hydrogen permeation to reaction. At the effective area of

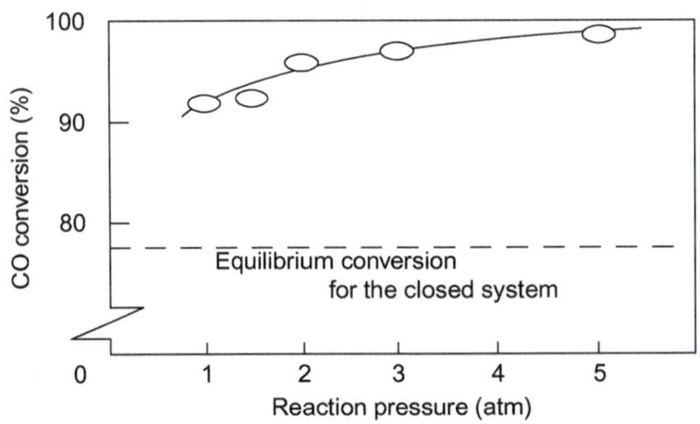

FIGURE 6.2 Effect of reaction pressure on conversion of carbon monoxide. Reaction conditions: catalyst (Girdler G-3), 3.0 g; temperature, 400 °C; H_2O/CO ratio, 1; feed rate of carbon monoxide, 25 ml/min. Sweep argon: flow rate, 400 ml/min; pressure, 1 atm. *(Taken from Figure 3 of E. Kikuchi, S. Uemiya, N. Sato, H. Inoue, H. Ando, T. Masuda, Chem. Lett. (1989) 489.)*

25.1 cm^2 the rate of hydrogen permeation is of almost the same order as that of the rate of reaction.

Then Uemiya *et al.* [2] reported another Pd membrane reactor for WGS reaction using Fe-Cr oxide catalysts. The model of flow in the palladium membrane reactor is illustrated in Figure 6.3. They proposed that hydrogen is permeated through palladium membrane via a solution diffusion transport mechanism, and the rate of hydrogen permeation, J, per unit area of membrane, is written in terms of Fick's first law as follows:

$$J = (Q/t)(P_d{}^n - P_d{}^n)$$

Q is the hydrogen permeation coefficient, t is the thickness of the palladium film and n is a constant indicating pressure dependency. P_d is the partial pressures of hydrogen in the high- and low-pressure sides. They studied the pressure dependence of hydrogen permeation rate in membrane having a thickness of 20 μm at 673 K. The rate of hydrogen permeation was found to be proportional to the difference between high and low pressures to their 0.76 power. Based on flow model they conducted simulation studies and compared with experimental results. The theoretical studies were well agreed with experimental results when they studied

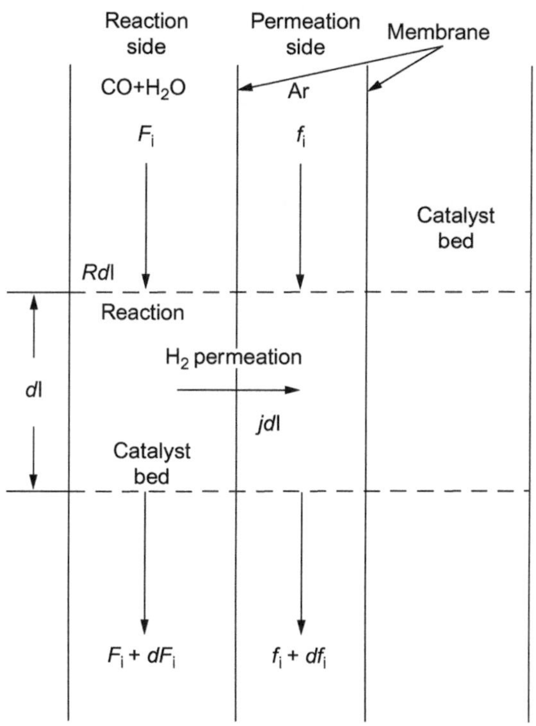

FIGURE 6.3 Flow model of reaction and permeation in palladium membrane reactor. *(Taken from Figure 1 of S. Uemiya, N. Sato, H. Ando, E. Kikuchi, Ind. Eng. Chem. Res. 30 (1991) 581.)*

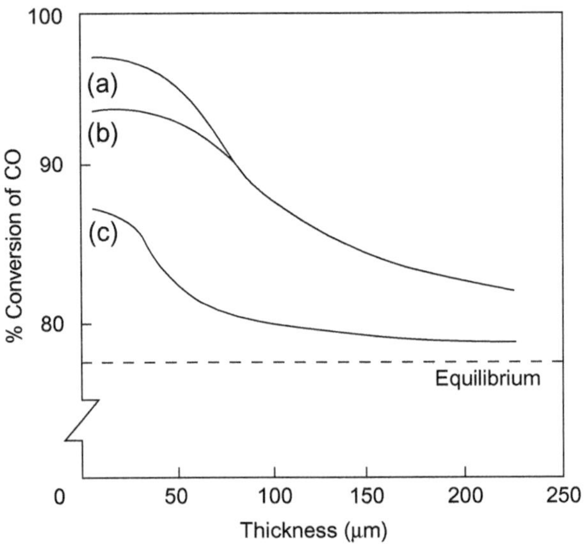

FIGURE 6.4 Conversion of carbon monoxide as a function of palladium thickness. Feed rate of CO: (a) 25, (b) 25 and (c) 100 cm^3(STP)/min. Flow rate of sweep argon: (a) 3200, (b) 400 and (c) 400 cm^3(STP)/min. *(Taken from Figure 1 of S. Uemiya, N. Sato, H. Ando, E. Kikuchi, Ind. Eng. Chem. Res. 30 (1991) 581.)*

the WGS reaction as a function of space velocity and steam to CO ratio. As expected, CO conversion exceeds equilibrium conversion by decreasing space velocity. They also conducted simulation studies on dependence of CO conversion on Pd thickness. The results are presented in Figure 6.4. The level of carbon monoxide conversion increased with decreasing thickness of palladium, as a result of improved rate of hydrogen permeation. The level of conversion was raised by an increase in the flow rate of argon, resulting from reduction of the partial pressure of hydrogen in the permeation side.

So many membrane reactors were reported in the literature for the WGS reaction. Along with Pd membranes, studies on silica membranes, zeolite membranes, hollow membrane reactors, electrochemical WGS reactor are also available in the literature. However, Pd-based membranes were investigated extensively for the WGS reaction. In the following sections we have given detailed description about various membrane reactors investigated for the WGS reaction. A detailed description about the theoretical and simulation studies of the membrane reactors for the WGS reaction was also discussed.

6.2 Pd-BASED MEMBRANE REACTORS

As explained earlier some experimental and computational studies were available in the literature before 1990 [1,2]. In 1996, Basile *et al.* [3] reported detailed experimental study, i.e., the influence of various reaction conditions on palladium membrane for WGS reaction obtained by coating an ultrathin double-layer palladium film on the inner surface of the support of a commercial tubular ceramic membrane by condensation technique. When they investigated the influence of steam to CO ratio for the WGS reaction, 0.96 H_2O/CO ratio was found to be the best. The CO conversion depends on the flow of the sweep gas utilized. Without sweep gas, the CO curve is below the equilibrium value for H_2O/CO molar ratio greater than 1.5. Considering a nitrogen sweep gas rate of 28.2 cm^3/min, the experimental curve is above the equilibrium curve for all the H_2O/CO molar ratios. Increasing pressure has positive effect on the CO conversion. They were able to get 99.9% CO conversion at 1.2 bar pressure. Increase in the temperature increases the CO conversion initially and then decreases the CO conversion because of thermodynamic constrains. They found that 600 K is the best temperature.

Basile *et al.* investigated the effect of method of deposition of Pd film on the surface of the ceramic membrane [4]. They used three different methods: magnetron sputtering technique, the physical vapour deposition technique and the co-condensation technique. The experimental results indicate that the physical vapour deposition and magnetron sputtering techniques do not allow an interesting homogeneous metallic film to be obtained on the outer surface of tubular supports. The water-gas reaction results show that the membrane reactor synthesized by co-condensation exhibits better WGS activity compared to the membranes synthesized by magnetron sputtering and the physical vapour deposition techniques. The presence of large pores results in conversion lower than equilibrium for membrane reactors synthesized by both magnetron sputtering and the physical vapour deposition technique.

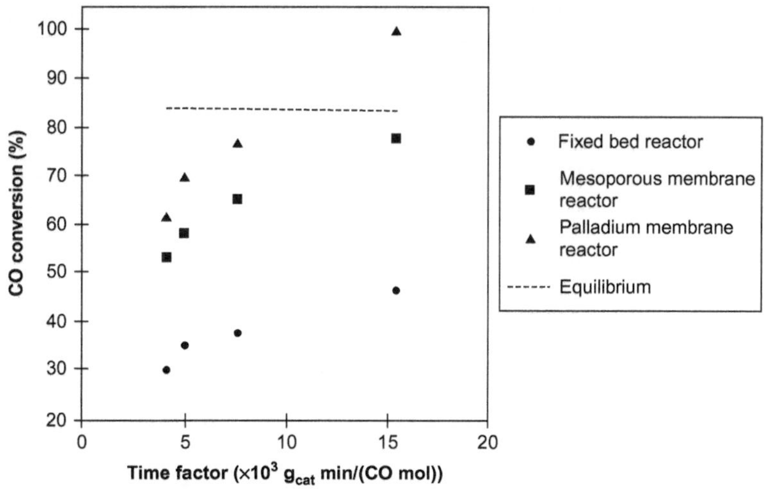

FIGURE 6.5 Effect of the time factor on the CO conversion for the three reaction systems and mix1. $T=505$ K; $H_2O/CO=1.1$; fixed-bed reactor: $P=1$ atm; mesoporous membrane reactor: $P_{lumen}=P_{shell}=1$ atm; no sweep gas; palladium membrane reactor: $P_{lumen}=P_{shell}=1$ atm; sweep gas flow rate $=43.0$ ml/min. *(Taken from Figure 2 of A. Criscuoli, A. Basile, E. Drioli, Catal. Today 56 (2000) 53.)*

Criscuoli *et al.* compared Pd membrane reactor with mesoporous membrane reactor and fixed-bed reactor [5]. Figure 6.5 shows the effect of space velocity on the CO conversion for the three reaction systems. As expected both membrane reactors exhibit better CO conversion than traditional reactor. Between the two membrane reactors Pd membrane reactor exhibits much better CO conversion compared to mesoporous membrane reactor. At the highest time factor, Pd membrane reactor exhibits 100% CO conversion. By increasing the Pd membrane thickness, the hydrogen permeation rate decreases and lower conversions of carbon monoxide are achieved. When they compared experimental results with simulation results the model fits well with the experimental points.

Basile *et al.* compared Pd membrane reactor with Pd/Ag membrane reactor. In this study, they used thin rolled membranes [6]. In rolled membranes, the main function of the ceramic support was to separate the Pd or Pd/Ag membrane from the catalyst bed of the MR. The catalyst is inside the ceramic support, while the permeating tubes are outside the ceramic support. Both the membranes exhibit excellent hydrogen permeation selectivity. The experimental data have shown that both membranes work well in terms of CO conversion. The maximum conversion of CO exceeds the value of 96.80%.

Tosti *et al.* evaluated Pd-Ag membrane reactors in pilot plant environment [7]. The tests carried out during several months of operation have simulated the WGS reaction. Operating conditions are temperature of 325 °C, feed pressure of 100 kPa, shell pressure of 100 kPa and nitrogen (purging gas) flow rate of 2.74×10^{-4} mol/s. They tested membrane reactor in different feed compositions with different CO feed flow rates. The results are presented in Figure 6.6. As expected increase in the CO feed flow rate decreases the CO conversion. Also increase in the water quantity increases the CO conversion. Introduction of CO_2 in the feed decreases the CO conversion.

Barbieria *et al.* [8] proposed innovative Pd-Ag membrane reactor for the WGS reaction. Their membrane design is presented in Figure 6.7.

The innovative membrane reactor exhibits 25% less volume index compared to the regular membrane reactor. This shows the clear advantage in the use of the innovative solution that, allowing the problems related to the good

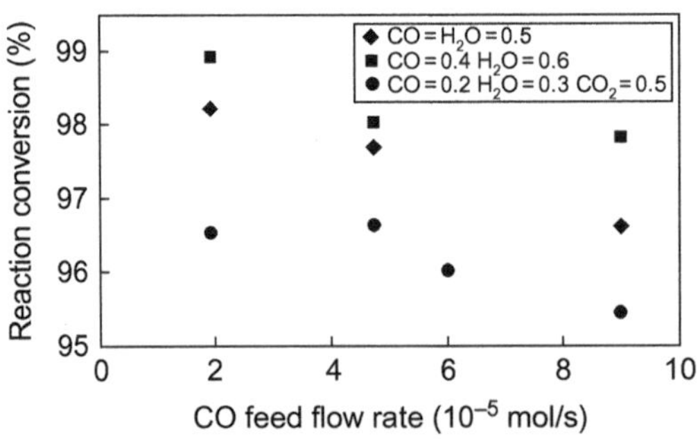

FIGURE 6.6 Measured conversion values for the WGS reaction. *(Taken from Figure 6 of S. Tosti, A. Basile, G. Chiappetta, C. Rizzello, V. Violante, Chem. Eng. J. 93 (2003) 23.)*

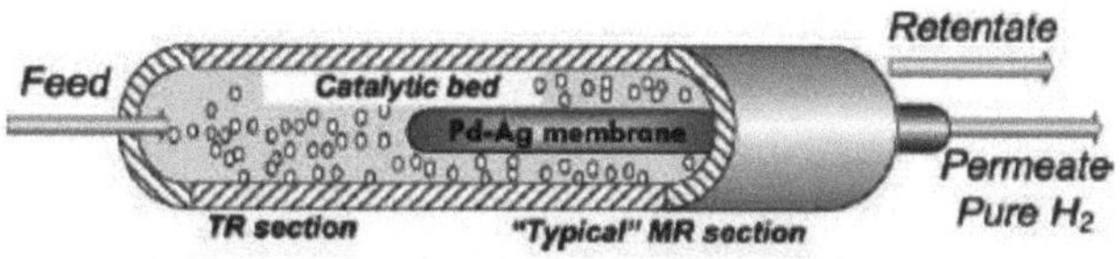

FIGURE 6.7 Configuration of the 'innovative' MR. *(Taken from Figure 2 of G. Barbieria, A. Brunetti, G. Tricoli, E. Drioli, J. Power Sources 182 (2008) 160.)*

exploitation of the membrane area to be overcome, shows further reaction volume reduction with respect to that achieved with the traditional MR.

Brunetti *et al.* tested this membrane in wet (33% H_2O, 33% CO, 29% H_2, 4% N_2) and dry syngas (45% CO, 50% H_2, 4% N_2) compositions with no sweep gas [9]. The presence of H_2 did not have any influence on CO conversion. The measured CO conversion of 90% was three to four times higher than that of a traditional reactor. The improvement of CO conversion and hydrogen recover is more effective at higher feed pressure and GHSV. They recovered 80% of the H_2 stream in the permeate side.

Then Brunetti *et al.* evaluated Pd-Ag membrane at elevated temperatures [10]. Figure 6.8 shows the outlet concentrations in traditional and membrane

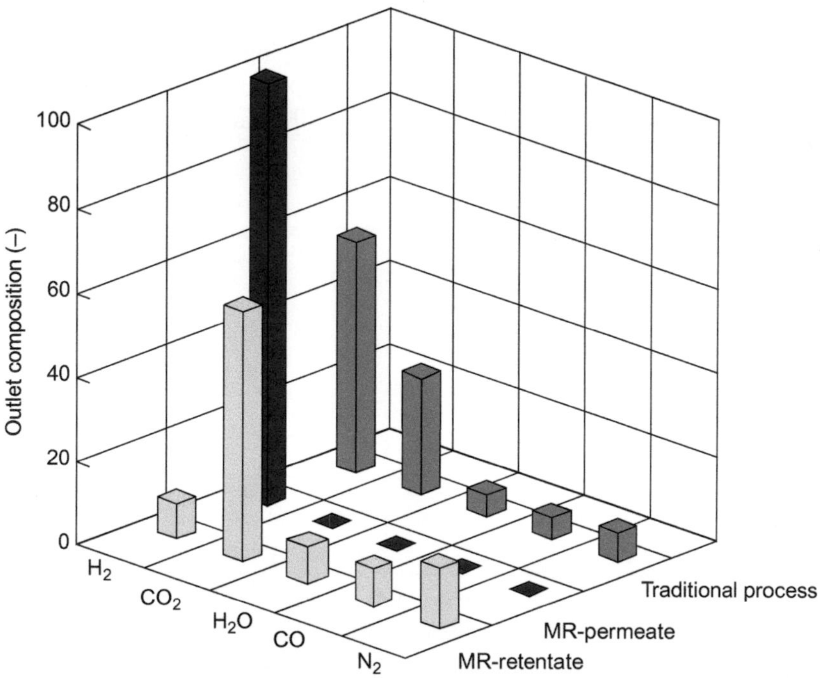

FIGURE 6.8 Outlet stream composition of the MR and the traditional process. Operating conditions as in Table 2 of the corresponding reference. *(Taken from Figure 5 of A. Brunetti, G. Barbieria, E. Drioli, A. Caravella, RSC Adv. 1 (2011) 651.)*

reactors. They achieved 95% conversion with syngas feed composition CO:CO_2:H_2:N_2=52:19:20.5:8.5% molar (dry basis) at 300 °C and 15 bar pressure. In addition, around 90% of the H_2 fed and produced by the reaction in the Pd-Ag MR was recovered in the permeate, when the MR operates at 15 bar and ca. 450 °C (outlet temperature). This stream, completely pure in H_2, does not require any separation/purification, contrarily to the one exiting from the traditional process.

Then Brunetti *et al.* evaluated Pd-Ag membrane at medium/higher temperatures using Ce-Cr catalysts and compared with low temperature results [11a]. The results are presented in Figure 6.9. The fast kinetics together with the high permeation rate offered by the high temperature allowed the thermodynamics and the further limitations due to the H_2 presence (50%) in the feed stream to be significantly exceeded. At the higher temperature of 375 °C and at 6021 h^{-1}, the catalyst volume required by membrane reactor was only 15-20% of the traditional reactor to achieve the same conversion.

Brunetti *et al.* also found that with Fe-Cr catalysts and at temperatures more than 400 °C and feed pressures 15 bar the membrane reactor only require 9% of the catalyst volume of the traditional reactor [11b].

Mendes *et al.* [12] designed 'finger-like' configuration of the self-supported Pd-Ag membrane and used as a packed-bed MR for producing ultra-pure hydrogen via WGS reaction. The design and picture was presented in Figure 6.10. They used Cu-Zu-Al_2O_3 as a catalyst. They proposed that H_2 recovery can be improved by increasing the operating temperature and/or applying a higher H_2 partial pressure. They achieved 100% CO conversion and complete H_2 recovery by operating the membrane reactor at 300 °C with a GSHV 1200kg_{cat}^{-1} h^{-1}, P_{feed} 4 bar, P_{perm} 3 bar and using 1000 m/min of sweep gas.

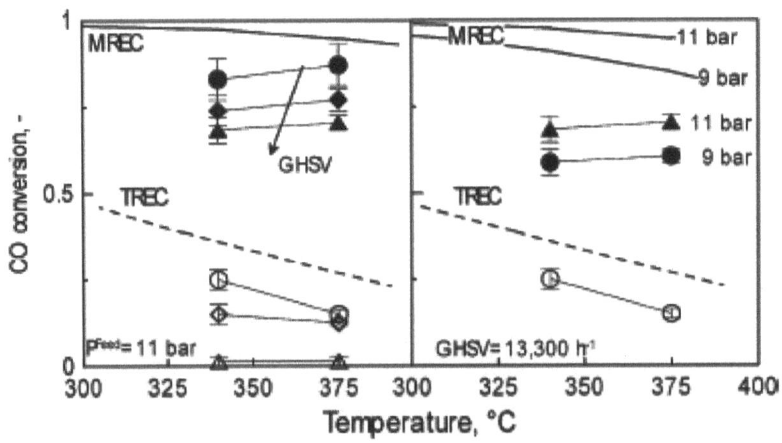

FIGURE 6.9 CO conversion as a function of temperature for different values for GHSV equal to 6021, 10,035, 13,300 h^{-1} (left side) and feed pressure (right side). *(Taken from Figure 8 of Brunetti, G. Barbieria, E. Drioli, A. Caravella, Chem. Eng. Technol. 35 (2012) 1238.)*

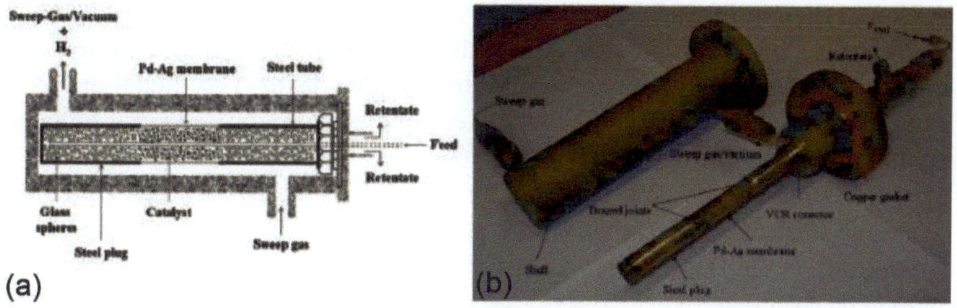

FIGURE 6.10 Scheme (a) and picture (b) of the Pd-Ag 'finger-like' configuration membrane reactor. *(Taken from Figure 1 of D. Mendes, V. Chibante, J. Zheng, S. Tosti, F. Borgognoni, A. Mendes, L.M. Madeira, Int. J. Hydrogen Energy 35 (2010) 12596.)*

Then Mendes *et al.* developed the pseudo-homogeneous 1-D model for the Pd-Ag finger-like shape membrane reactor and compared with the experimental results [13]. The simulation results are well agreed with the experimental results. Then they investigated this membrane reactor for parameter space such as temperature, pressure and sweep gas. Increasing reaction temperature increases the CO conversion up to 250 °C and further increase in reaction temperature decreases the CO conversion. However, H_2 recovery increases with increasing temperature. Their simulation results suggest that it is possible to achieve 100% CO conversion both in sweep gas mode as well as vacuum mode.

Tosti *et al.* tested Pd-Ag membrane reactor for 12 months for H_2 permeation [14]. Excellent stability was observed for 12 months of operation. In fact, the complete hydrogen selectivity and none failure (formation of cracks, holes) were observed. They proposed that the reliability is a result of both the tube manufacturing procedure and the reactor design configuration (finger-like). Figure 6.11 shows the picture of membrane reactor before and after the 12 months of operation.

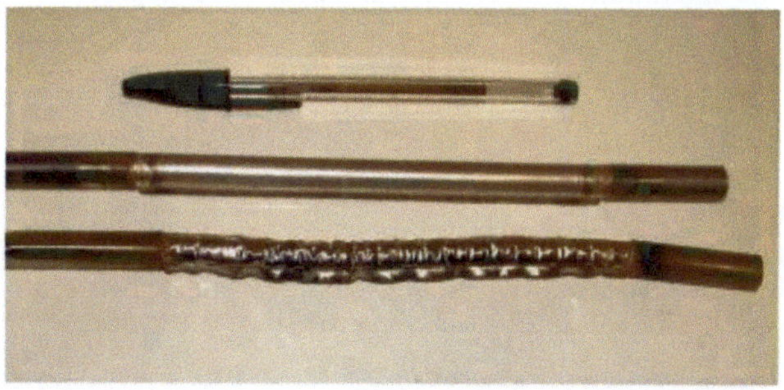

FIGURE 6.11 Pd-Ag thin wall permeator tubes: as produced (above) and after testing (below). *(Taken from Figure 6 of S. Tosti, A. Basile, L. Bettinali, F. Borgognoni, F. Chiaravalloti, F. Gallucci, J. Membr. Sci. 284 (2006) 393.)*

Cornagliaa *et al.* [15] reported Pd-Ag membranes using Pt/La-Si catalyst. They conducted WGS reaction for 100 h. The WGS selectivity remained constant at almost 100%. They did simulation studies using 1 D model and concluded that 100% CO conversion and complete hydrogen recovery with the membrane can be achieved.

Cornagliaa *et al.* [16] evaluated the membrane reactor at higher pressure (100 and 400 kPa) and higher temperature (673 and 723 K). They found that 500 ml/min is the optimum sweep gas flow rate for the higher CO conversion and H_2 recovery.

Iyoha *et al.* [17] reported Pd-Cu membrane reactor for WGS reaction. The 3.175 mm OD, 125-μm thick Pd and Pd 80 wt%-Cu 20 wt% alloy tubes were used in their study. A four-tube MR, rather than a single tube of larger diameter, was designed to increase the catalytic surface area to reaction volume ratio of the MR. The picture of the membrane reactor is shown in Figure 6.12. They conducted WGS reaction in counter current mode at 1173 K. Interestingly, Pd-Cu membrane reactor exhibits lesser CO conversion compared to Pd membrane. This is probably the result of the decrease in H_2 permeance with increasing Cu content for FCC alloys. Both membrane systems resulted in high H_2 recovery at residence times as low as 1 s. At 1173 K, the H_2 permeances of Pd-Cu and Pd were determined to be 1.42×10^{-4} and 3.10×10^{-4} mol H_2, respectively. The CO conversions remained below the equilibrium conversion value for Pd-Cu membrane reactor increasing steam to CO ratios.

Then Iyoha *et al.* evaluated [18] Pd and Pd-Cu membrane reactors in simulated coal syngas containing H_2S. For both the membrane reactors introduction of H_2S in the feed by switching the feed mixture to 90% H_2-1000 ppm H_2S-He resulted in no discernible change in H_2 permeance. 99.7% conversion of CO contained in a simulated syngas feed consisting of 53% CO, 35% H_2 and 12% CO_2 (dry basis) was attained in a Pd four-tube membrane reactor at 1173 K with a steam to CO ratio of 1.5. The Pd-Cu membrane reactor also effectively enhanced CO conversions above the equilibrium value of 32% (associated with non-membrane reactors) over the conditions of the study. However, the maximum conversions attained were appreciably lower than

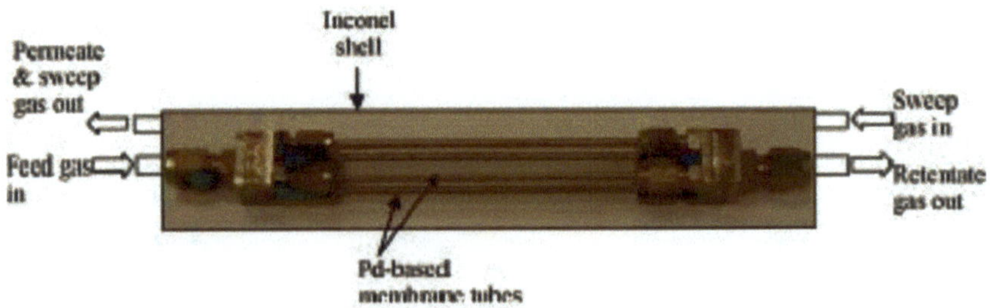

FIGURE 6.12 Detail of the NETL four-tube Pd-based membrane reactor. *(Taken from Figure 1 of O. Iyoha, R. Enick, R. Killmeyer, B. Howard, B. Morreale, M. Ciocco, J. Membr. Sci. 298 (2007) 14.)*

those obtained in the Pd membrane reactor, reaching maximum CO conversions of 62% and 78% at 2 and 2.8 s residence times, respectively. This was primarily attributed to the lower H_2 membrane permeance compared to that of Pd. Exposure of both membrane reactors to syngas mixtures containing H_2S levels such that the H_2S-to-H_2 ratio was less than 0.0011 did not appear to affect the H_2 permeance, mechanical integrity and H_2 selectivity of the membrane reactors at 1173 K. However, a steep drop in CO conversion was observed. When the H_2S-to-H_2 feed ratio was increased to 50 and 90 ppm H_2S for the Pd and Pd-Cu membrane reactor systems, the membrane reactors were observed to fail within minutes.

Hwang *et al.* [19] prepared defect free Pd-Cu-Ni membrane and evaluated for WGS reaction using Pt/CeO_2 catalyst. They conducted WGS reaction using 7.0 vol% CO, 8.5 vol% CO_2, 22.0 vol% H_2O, 37 vol% H_2 and 25.5 vol% N_2 feed gas mixture. The membrane reactor exhibits 3000 min of long-term stability with 98.2% CO conversion.

Hwang *et al.* [20] prepared Pd-Au membrane reactor for the WGS reaction by using Ni catalyst. They compared their hydrogen flux results with the literature reports. Their membrane reactor exhibits excellent hydrogen selectivity and permeance. They studied the WGS reaction at 340 °C with a steam to CO ratio of 3. The CO conversion increased from 98.6% to 99.1% with increasing system pressure from 0 to 25 bar. Also 85.4% of hydrogen was recovered through the Pd-based membrane. Also the CO conversion was above 99% and the H_2 recovery was above 94% at pressure 30 bar in the membrane reactor. The best result for the concentration of the enriched CO_2 in the retanate side was 85.3% under the conditions of 350 °C, pressure 30 bar and a steam to carbon ratio of 2.0.

Then Hwang *et al.* [21] prepared plat type Pd membrane reactor using the magnetron sputtering method over a nickel metal support. They conducted WGS reaction using nickel catalyst. The nickel metal catalyst with a disc shape was placed on a membrane without a metal cage or mesh to hold the catalyst in the reactor. However the membrane did not work very well.

Tosti *et al.* [22] compared a thin-walled dense tube Pd membrane and composite Pd-ceramic tube membranes. They developed a finite elemental model for membrane reactors. Table 6.1 shows the WGS reaction conversion values calculated for the dense and the composite membrane reactors both by taking into account the wall effects ('WE' case) resistance and by neglecting such a resistance ('no WE' case). The main result is that by increasing the temperature the reaction conversion increases. Table 6.2 shows in co-current mode. Both the membranes show similar activities.

Bi *et al.* [23] reported WGS reaction in Pd membrane reactor using Pt/Ce-Zr catalyst. They prepared Pd membrane on outer surface of porous ceramic tubes. Figure 6.13 shows WGS CO conversion and H_2 recovery in the Pd membrane reactor charged with the Pt/$Ce_{0.6}Zr_{0.4}O_2$ catalyst as a function of reaction pressure at 623 K, GHSV = 4050 kg^{-1} h^{-1} and steam/CO = 3. In the Pd membrane

TABLE 6.1 Reaction Conversion Percent: Counter-Current Sweep Mode

	Dense membrane		Composite membrane	
	WE (%)	*No WE (%)*	*WE (%)*	*No WE (%)*
T=300 °C	64.47	64.65	64.66	64.80
T=350 °C	97.55	97.69	97.64	97.73
T=400 °C	99.23	99.31	99.21	99.28

TABLE 6.2 Reaction Conversion Percent: Co-Current Sweep Mode

	Dense membrane		Composite membrane	
	WE (%)	*No WE (%)*	*WE (%)*	*No WE (%)*
T=300 °C	63.68	63.78	63.74	63.82
T=350 °C	92.12	92.10	92.09	92.07
T=400 °C	89.99	89.99	89.99	89.99

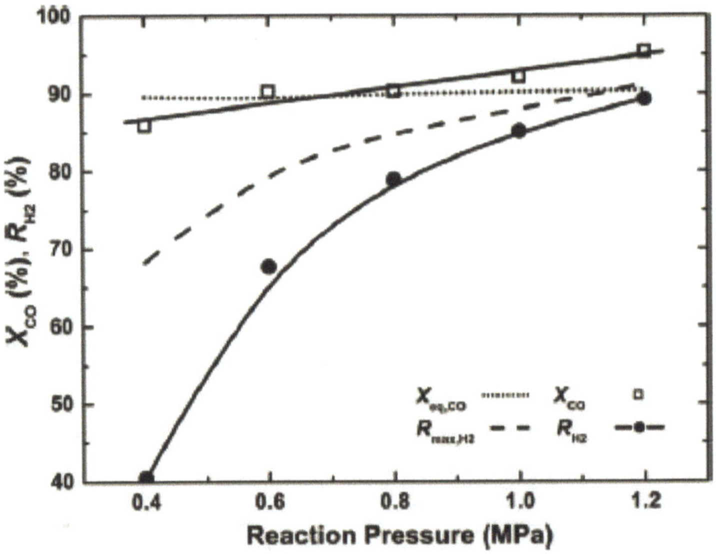

FIGURE 6.13 Influence of pressure difference on CO conversion and H_2 recovery in the membrane reactor at GHSV=40501 $kg^{-1}\,h^{-1}$, T=623 K and S/C=3. *(Taken from Figure 4 of Y. Bi, H. Xu, W. Li, A. Goldbach, Int. J. Hydrogen Energy 34 (2009) 2965.)*

reactor CO conversion improves with increasing reaction pressure because H_2 is removed faster from the reaction zone. Here the increase of reaction pressure from 0.4 to 1.2 MPa raised R_{H2} from 40.5% to 89.2%, driving up CO conversion from 86.0% to 95.5% in parallel. The purity of the permeated H_2 declined slightly from 99.7% to 99.2% with increasing H_2 recovery. Increase in the temperature does not have much effect on CO conversion; however, H_2 recovery increased from 70% to 86.4% with increasing temperature from 598 to 673 K. Increase in the steam/CO ratio has negative effect on H_2 recovery. H_2 recovery decayed rapidly from 84.8% to 48.7% with increasing space velocity from 4000 to 9000 $kg^{-1}\ h^{-1}$, while the permeate purity increased from 99.4% to 99.7%. This is probably because the residence time of the reaction mixture in the membrane zone became too short for H_2 diffusion from the reactor wall through the catalyst bed to the membrane surface.

Piemonte *et al.* [24] investigated the influence of reactor length/diameter ratio, position of membrane inside the reactor on the membrane reactor performance. They investigated the effect of slenderness ratio (reactor length to reactor radius ratio). They found that the CO conversion decreases with increasing slenderness ratio. Figure 6.14 shows the different CO conversion curves obtained by varying the membrane length with respect to the reactor length.

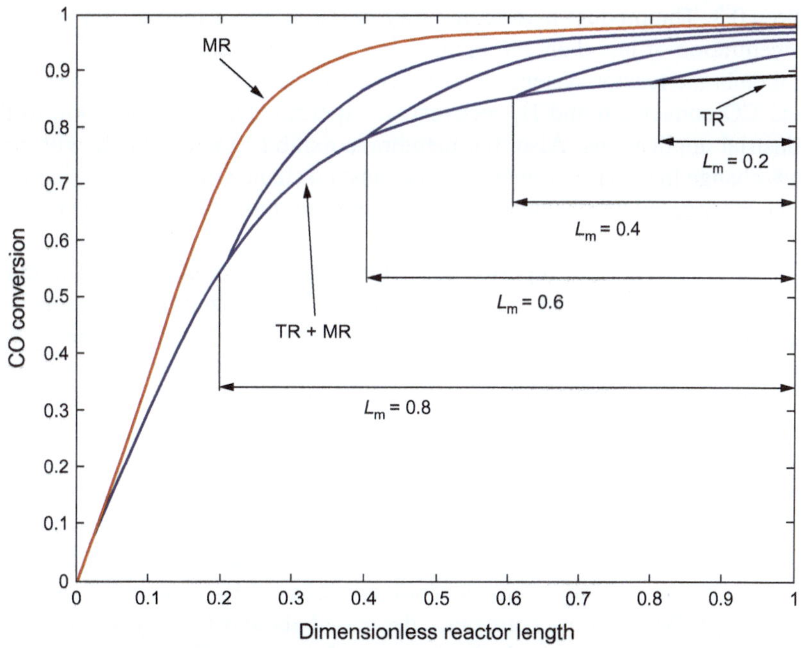

FIGURE 6.14 CO conversion for the 'TR+MR' configuration for different L_m values. *(Taken from Figure 9 of V. Piemonte, M. De Falco, B. Favetta, A. Basile, Int. J. Hydrogen Energy 35 (2010) 12609.)*

The presence of the membrane at the initial reaction stage (reactor entrance) increases cost without giving a significant performance enhancement. The membrane operates only in the second part of the catalytic bed where the hydrogen partial pressure is high enough to promote the permeation in the desired direction, from the reaction to permeate side.

Pinaccia *et al.* [25] investigated the tubular Pd membrane reactor at higher temperatures, i.e., above 400 °C and higher pressures (100-800 kPa). They did permeation tests for 1200 h at 400 °C and the membrane exhibits excellent stability. After permeation they evaluated membrane reactor for WGS reaction using commercial Fe-Cr catalyst in the syngas mixture. They are able to achieve 85% CO conversion and 82% H_2 recovery with 97% purity.

Augustine *et al.* [26] reported Pd membrane reactors at elevated temperatures. Figure 6.15 shows the effect of temperature on the WGS activity and H_2 recovery of Pd membrane reactor. The CO conversion increases with increasing temperature up to 450 °C and further increase in temperature has no effect on the CO conversion. Similar trends were observed with H_2 recovery.

Abdollahi *et al.* [27] evaluated commercial scale Pd membrane reactor for the WGS reaction. A tubular Pd membrane reactor of length = 762 mm, ID = 3.5 mm, OD = 5.7 mm was synthesized and commercial Cu-Zn-Al_2O_3 catalyst was used. Almost complete CO conversion and 90% hydrogen recovery were achieved with $T = 300$ °C, $P = 4.46$ bar and the permeate sweep gas ratio = 0.3. The product hydrogen purity is always more than 99.9% with CO concentration of less than 100 ppm. Their simulation results also suggest that Pd membrane reactor system under study is capable of delivering almost complete CO conversion and H_2 recovery at experimental conditions akin to the industrial applications. Also, the membrane exhibits good stability with only a 6% change in the H_2 permeance and almost no change in the permeation rates of the other gases after being used in the reactor for more than a month under the WGS environment.

Augustine *et al.* [28] reported durability of porous stainless steel supported Pd tubular membranes for WGS reaction. The synthesized membrane is very stable in the presence of H_2/H_2O mixture. No significant change in the H_2 permeance is observed for 2000 h. However, under WGS conditions when the H_2O:CO ratio was 2:1, a reduction in H_2 recovery was observed over 65 h due to coke formation on the membrane surface. They conducted another WGS experiment for 1000 h with a higher H_2O:CO ratio of 3:1 and stable behaviour was observed. They achieved 97% CO conversion and 85% H_2 recovery from a simulated syngas mixture for 900 h.

Liguori *et al.* [29] evaluated porous stainless steel supported Pd tubular membranes for more than 4000 h. They investigated H_2 permeation measurements for 1200 h. The H_2 permeance decreased about 8%. They also observed that H_2 permeance decreases in the presence of steam and CO. In the presence of syngas MR was able to achieve up to 76% of CO conversion and 75% of hydrogen recovery with a H_2 permeate purity exceeding 97%.

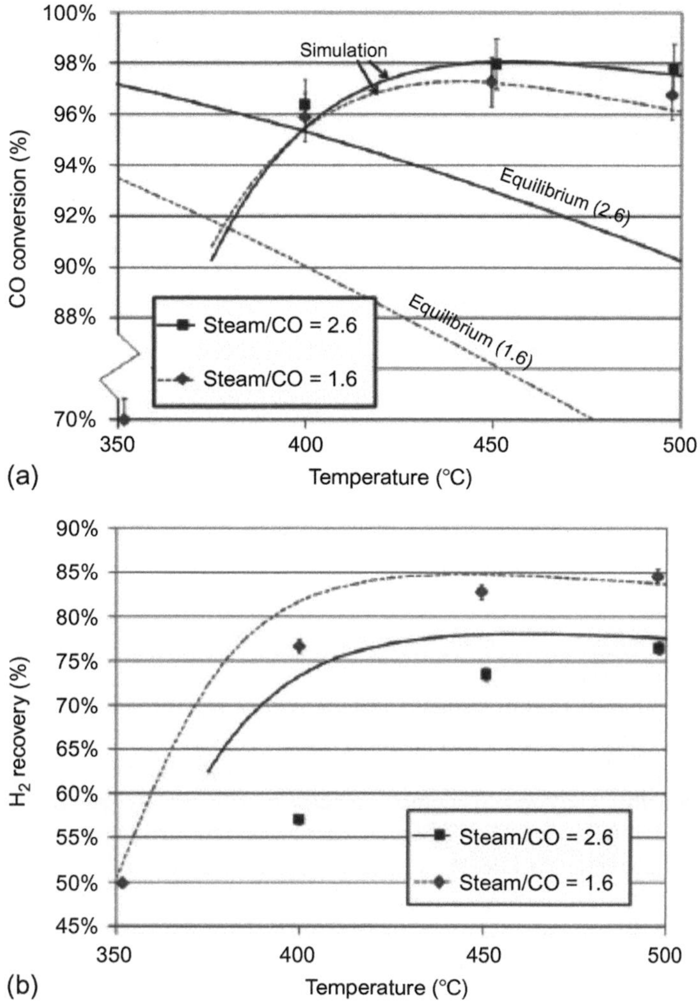

FIGURE 6.15 CO conversion (a) and H_2 recovery (b) as a function of temperature for a constant CO feed rate of 7.95 mmol/min (membrane: AA-6R; $P=14.4$ bar; ◊: $H_2O/CO=1.6$, $GHSV=2100\ h^{-1}$; □: $H_2O/CO=2.6$, $GHSV=2900\ h^{-1}$). *(Taken from Figure 5 of A.S. Augustine, Y.H. Ma, N.K. Kazantzis, Int. J. Hydrogen Energy 36 (2011) 5359.)*

Li *et al.* [30] synthesized membrane reactor with three Pd tubes by varying the thickness from 5.6 to 6.1 μm. They evaluated membrane reactor for WGS reaction for 30 days. The H_2 permeance results at two pressures are presented in Figure 6.16. It can be seen that the pure H_2 permeance remained stable during the reaction test of 27 days, showing a good chemical and mechanical stability of the Pd membranes used in this study. It is implied that there was no degradation of the membrane performance due to, e.g., carbon formation on the

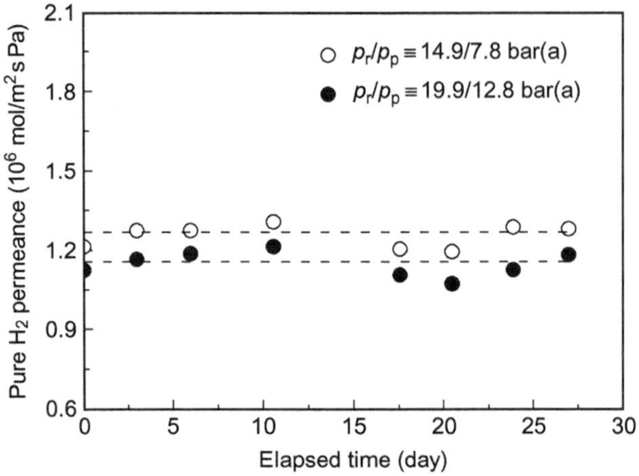

FIGURE 6.16 Stability evaluation of the three membranes during the WGS reaction test: The pure H_2 permeance of the three membranes measured at 673 K and a feed/permeate pressure of 14.9/7.8 and 19.9/12.8 bar(a). *(Taken from Figure 2 of H. Li, J.A.Z. Pieterse, J.W. Dijkstra, J. Boon, R.W. van den Brink, D. Jansen, Int. J. Hydrogen Energy 37 (2012) 4139.)*

membrane surface. The membrane reactor exhibits 95% of CO conversion and above 90% of H_2 recovery.

Catalano *et al.* [31] investigated Pd membrane reactor at larger scale. They prepared Pd membrane with 25 cm length and 2.54 cm o.d. and tested at higher temperatures (440 °C), pressures (20 bar) and relatively high feed flow rates (up to 1.5 Nm^3/h). They tested membrane reactor with real syngas mixture, i.e., 42.2% CO, 40% H_2 and 17.8% CO_2 on a dry basis. The membrane exhibits maximum CO conversion of 98.1% and hydrogen recovery up to 81.5%. The purity of the hydrogen produced was consistently over 99.97% for membrane at pressures up to 20 bar and a temperature of 420 °C. However, after several days of testing a decline in the membrane selectivity was observed giving a lower hydrogen purity of 99.2% at 20 bar of pressure. This decline in selectivity was due to the high temperature at which the membrane was exposed: in this temperature region, indeed, a higher leak growth is present.

All the above literature reports suggest that Pd membrane reactors can replace traditional WGS reactors in the coal gasification plants. However, because of their cost and poor thermal stability at higher temperatures we cannot use these reactors in the industrial sector.

6.3 SILICA MEMBRANES

Recently, investigation of silica membranes was also reported in the literature. Giessler *et al.* [32] were the first who investigated molecular sieve silica membrane for WGS reaction. They prepared membrane by using tetra ethyl ortho silicate and methyl trimethyl silane over porous alumina support using sol-gel method. They used $CuO/ZnO/Al_2O_3$ as catalyst and evaluated for low temperature WGS reaction. They synthesized both hydrophilic and hydrophobic membranes. They proposed that hydrophobic molecular sieve silica membrane is suitable for WGS reaction compared to hydrophilic membrane reactors. They suggested that water vapour causes the unwanted structural changes in the hydrophobic membrane reactors. They are able to achieve 99% CO conversion at 280 °C. The use of sweep gas increases the CO conversion up to 80 ml/min and further increase in the CO conversion decreases the CO conversion for hydrophobic membrane reactor. The optimal operation conditions for hydrophobic membrane are a sweep gas flow of 80 cm^3/min, feed flow rate of 50 cm^3/min and a H_2O/CO molar ratio of 1.0.

In 2007, Brunetti *et al.* [33] reported porous stainless steel supported silica membrane reactors for WGS reaction using CuO/CeO_2 catalyst. They prepared membrane with soaking-rolling procedure, modifying the macro-pores of the support disc by packing silica xerogel (500 nm) and coating the intermediate layer of alumina in order to improve the H_2 membrane selectivity. They measured CO content in the permeate stream ranging from 1% to 10% depending on operating conditions. The membrane exhibits CO conversion more that equilibrium CO conversion only at 280 °C reaction temperature. Also the membrane exhibits higher CO conversion at 4 bar compared to at 6 bar. A significant difference in permeances measured during and after the reaction tests with respect to those measured before the reaction was observed. The results are presented in Figure 6.17. The membrane exhibits higher H_2 permeance after the reaction

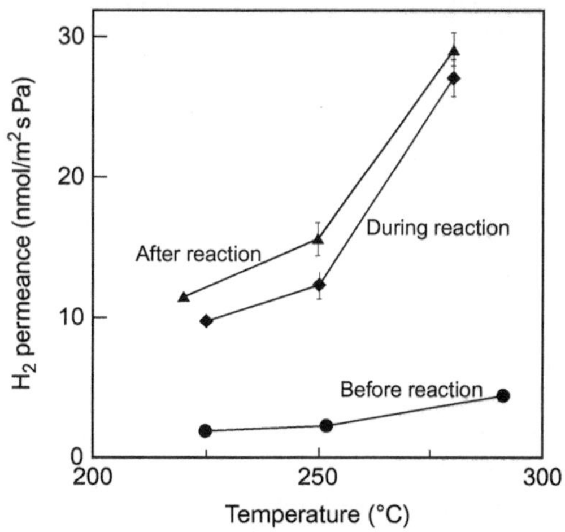

FIGURE 6.17 Hydrogen experimental permeance as a function of the temperature. *(Taken from Figure 10 of A. Brunetti, G. Barbieri, E. Drioli, K.-H. Lee, B. Seac, D.-W. Lee, Chem. Eng. Process. 46 (2007) 119.)*

compared to that before the reaction. This is probably due to an incomplete stabilization of the membrane properties, achieved progressively during the reaction, as confirmed also by the agreement of the results obtained during the reaction with those obtained after the reaction tests, and also to structural changes in the separative layer for the exposure to water vapour and to the modification of the support external area that changes its morphology (superficial metal oxide reduction) owing to a continuous contact with the H_2 stream, producing micro-fractures in the separating layer.

Brunetti *et al.* evaluated the membrane reactor at much higher pressures, i.e., up to 31 bar [34]. They observed optimum operating conditions at 280 °C and 29 bar pressure. The membrane showed a higher permeance and no significant variation in selectivity after the reaction testing. In all cases, the permeating flux is a linear function of the driving force and no inhibition effect of other gases on the hydrogen flux was observed.

Battersby *et al.* [35] investigated hydrostable cobalt-doped silica membrane reactors for WGS reaction. They used metal for hydro stabilizes the membrane. The H_2/CO separation increases three times when reaction temperature increased from 150 to 250 °C. However, the reaction temperature has less influence on H_2/CO_2 separation. And also, increases in steam to CO ratio decrease the H_2/CO and H_2/CO_2 separations. However, they observed the maximum CO conversion of 95%.

Then Battersby *et al.* investigated [36] the effect of water content on the membrane reactor performance. The excess water in the reaction was shown to have a positive effect on conversion and H_2 separation, and it was led to greater densification of the silica structure over time. The membrane delivered good hydrothermal stability, operating under harsh thermal and chemical conditions for over 200 h. Then they performed the simulation studies using Matlab Simulink model for both high and low temperature WGS reactions. The model provided a very good fit against the experimental Co conversions, while there were a couple of points for both temperatures with slightly higher deviation of around 8-10%. Using the model parameters determined from the tube scale permeation and reaction testing, they developed full model to predict operation of the MR conversion and H_2 recovery based on the operational variables such as feed rate, pressure and permeate sweep rate. The model results reveal that at higher reaction temperatures the CO conversion improved much better compared to lower reaction temperature because CO conversion enhancement is a factor of the H_2 diffusion through the membrane. H_2 recovery increases 50% with increasing feed pressure from 4 to 15 bar. They observed that the best performance of the MR was achieved at lower feed H_2O:CO ratios and higher temperatures. However, they are able to achieve only maximum conversion of 93% conversion with a H_2 recovery rate of 95%.

Then Araki *et al.* [37] synthesized the silica membrane by chemical vapour deposition method on alumina support. They used commercial Pt catalysts for WGS reaction. Increase in reaction temperature from 525 to 575 K increases the

CO conversion and H_2 recovery, and further increase in reaction temperature decreases the CO conversion and H_2 recovery slightly. As expected increase in steam/CO ratio increases the CO conversion and reaches 100% at a steam to CO ratio of 2.

6.4 PROTON-CONDUCTING MEMBRANES

Li *et al.* [38] first time ever reported mixed protonic-electronic conducting $SrCe_{0.9}Eu_{0.1}O_{3-\delta}$ membrane coated on a Ni-$SrCeO_{3-\delta}$ support. They prepared membrane by conventional solid-state reaction by mixing stoichiometric amounts of $SrCO_3$, CeO_2 and Eu_2O_3 powders, followed by ball milling and calcining at 1300 °C. A Ni-$SrCeO_{3-\delta}$ tubular support was fabricated using tape-casting followed by a rolling process. The tubular support was sealed at one end and presintered. $SrCe_{0.9}Eu_{0.1}O_{3-\delta}$ was coated on the inner side of the pre-sintered support. The tubular membranes were finally sintered at 1450 °C. They used Ni catalyst to perform WGS reaction. A 46% increase in CO conversion and total H_2 yield was achieved at 900 °C under 3% CO and 6% H_2O, resulting in a 92% single pass H_2 production yield and 32% single pass yield of pure permeated H_2.

Then Li *et al.* synthesized $SrCe_{0.7}Zr_{0.2}Eu_{0.1}O_{3-\delta}$ membranes on tubular Nie$SrCe_{0.8}Zr_{0.2}O_{3-\delta}$ supports [39]. They achieved CO conversions of 84% and 90% at steam to CO ratios of 1 and 2 at 900 °C. They proposed that the membrane stability increases under WGS conditions with Zr substitution.

6.5 CO_2-SELECTIVE MEMBRANE REACTORS

Reports are also available on CO_2 selective membrane reactors for WGS reaction. Zou *et al.* [40] first time synthesized polymeric CO_2-selective membrane by incorporating fixed and mobile carriers in cross-linked poly vinyl alcohol. Micro-porous Teflon was used as support. They used $CuO/ZnO/Al_2O_3$ catalyst for low temperature WGS reaction. They investigated the effect of water content on the CO_2 selectivity and CO_2/H_2 selectivity. As the water concentration in the sweep gas increased, both CO_2 permeability and CO_2/H_2 selectivity increased significantly. Figure 6.18 shows the influence of temperature on CO_2 permeability and CO_2/H_2 selectivity. Both CO_2 permeability and CO_2/H_2 selectivity decrease with increasing reaction temperature. After the catalyst activation, the synthesis gas feed containing 1% CO, 17% CO_2, 45% H_2 and 37% N_2 was pumped into the membrane reactor. They are able to achieve almost 100% CO conversion. They also developed a one-dimensional non-isothermal model to simulate the simultaneous reaction and transport process and verified the model experimentally under an isothermal condition.

Then they investigated the influence of system parameters on the CO_2 membrane reactor performance [41]. They investigated WGS reaction with auto-thermal reforming and steam reforming feeds. The required membrane

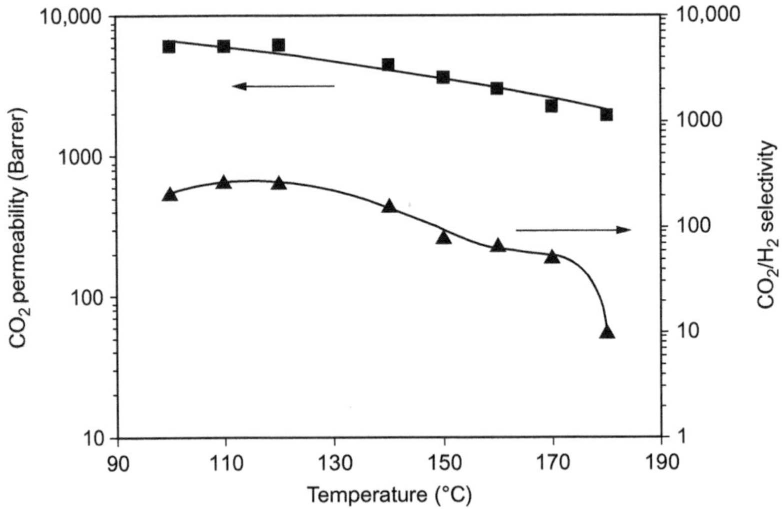

FIGURE 6.18 CO_2 permeability (■) and CO_2/H_2 selectivity (▲) of membrane versus temperature. Circular cell; feed gas 20% CO_2, 40% H_2 and 40% N_2, with increasing water rates at elevated temperatures; $p_f = 2.1$ atm, $p_s = 1.0$ atm; membrane thickness = 26 μm in. *(Taken from Figure 6 of J. Zou, J. Huang, W.S.W. Ho, Ind. Eng. Chem. Res. 46 (2007) 2272.)*

area decreases with increasing reaction temperature because of increase in WGS reaction rate. Similar trend was observed with increasing feed pressure. Increase in catalyst activity also decreases the required membrane area. The modelling results have shown that a CO concentration of less than 10 ppm is achievable from syngas containing up to 10% CO.

Then Ramasubramanian *et al.* [42] reported spiral wound CO_2-selective membrane reactor for GS reaction by using $CuO/ZnO/Al_2O_3$. Figure 6.19 shows the configuration of spiral wound model membrane reactor.

Fukuda *et al.* [43] reported plate type CO_2 membrane reactor with micro-channels and did simulation studies.

6.6 ZEOLITE MEMBRANE REACTORS

Dong and coworkers synthesized zeolite membrane reactor for the ultra-high temperature WGS reaction. Initially, porous alumina supported MFI zeolite tubular membranes have been prepared and evaluated for ultra-high temperature WGS reaction [44]. Then the zeolite membrane was modified by selectively depositing molecular silica at small number of active sites by using catalytic cracking deposition. The H_2 permeance and H_2/CO_2 separation factor at various modification stages are presented in Figure 6.20.

After the first modification, the H_2 permeance decreased from 3.75×10^{-7} to 2.7×10^{-7} mol/m^2 s Pa, whereas the H_2/CO_2 separation factor increased from 3.4 to 68. The H_2 permeance increased to 3.6×10^{-7} mol/m^2 s Pa, and the separation factor stabilized at a value of 57 after 10 h of annealing in the

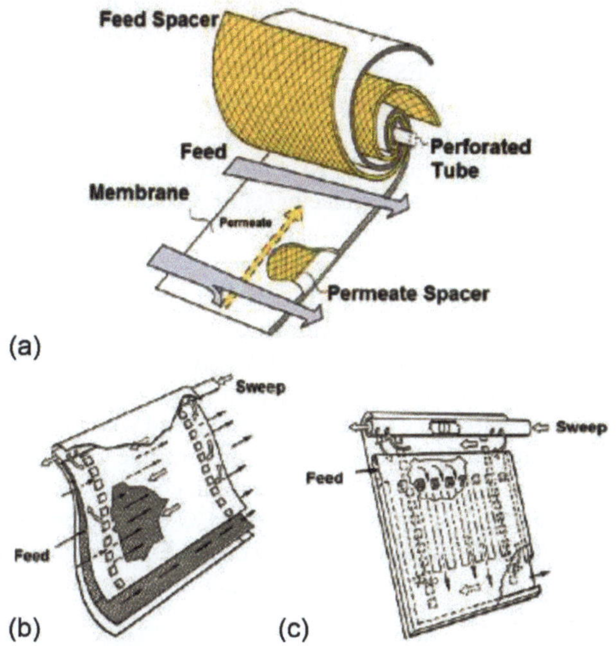

FIGURE 6.19 Spiral wound module configurations: (a) without sweep gas (reprinted with permission from MTR, Inc.[26]), (b) counter current flow with sweep and (c) crossflow with sweep. Both (b) and (c) reprinted from D. Reddy, T.Y. Moon, C.E. Reineke Counter current Dual flow spiral wound dual pipe membrane separation U.S. Patent 5,034,126, July 23, 1991. *(Taken from Figure 1 of the T. Fukuda, T. Maki, K. Mae, Chem. Eng. Tech. 35 (2012) 1205.)*

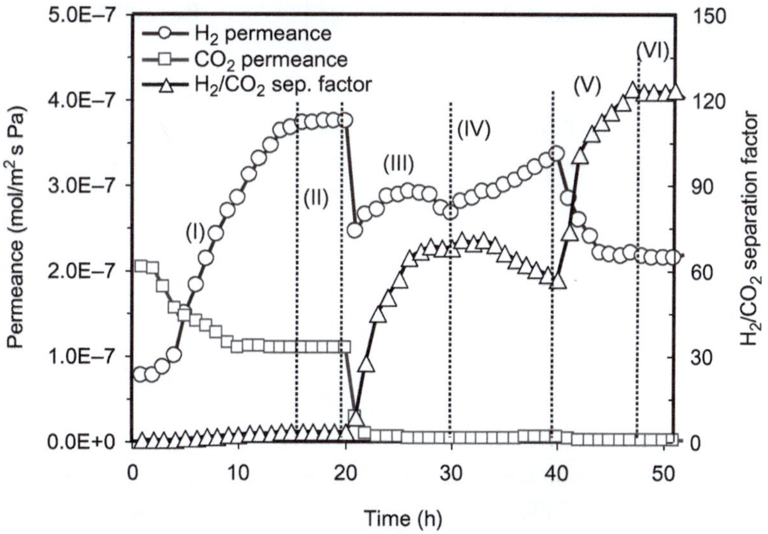

FIGURE 6.20 Separation results for an equimolar H_2/CO_2 mixture during the CCD modification. (I) Heating from 298 to 723 K; (II) dwelling at 723 K; (III) first MDES CCD; (IV) annealing in H_2/CO_2 feed without MDES; (V) second MDES CCD; and (VI) H_2/CO_2 feed without MDES.

H_2/CO_2 stream without MDES vapour. The second modification of the membrane resulted in a further increase in the H_2 separation factor from 57 to a stable value of 123 under the feed containing MDES vapour, with only a small decrease in H_2 permeance to 2.2×10^{-7} mol/m^2 s Pa. After the termination of feeding MDES, the H_2 separation factor stabilized at ~108 in about 24 h with a virtually unchanged H_2 permeance and a slightly increased CO_2 permeance.

Then, membrane WGS reactor were designed by using a novel catalyst Fe/Ce which is developed in our lab and performed high temperature WGS reaction [45]. As expected our membrane reactor exhibits much higher CO conversion compared to the traditional reactor at all temperatures and also at 500 and 550 °C, the CO conversion in the membrane reactor exceeded the equilibrium CO conversions. As expected the CO conversion increased with increasing reaction temperature from 400 to 550 °C. Figure 6.21 presents the CO conversion in the membrane reactor as a function of temperature in comparison with the results obtained from the TR. The membrane reactor CO conversion increased with increasing steam/CO ratio from 1 to 3.5. However, CO conversion enhancement in the membrane reactor was more significant at lower steam to CO ratio

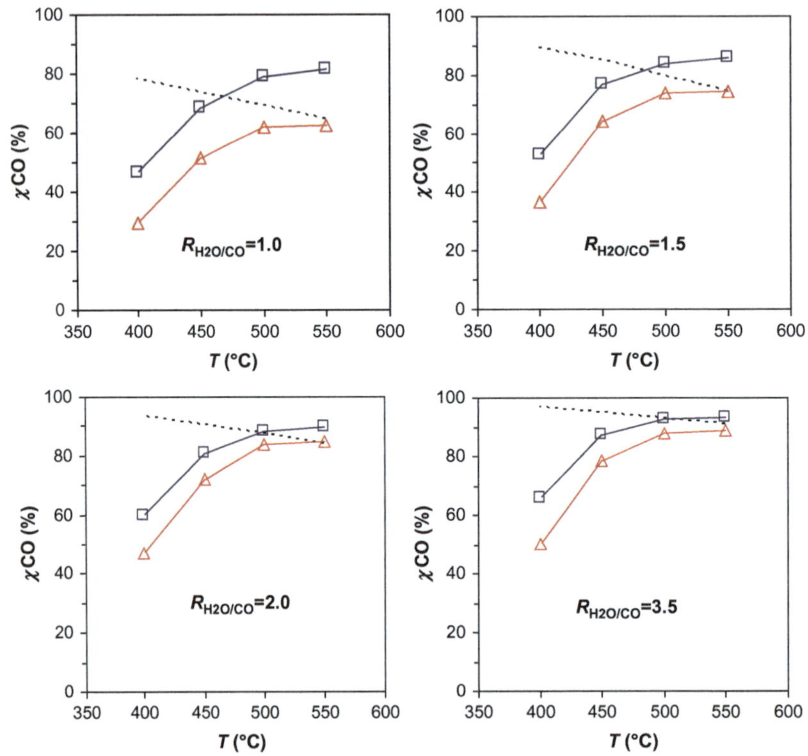

FIGURE 6.21 WGS activity profiles of membrane WGS reactor at various temperatures and various steam to CO ratios (GHSV = 60,000 h^{-1}).

compared to higher steam to CO ratio. The CO conversion in membrane reactor increased with decreasing WHSV because of increased time for reaction and H_2 transport. The enhancement of CO conversion by reducing WHSV is more significant at low temperatures where kinetic and mass transport limits are severe.

Also the CO conversion increased with increasing sweeping gas flow rate. The modified membrane was also tested for the separation of an equimolar H_2/CO_2 mixture before and after the WGS reactions. Interestingly only a 16.9% decrease in separation factor (to 37.3) and a 17.7% decrease in H_2 permeance (to 1.21×10^{-7} mol/m^2 s Pa) were observed after the WGS operations as compared to the as-modified membrane. These demonstrate that the CCD modified MFI-type zeolite membrane has good stability at high temperature in the WGS reaction condition. However, CO conversion in the MR did not come much closer to 100% even at 550 °C and steam to CO ratio of 3.5 and WHVS decreased to 7500 h^{-1}. The inefficiency of H_2 removal rather than the reaction rate is more likely the factor limiting the CO conversion in membrane reactor. The completion of CO conversion in the MR requires nearly complete removal of H_2 through the membrane. Overcoming the hindrance of mass transfer inefficiency on the enhancement of CO conversion requires further membrane improvement for higher H_2 selectivity and high permeance.

WGS measurements a disc membrane was prepared instead of tubular membrane [46]. The WGS reaction experiments were performed with feed side pressures varying from 2 to 6 atm. The permeate side was swept by a N_2 flow at atmospheric pressure, and the WHSV was fixed at 7500 h^{-1} and steam to CO ratio fixed at 3.5. The results are presented in Figure 6.22. The CO conversion of the membrane reactor is much higher than that of the traditional reactor

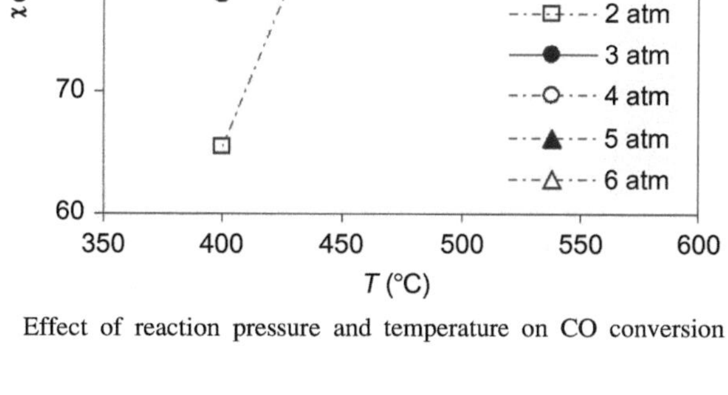

FIGURE 6.22 Effect of reaction pressure and temperature on CO conversion of membrane WGS reactor.

and increases with reaction pressure at all temperatures. However, it is observed that CO conversion tends to level off after certain points in both cases of increasing pressure at a fixed temperature and increasing temperature under a fixed pressure. The level-off behaviour indicates that adjusting or improving one individual factor alone is insufficient for achieving nearly complete CO conversion. It is seen that the CO conversion increases with increasing steam to CO ratio at both low and high reaction pressures. The enhancement of CO conversion is more pronounced for a steam to CO ratio increase from 1 to 1.5 than for an increase from 2 to 3.5. CO conversion in the membrane reactor increased with decreasing WHSV because of longer residence time for reaction and H_2 permeation at smaller WHSV. The CO conversion also enhanced by simultaneously increasing feed pressure and lowering the WHSV. On the whole, the zeolite membrane showed good stability after operating for more than 2000 h in WGS feed streams with and without the presence of H_2S. However, even at high pressures the membrane reactor exhibits a maximum conversion of 98%.

Then, the experimental results are compared with the 1-D PFR model calculations. Interestingly, our calculated results agree well with the experimental data especially at temperatures above 450 °C [47]. To investigate the possibility for the current membranes to achieve near-completion CO conversion of >99.5% under practically meaningful conditions, the 1-D PFR model was used to simulate the MR performance for operations beyond the experimental conditions used in this study. The operating conditions used in the calculations include catalyst load (m) of 78.7 mg_{cat}/cm^2 membrane, WHSV of 7500 h^{-1} and steam to CO ratio of 3.5, which were same as those used in the experiments. The results show that increasing both temperature and reaction pressure enhances the CO conversion in the membrane reactors. However, CO conversion tends to plateau above certain temperature and pressure. The highest CO conversion value achieved is 99.2% at temperature >500 °C and pressure >30 atm. The results are presented in Figure 6.23. Then, more simulations were carried out to further investigate the feasibility of the current membrane for achieving of CO conversion more than 99.5%, which is the final conversion level of the multiple reactor systems used in the industry.

Zhang *et al.* [48] investigated MFI zeolite membrane reactor for low temperature WGS reaction by using $Cu/Zn/Al_2O_3$ catalyst. They prepared membrane in disc shape. They also performed modification by catalytic cracking deposition. As expected the CO conversion increases with increasing reaction temperature. However only at 300 °C and GHSV 1500 h^{-1} the membrane exhibits higher CO conversion than equilibrium conversion. The steam to CO ratio also has positive effect on the WGS activity of membrane reactor. Their modified MFI zeolite membrane showed excellent CO tolerance during the experiment lasting for 20 days.

Lin's group [49] synthesized bilayer MFI zeolite membrane reactor for WGS reaction. They investigated the membrane reactor for long-term

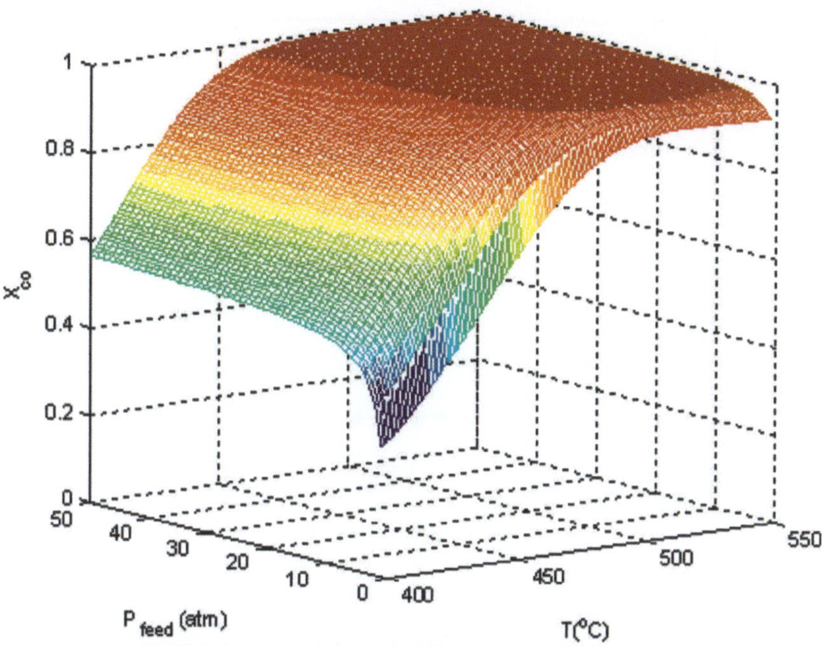

FIGURE 6.23 Calculated CO conversion as a function of reaction temperature and pressure.

time-on-stream experiments. The results are presented in Figure 6.24. The CO conversion in the membrane reactor increases first and then maintained at about 84%. The initial enhancement of CO conversion may be caused by the fact that WGS catalyst has not been fully activated by the process gas. During the first 22 days of WGS reaction, no obvious change in hydrogen recovery was observed. From the 23rd day, the hydrogen recovery increased slightly from 22% to about 23.2%. The stable CO conversion and H_2 recovery during the long-term WGS reaction in the modified MFI zeolite membrane reactor indicates that both the modified bilayer MFI zeolite membrane reactor and the WGS catalyst are extremely stable under WGS reaction environment. The bilayer MFI zeolite membrane is also highly stable at 500 °C for at least 24 days under industrially relevant conditions for WGS reaction (equimolar mixture of CO, CO_2, H_2 and H_2O plus 400 ppm H_2S).

6.7 THEORETICAL INVESTIGATION OF MEMBRANE REACTORS

Damle *et al.* [50] in 1992 developed a simplified process model to simulate catalytic membrane WGS reactor. They assumed the permeability ratios of different gases to be constant during membrane separation. The model further assumes that the WGS reaction is not limited by chemical kinetics and thus

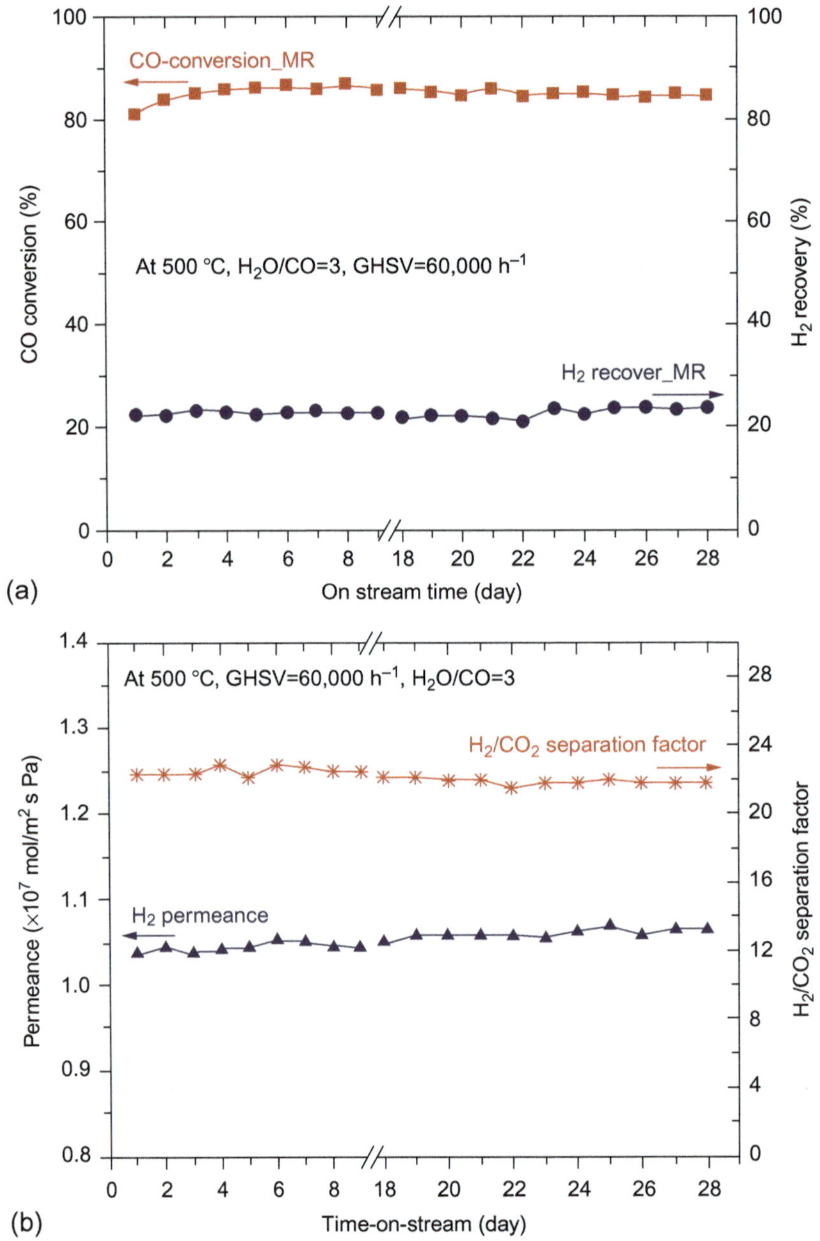

FIGURE 6.24 CO conversion and H_2 recovery (a), and H_2 permeance and H_2/CO_2 reparation factor (b) versus time-on-stream for WGS reaction in the modified bilayer MFI zeolite membrane (total feed gas flow rate: 100 ml/min (STP), feed side pressure: 2 atm, helium sweeping: 20 ml/min (STP), permeate side pressure: 1 atm). *(Taken from Figure 6 of H. Wang, X. Dong, Y.S. Lin, J. Membr. Sci. 450 (2014) 425.)*

the feed gas stream is assumed to be continuously at equilibrium throughout the membrane reactor. They applied the model for possible application of membrane reactors in coal gasification. The feed composition in exit of gasification chamber is 48.6% H_2, 21.0% CO_2, 17.3% CO and 13.1% N_2 on a dry basis. They found that higher pressures are required for the concurrent mode and lower pressures are adequate for counter-current mode. Also, simulations indicated that a large number of stages will be required to increase the hydrogen concentration to more than 80%. A two-stage process provides 80-85% hydrogen and a three-stage process provides >90% hydrogen.

Violante *et al.* [51] simulated multi-layer membrane reactor for WGS reaction. They proposed that the permeability of the multi-layer metallic/ceramic membrane is higher than that of a traditional metallic membrane. With same pressure difference across the membrane and with the same exchange area, the hydrogen flux through the multi-layer membrane is five times larger than the flux through a Pd-Ag membrane. They also proposed that the membrane should operate at suitable temperature and pressure.

Adrover *et al.* [52] discussed heat effects in membrane WGS reactor. They proposed that for non-adiabatic operation the proper selection of operating conditions is important to avoid the undesired temperature raises. They also proposed that heat effects are negligible in small-scale laboratory designs. However, for intermediate or larger scale applications the temperature variations have significant effects on chemical kinetics and equilibrium.

Then Adrover *et al.* [53] simulated multi-tubular membrane reactor without a sweep gas. The scheme of the multi-tubular membrane reactor is presented in Figure 6.25. They found that flow configuration has significant influence on thermal behaviour of membrane reactor. Operation of membrane reactor in concurrent mode reduces the thermal effects by minimizing the temperature rise in the catalyst bed. This is due to the heat exchange with the permeate stream, which acts as a cooling medium along the reactor length. On the other hand, counter-current configuration facilitates the reaction ignition due to the preheating effect caused by the permeate stream, and this flow scheme can lead to high temperature rises along the axial position.

Chein *et al.* [54] did simulation studies by using pre-exponential factor. CO conversion and H_2 recovery increase with increasing pre-exponential factor up

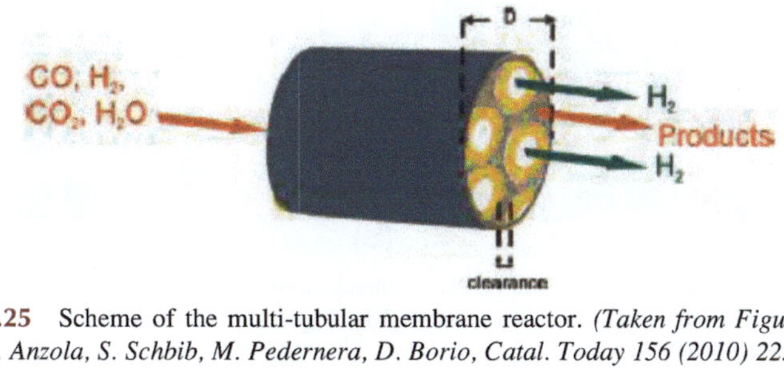

FIGURE 6.25 Scheme of the multi-tubular membrane reactor. *(Taken from Figure 1 of M.E. Adrover, A. Anzola, S. Schbib, M. Pedernera, D. Borio, Catal. Today 156 (2010) 223.)*

to 10^{-8} and further increase has no effect on the CO conversion and H_2 recovery. These results reveal that high membrane permeance is required to have high CO conversion and H_2 recovery. Increasing the H_2O/CO molar ratio increases the CO conversion but causes a decrease in H_2 recovery because of the reduction in hydrogen permeation driving force. For high membrane permeance CO conversion and H_2 recovery approach to limiting values when the operating pressure is increased. Lowering the sweep gas flow has the effect of decreasing the CO conversion and H_2 recovery. In the high membrane permeance case the CO conversion and H_2 recovery approach to limiting values as the sweep gas flow rate is increased.

Boutikosa and Nikolakisb [55] did simulation studies over Pd-Ag isothermal tube-shell reactor. They investigated the effect of H_2/CO and H_2/CO_2 perm-selectivity on the membrane reactor performance. The results are presented in Figure 6.26. The CO conversion increased with membrane perm-selectivity, while the opposite was observed for H_2 recovery. When H_2/CO perm-selectivity was higher than H_2/CO_2 perm-selectivity, CO conversion was enhanced due to the selective removal of both reaction products. On the other hand, when H_2/CO perm-selectivity was smaller than H_2/CO_2 perm-selectivity, CO conversion decreased because a larger fraction of CO permeated through the membrane as well as because a larger fraction of the CO_2 produced remained on the tube side of the reactor. They proposed that utilization of CO_2 selective, instead of H_2 selective, membranes could improve CO conversion only if the CO_2 content of the feed is higher than that of H_2. They also found that the reaction mechanism only has slight influence on the CO conversion.

Dijkstra *et al.* [56] compared membrane WGS reaction with membrane reforming in the natural gas combined cycles with CO_2 capture. Their simulation results indicate that membrane WGS reaction suits well for CO_2 capture compared to the membrane reforming. The lower hydrogen partial pressure in a membrane reformer compared to membrane WGS causes high investment

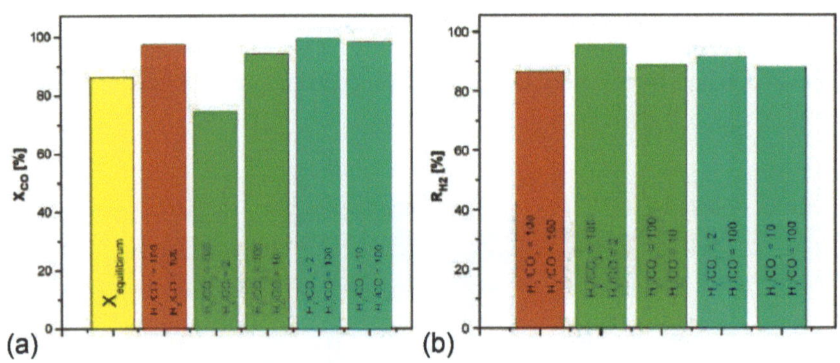

FIGURE 6.26 Effect of H_2/CO_2 and H_2CO perm-selectivity on membrane reactor performance: (a) CO conversion and (b) H_2 recovery. *(Taken from Figure 6 of P. Boutikosa, V. Nikolakisb, J. Membr. Sci. 350 (2010) 378.)*

costs because the membrane area is high and compression of the hydrogen fuel before entering the gas turbine is necessary.

Lotric *et al.* [57] investigated suitability of Pd membrane reactor for the integrated gasification combined cycle. They did modelling of WGSR kinetics based on Bradford mechanism which was used to develop a code within Mathematica programming language to simulate the chemical reactions. Based on their studies more than 89.1% of CO conversion can be achieved in membrane reactor. They also proposed that reducing the molar fraction of H_2 in the permeate stream would favour the CO conversion. Also increasing the reactor surface area by increasing the reactor length favours CO conversion. Their simulation results suggest that using high temperature catalysts and high process temperatures increases the CO conversion in membrane reactor. They suggested that the WGS membrane reactors are suitable for IGCC application; however, so many issues are still to be addressed. The main issue is the stability of membrane reactor during various environments like acid.

Falcoa *et al.* [58] developed two-dimensional non-isothermal models for Pd membrane reactors. They proposed that the maximum CO conversion is achieved at 600 K. Lima *et al.* developed one-dimensional isothermal model for membrane reactors in IGCC plant. Their simulation results show that counter-current configuration can achieve the specific targets for the IGCC plant. Their optimization experiments show that performing WGS reaction before the membrane reactor saves the energy as high as 25%. For industrial-scale reactors, these savings represent an amount of as high as $5,000,000 per year. Romero *et al.* show that use of Pd films combined with composite catalytic membrane showed significant improvement in overall H_2 recoveries.

Reddy and Wilhite [59] investigated application of membrane reactors in diesel reformate mixture purification isothermal two-dimensional model. The typical reformate mixture contains 9% CO, 3% CO_2, 28% H_2 and 15% H_2O. Simulations indicate that apparent CO:H_2 selectivities of 90:1 to >200:1 at H_2 recoveries of 20% to upwards of 40% may be achieved through appropriate design of the catalytic membrane and selection of operating conditions. Comparison of adiabatic and isothermal simulations indicates that accumulation of reaction heat reduces apparent perm-selectivities; however, this may be mitigated by external imposition of a countering thermal gradient.

REFERENCES

[1] E. Kikuchi, S. Uemiya, N. Sato, H. Inoue, H. Ando, T. Masuda, Membrane reactor using microporous glass supported thin film of Pd – application to the WGS reaction, Chem. Lett. (1989) 489–492.

[2] S. Uemiya, N. Sato, H. Ando, E. Kikuchi, The water gas shift reaction assisted by a palladium membrane reactor, Ind. Eng. Chem. Res. 30 (1991) 585–589.

[3] A. Basile, A. Criscuoli, F. Santellab, E. Driol, Membrane reactor for water gas shift reaction, Gas Sep. Purif. 10 (1996) 243–254.

[4] A. Basile, E. DriolL, G. Vkullie, F. Santella, V. Violante, G. Capannellid, A study on catalytic membrane reactors for water gas shift reaction, Gas Sep. Purif. 10 (1996) 53–61.

[5] A. Criscuoli, A. Basile, E. Drioli, An analysis of the performance of membrane reactors for the water-gas shift reaction using gas feed mixtures, Catal. Today 56 (2000) 53–64.

[6] A. Basile, G. Chiappetta, S. Tosti, V. Violante, Experimental and simulation of both Pd and Pd/Ag for awater gas shift membrane reactor, Sep. Purif. Technol. 25 (2001) 549–571.

[7] S. Tosti, A. Basile, G. Chiappetta, C. Rizzello, V. Violante, Pd-Ag membrane reactors for water gas shift reaction, Chem. Eng. J. 93 (2003) 23–30.

[8] G. Barbieria, A. Brunetti, G. Tricoli, E. Drioli, An innovative configuration of a Pd-based membrane reactor for the production of pure hydrogen experimental analysis of water gas shift, J. Power Sources 182 (2008) 160–167.

[9] A. Brunetti, G. Barbieria, E. Drioli, Upgrading of a syngas mixture for pure hydrogen production in a Pd-Ag membrane reactor, Chem. Eng. Sci. 64 (2009) 3448–3454.

[10] A. Brunetti, G. Barbieria, E. Drioli, A. Caravella, Pd-based membrane reactors for one-stage process of water gas shift, RSC Adv. 1 (2011) 651–661.

[11] (a) A. Brunetti, G. Barbieria, E. Drioli, A. Caravella, Process intensification by membrane reactors: high-temperature water gas shift reaction as single stage for syngas upgrading, Chem. Eng. Technol. 35 (2012) 1238–1248. (b) A. Brunetti, G. Barbieria, E. Drioli, Medium/high temperature water gas shift reaction in a Pd-Ag membrane reactor: an experimental investigation, RSC Adv. 2 (2011) 226–233.

[12] D. Mendes, V. Chibante, J. Zheng, S. Tosti, F. Borgognoni, A. Mendes, L. M. Madeira, Enhancing the production of hydrogen via wateregas shift reaction using Pd-based membrane reactors, Int. J. Hydrogen Energy 35 (2010) 12596–12608.

[13] D. Mendes, S. Tosti, A. Mendes, L. M. Madeira, S. Sa, J. M. Sousa, Experimental and modeling studies on the low-temperature water-gas shift reaction in a dense Pd-Ag packed-bed membrane reactor, Chem. Eng. Sci. 66 (2011) 2356–2367.

[14] S. Tosti, A. Basile, L. Bettinali, F. Borgognoni, F. Chiaravalloti, F. Gallucci, Long-term tests of Pd-Ag thin wall permeator tube, J. Membr. Sci. 284 (2006) 393–397.

[15] C. A. Cornagliaa, S. Tosti, M. Sansovinib, J. Múneraa, E. A. Lombardoa, Novel catalyst for the WGS reaction in a Pd-membrane reactor, Appl. Catal. A Gen. 462–463 (2013) 278–286.

[16] C. A. Cornagliaa, S. Tosti, M. Adrover, D. Borio, N. Pedernera, J. Múneraa, E. A. Lombardoa, Production of ultrapure hydrogen in a Pd-Ag membrane reactor using noble metals supported on La-Si oxides. Heterogeneous modeling for the water gas shift reaction, Int. J. Hydrogen Energy 38 (2013) 10485–10493.

[17] O. Iyoha, R. Enick, R. Killmeyer, B. Howard, B. Morreale, M. Ciocco, Wall-catalyzed water-gas shift reaction in multi-tubular Pd and 80 wt%Pd-20 wt%Cu membrane reactors at 1173 K, J. Membr. Sci. 298 (2007) 14–23.

[18] O. Iyoha, R. Enick, R. Killmeyer, B. Howard, B. Morreale, M. Ciocco, H_2 production from simulated coal syngas containing H_2S in multi-tubular Pd and 80 wt% Pd-20 wt% Cu membrane reactors at 1173 K, J. Membr. Sci. 306 (2007) 113–115.

[19] K. Hwang, S. J. King, J. S. Park, A catalytic membrane reactor for water-gas shift reaction, Korean J. Chem. Eng. 27 (2010) 816–821.

[20] K. Hwang, C. B. Lee, S. K. Ryi, J. S. Park, Hydrogen production and carbon dioxide enrichment using a catalytic membrane reactor with Ni metal catalyst and Pd-based membrane, Int. J. Hydrogen Energy 37 (2012) 6626–6634.

[21] K. Hwang, S. W. Lee, S. K. Ryi, D. K. Kim, T. H. Kim, J. S. Park, Water-gas shift reaction in a plate-type Pd-membrane reactor over a nickel metal catalyst, Fuel Process. Technol. 106 (2013) 133–140.

[22] S. Tosti, F. Borgognoni, C. Rizzello, V. Violante, Water gas shift reaction via Pd-based membranes, Asia-Pac. J. Chem. Eng. 4 (2009) 369–379.

[23] Y. Bi, H. Xu, W. Li, A. Goldbach, Water-gas shift reaction in a Pd membrane reactor over Pt/$Ce_{0.6}Zr_{0.4}O_2$ catalyst, Int. J. Hydrogen Energy 34 (2009) 2965–2971.

[24] V. Piemonte, M. De Falco, B. Favetta, A. Basile, Counter-current membrane reactor for WGS process: membrane design, Int. J. Hydrogen Energy 35 (2010) 12609–12617.

[25] P. Pinaccia, M. Brogliaa, C. Valli, G. Capannellib, A. Comite, Evaluation of the water gas shift reaction in a palladium membrane reactor, Catal. Today 156 (2010) 165–172.

[26] A. S. Augustine, Y. H. Ma, N. K. Kazantzis, High pressure palladium membrane reactor for the high temperature wateregas shift reaction, Int. J. Hydrogen Energy 36 (2011) 5359–5360.

[27] M. Abdollahia, J. Yua, P. K. T. Liub, R. Ciorab, M. Ahimia, T. T. Tsotsis, Ultra-pure hydrogen production from reformate mixtures using a palladium membrane reactor system, J. Membr. Sci. 390–391 (2012) 32–42.

[28] A. S. Augustine, I. P. Mardilovich, N. K. Kazantzis, Y. H. Ma, Durability of PSS-supported Pd-membranes under mixed gas and water-gas shift conditions, J. Membr. Sci. 415–416 (2012) 213–220.

[29] S. Liguoria, P. Pinaccic, P. K. Seelamd, R. Keiskid, F. Dragoc, V. Calabròa, A. Basile, A. Iulianelli, Performance of a Pd/PSS membrane reactor to produce high purity hydrogen via WGS reaction, Catal. Today 193 (2012) 87–94.

[30] H. Li, J. A. Z. Pieterse, J. W. Dijkstra, J. Boon, R. W. van den Brink, D. Jansen, Bench-scale WGS membrane reactor for CO_2 capture with co-production of H_2, Int. J. Hydrogen Energy 37 (2012) 4139–4145.

[31] J. Catalano, F. Guazzone, I. P. Mardilovich, N. K. Kazantzis, Y. H. Ma, Hydrogen production in a large scale water-gas shift Pd-based catalytic membrane reactor, Ind. Eng. Chem. Res. 52 (2013) 1042–1055.

[32] S. Giessler, L. Jordan, J. C. Dinizda Costa, G. Q. Lu, Performance of hydrophobic and hydrophilic silica membrane reactors for the water gas shift reaction, Sep. Purif. Technol. 32 (2003) 255–264.

[33] A. Brunetti, G. Barbieri, E. Drioli, K.-H. Lee, B. Seac, D.-W. Lee, WGS reaction in a membrane reactor using a porous stainless steel supported silica membrane, Chem. Eng. Process. 46 (2007) 119–126.

[34] A. Brunetti, G. Barbieri, E. Drioli, T. Granato, K. H. Lee, Aporous stainless steel supported silica membrane for WGS reaction in a catalytic membrane reactor, Chem. Eng. Sci. 62 (2007) 5621–5626.

[35] S. Battersby, M. C. Duke, S. Liu, V. Rudolph, J. C. Diniz da Costa, Metal doped silica membrane reactor: operational effects of reaction and permeation for the water gas shift reaction, J. Membr. Sci. 316 (2008) 46–52.

[36] S. Battersby, S. Smarta, B. Ladewigb, S. Liua, M. C. Dukec, V. Rudolpha, J. D. Costa, Hydrothermal stability of cobalt silica membranes in a water gas shift membrane reactor, Sep. Purif. Technol. 66 (2009) 299–305.

[37] S. Araki, H. Miyanishi, H. Yano, S. Tanaka, Y. Miyake, Water gas shift reaction in a membrane reactor using a high hydrogen permselective silica membrane, Sep. Sci. Technol. 48 (2013) 76–83.

[38] J. Li, H. Yoon, T. Oh, E. D. Wachsman, High temperature $SrCe_{0.9}Eu_{0.1}O_{3-\delta}$ proton conducting membrane reactor for H_2 production using the water-gas shift reaction, Appl. Catal. B Environ. 92 (2009) 234–239.

[39] J. Li, H. Yoon, T. Oh, E. D. Wachsman, $SrCe_{0.7}Zr_{0.2}Eu_{0.1}O_3$-based hydrogen transport water gas shift reactor, Int. J. Hydrogen Energy 37 (2012) 16006–16012.

[40] J. Zou, J. Huang, W. S. W. Ho, CO_2-selective water gas shift membrane reactor for fuel cell hydrogen processing, Ind. Eng. Chem. Res. 46 (2007) 2272–2279.

[41] J. Huang, W. S. W. Ho, Effects of system parameters on the performance of CO_2-selective WGS membrane reactor for fuel cells, J. Chin. Inst. Chem. Eng. 39 (2008) 129–136.
[42] K. Ramasubramanian, M. Song, W. S. W. Ho, Spiral-wound water-gas-shift membrane reactor for hydrogen purification, Ind. Eng. Chem. Res. 52 (2013) 8829–8842.
[43] T. Fukuda, T. Maki, K. Mae, Design of a plate-type catalytic microreactor with CO_2 permeation membrane for water-gas shift reaction, Chem. Eng. Technol. 35 (2012) 1205–1213.
[44] Z. Tang, T. M. Nenoff, J. Dong, Internal surface modification of MFI-type zeolite membranes for high selectivity and high flux for hydrogen, Langmuir 25 (2009) 4848–4852.
[45] Z. Tang, S. J. Kim, G. K. Reddy, J. Dong, P. Smirniotis, Modified zeolite membrane reactor for high temperature water gas shift reaction, J. Membr. Sci. 354 (2010) 114–122.
[46] S. J. Kim, Z. Xu, G. K. Reddy, J. Dong, P. Smirniotis, Effect of pressure on high-temperature water gas shift reaction in microporous zeolite membrane reactor, Ind. Eng. Chem. Res. 51 (2012) 1364–1375.
[47] S. J. Kim, S. Wang, G. K. Reddy, J. Dong, P. Smirniotis, Zeolite membrane reactor for high-temperature water-gas shift reaction: effects of membrane properties and operating conditions, Energy Fuel 27 (2013) 4471–4480.
[48] Y. Zhang, Z. Wu, Z. Hong, X. Gu, N. Xu, Hydrogen-selective zeolite membrane reactor for low temperature water gas shift reaction, Chem. Eng. J. 197 (2012) 314–321.
[49] H. Wang, X. Dong, Y. S. Lin, Highly stable bilayer MFI zeolite membranes for high temperature hydrogen separation, J. Membr. Sci. 450 (2014) 425–432.
[50] A. S. Damle, S. K. Gangwal, V. K. Venkataraman, A simple model for a water gas shift membrane reactor, Gas Sep. Purif. 8 (1994) 101–106.
[51] V. Violante, A. Basile, E. Drioli, Composite catalytic membrane reactor analysis for the water gas shift reaction in the tritium fusion fuel cycle, Fusion Eng. Des. 30 (1995) 217–223.
[52] M. E. Adrover, E. López, D. O. Borio, M. N. Pedernera, Heat effects in a membrane reactor for the water gas shift reaction, Nat. Gas Convers. 167 (2007) 184–188.
[53] M. E. Adrover, A. Anzola, S. Schbib, M. Pedernera, D. Borio, Effect of flow configuration on the behavior of a membrane reactor operating without sweep gas, Catal. Today 156 (2010) 223–228.
[54] R. Chein, Y. C. Chen, J. N. Chung, Parametric study of membrane reactors for hydrogen production via high-temperature water gas shift reaction, Int. J. Hydrogen Energy 38 (2013) 2292–2305.
[55] P. Boutikosa, V. Nikolakisb, A simulation study of the effect of operating and design parameters on the performance of a water gas shift membrane reactor, J. Membr. Sci. 350 (2010) 378–386.
[56] J. W. Dijkstra, G. Raju, G. Peppink, D. Jansen, Techno-economic evaluation of membrane technology for pre-combustion decarbonisation: water-gas shift versus reforming, Energy Procedia 4 (2011) 723–730.
[57] A. Lotric, M. Sevavcnik, S. C. Kunze, H. Spliethoff, Simulation of water-gas shift membrane reactor for integrated gasification combined cycle plant with CO_2 capture, J. Mech. Eng. 57 (2011) 911–926.
[58] M. Falcoa, V. Piemontea, A. Basile, Performance assessment of water gas shift membrane reactors by a two-dimensional model, in: Proceedings of the 11th International Symposium on Process Systems Engineering, Singapore, 2012, , pp. 611–614.
[59] K. B. V. Reddy, B. A. Wilhite, Theoretical investigation of a water-gas-shift catalytic membrane for diesel reformate purification, AIChE J. 59 (2013) 4334–4344.

Chapter 7

Homogeneous WGS Reaction

7.1 HOMOGENEOUS CATALYTIC WGS REACTION

If the catalyst and reactant are in the same state (in this case both are in liquid state) then the reaction is called homogeneous WGS reaction. In 1953, Reppe was the first researcher who discussed the concept of homogeneous WGS reaction [1]. Although he published a number of catalytic reactions of CO and H_2O with organic substrates, Fenton was the first person who patented the homogeneous catalysis of the water gas shift reaction (WGSR). He filed three patents on homogeneous WGS reaction by using group VIII metals in conjunction with phosphine, arsine or stibine ligands and amine or inorganic bases [2–4].

Metal carbonyl complexes are mainly investigated for homogeneous WGS reaction. In the later stages, pyridine complexes are also investigated for this reaction. The homogeneous WGS reaction can be conducted in acidic, basic or neutral media.

Then, in 1977, reports were published in the open literature about the homogeneous WGS reaction. Four groups published four different catalysts in the same year. Ruthenium carbonyl complexes were reported by Laine *et al.* [5] in the basic solution KOH as active catalysts for homogeneous WGS reaction. Kang *et al.* [6] reported that most of the group VIII metals act as active catalysts for the homogeneous WGS reaction in THF containing trimethyl amine as base. On the other hand, Cheng *et al.* [7] reported Rh carbonyl iodide as active catalyst in the acidic medium. Zudin *et al.* [8] also reported palladium phosphine as an active catalyst in the aq trifluoroacetic acid medium.

Various transition metal complexes have been investigated for homogeneous WGS reaction in either basic or acidic or neutral media. The metals mainly investigated are Rh, Ru Fe, Ir and Pt group metals or mixer of these metals. In the following section, we are discussing about the WGS activity of these metals.

7.2 Rh-BASED COMPLEXES FOR HOMOGENEOUS WGS REACTION

In 1977, Cheng *et al.* [7] reported first ever reported the rhodium carbonyl systems for the homogeneous WGS reaction.

Water Gas Shift Reaction. http://dx.doi.org/10.1016/B978-0-12-420154-5.00007-3

7.2.1 Catalyst Synthesis

The catalyst is prepared under N_2 from $[Rh(CO)_2Cl]_2$ (97 mg, 0.5 mmol of Rh), glacial acetic acid (26 mL), concentrated HCl (12 mL), NaI (16.8 g) and H_2O (12 mL); the solution is filtered to remove precipitated NaCl and any undissolved NaI.

7.2.2 Catalytic Measurements

The catalyst solution was placed in a 1-L flask which was evacuated, flushed once with CO and then charged at room temperature to an initial CO pressure of 250-400 Torr. Approximately 50 Torr of N_2 was then added as an internal calibrant for GC analysis. They did total three runs. Run I was conducted at 90 °C and run II at 80 °C. After the first run the catalyst was recharged with CO and run III was conducted at 90 °C.

The amount of CO_2 produced is almost same as the amount of H_2 produced which clearly shows the stoichiometric nature of the reaction. By conducting different experiments they concluded the following observations: (i) substitution of HBF_4 for HCl yielded a dark brown, catalytically active solution, while omission of a strong acid produced only a trace of CO_2 and a yellow solution; (ii) reaction solutions with acetic acid present were three to five times more reactive than in its absence; (iii) ^{13}CO isotopic labelling experiments suggest that CO_2 is formed only from the reactant CO and not from the $Rh(CO)_5$.

In 1979, the same group prepared [9] binuclear Rh(0) molecular A frames and evaluated for homogeneous WGS reaction. They prepared $[Rh_2(\mu\text{-}H)(\mu\text{-}CO)(CO)_2(dpm)_2]^+A^-$ by protonation of $[Rh_2(\mu\text{-}H)(\mu\text{-}CO)(CO)_2(dpm)_2]$ complex with HA. The detailed synthesis procedure is presented elsewhere. Then this complex has been evaluated for homogeneous WGS reaction in the presence of THF. 0.05 g of $[Rh_2(\mu\text{-}H)(\mu\text{-}CO)(CO)_2(dpm)_2]^+A^-$ is dissolved in 30 mL of *n*-PrOH with 2 molar equiv. of LiCl in 1 mL of H_2O added, and the system is placed under 1 atm of CO at 90 °C. The rate of catalysis corresponds to 2.5 turnovers/h, which is one of the most active compared to their previous catalysts under these very mild temperature and pressure conditions. The catalysis proceeds much more slowly in the absence of Cl^- and is pH sensitive.

The same group investigated [10] rhodium carbonyl system at different reaction conditions such as effect of temperature, Rh concentration and reaction temperature. The influence of rhodium concentration on the rate of WGS reaction was determined by adding an additional weighed portion of $[Rh(CO)_2Cl]_r$ to the catalytic solution between runs, followed by degassing and recharging with CO. The effect of Rh concentration on the WGS activity of rhodium carbonyl catalysts is presented in Table 7.1. As shown in Table 7.1, increases in Rh concentration increase the rate of WGS reaction.

They also investigated the effect of CO partial pressure on the WGS activity. The results are presented in Table 7.2. To avoid changes in concentration

TABLE 7.1 The Effect of Rh Concentration on the WGS Activity of Rhodium Carbonyl Catalysts

Run	Rh concentration (M)	Temperature (°C)	Rate (Torr/h)
1A	0.0051	80	3.3
1B	0.0102	80	6.1
1C	0.0203	80	11.7
1A	0.0102	60	4.3
1B	0.0204	60	8.6

TABLE 7.2 The Effect of CO Partial Pressure on the WGS Activity

Run	P_{CO} (Torr)	Temperature (°C)	Rate (Torr/h)
1A	449	80	16.1
1B	225	80	8.6
1C	145	80	5.4
1D	68	80	2.5
2A	381	100	27.2
2B	207	100	15
2C	168	100	12.3
2D	66	100	4.0

between runs, a single solution was used which was 0.01 M in rhodium, 2 M in H^+ and I^- and 20 M in H_2O. The solution was degassed and recharged between runs, and the initial run was repeated at the end to be sure that no significant changes had occurred. Runs 3C and 3D contained nitrogen in addition to CO to keep the total pressure high enough to avoid sampling errors. As expected, decrease in the CO partial pressure decreases the WGS activity.

They also investigated the effect of reaction temperature on the WGS activity. As expected increasing reaction temperature increases the WGS activity. Increasing reaction temperature from 55 to 100 °C increases the rate from 2.4 to 27.7 Torr/h.

In 1979, Kaneda *et al.* [11] reported rhodium carbonyl system for homogeneous WGS reaction using diamine systems. The activity results are presented in Table 7.3. They made the following conclusions from the activity results:

TABLE 7.3 Water Gas Shift Reaction Using $Rh_6(CO)_{16}$-Amine Homogeneous Catalysts[a]

Run	$C_2H_5OC_2H_4OH$ (mL)	Amine (pK_a)[b]	Yield of hydrogen[c] (mmol)	Activity[d]
1[e]	0	$NH_2CH_2CH_2NH_2$ (9.97)	4.9	98
2	4.4	$NH_2CH_2CH_2NH_2$ (9.97)	3.7	75
3	3.0	$NH_2(CH_2)_3NH_2$ (10.65)	1.6	32
4	1.6	$NH_2(CH_2)_4NH_2$ (10.84)	0.3	6.4
5	0	$NH_2(CH_2)_6NH_2$ (11.0)[f]	0.4	8.6
6	2.8	$NH_2CH_2C(CH_3)HNH_2$	4.3	86
7	0	$N(CH_3)HCH_2CH_2N(CH_3)H$ (10.16)	3.5	69
8	0	$NH_2CH_2CH_2N(CH_3)_2$ (9.53)	2.3	46
9	0	$N(CH_3)_2CH_2CH_2N(CH_3)_2$ (8.97)	0.1	2.5
10	0	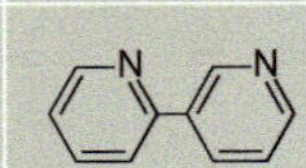 (4.14)	0	0
11	3.3	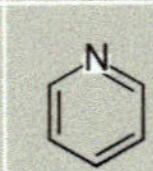 (5.17)	1.1	22

12	1.7	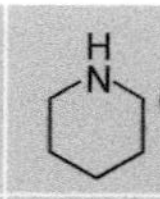 (11.2)	0.08	1.6
13	0	$(n\text{-}C_3H_7)_2NH$ (11.0)	Trace	0
14[g]	0	$(C_2H_5)_3N$ (10.9)	Trace	0
15[g,h]	0	$(C_2H_5)_3N$ (10.9)	Trace	0

[a]*Except where noted. 0.05 mmol of $Rh_6(CO)_{16}$, 85 mmol of amine, 40 mmol of H_2O, [10-amine] mL of ethoxyethanol and 700 mmHg of initial CO pressure were used; the reaction temperature was 100 °C; the total volume of the reaction system was 230 mL.*
[b]*Cited from M. Kotake, Constants of organic compounds, Asakura Publishing, Tokyo, 1963.*
[c]*The values were reproducible to about ±7%.*
[d]*Turnover number = mole of hydrogen/mole of rhodium complex per 4 h.*
[e]*10 mL of ethylenediamine (150 mol) was used.*
[f]*Cited from R. Barbucci, P. Paoletti and A. Vacca Che. Soc. (A) (1970) 2202–2207.*
[g]*10 mL of $(C_2H_5)_3N$ (72 mmol) was used.*
[h]*Reaction temperature was 80 °C.*
Taken From Table 1 of K. Kaneda, M. Hiraki, K. Sano, T. Imanaka, S. Teranishi, J. Mol. Catal. 9 (1980) 227.

(1) Increasing the volume of the amine increases the yield of hydrogen, and in the absence of the amine, only a trace of hydrogen is produced. (2) Increasing the number of methylene groups in $NH_2(CH_2)_nNH_2$ compounds decreases the yield. (3) Among ethylenediamine derivatives, methyl substituents on nitrogen depress the catalytic activity, while the methyl substituent on methylene carbon increases the activity (Table 7.3).

Then in 1981, Yoshida *et al.* [12] reported rhodium hydride catalysts for the homogeneous WGS reaction. The catalysts analyzed are presented in Table 7.4. The WGS reaction was carried out at 100 °C under CO pressure (20 kg/cm^2) for 18 h with rhodium(1)-hydrido compounds as catalyst precursors. The catalytic activity of the rhodium(1)-hydrido complexes depends on the nature of the auxiliary phosphine ligand and decreases in the order **2**>**3**>**1**≫**4**>**5**. The complex $RhCl(PPh_3)_3$ is virtually inactive as catalyst under similar conditions, and *trans*-$RhCl(CO)(PPh_3)_2$ was also found to be inactive as the catalyst even at higher temperatures (150 °C). The reaction rate is greatly influenced by the reaction medium. The rate was faster in strongly coordinating pyridine than in acetone. With **2**, the conversion of CO into CO_2 reaches 90% after 18 h in pyridine at 150 °C. Remarkably, **2** in pyridine is active at a temperature as low as 50 °C (Table 7.4).

In 1985, Mahajan *et al.* [13] reported polypyridine Rh complexes for the homogeneous WGS reaction. They prepared $[Rh(CO)_2Cl]_2$, $[Rh(bpy)_2]Cl$, $[Rh\text{-}(phen)_2]Cl$, $[Rh(bpy)_2]ClO_4$, $[Rh(CO)_2Cl]_2(\mu\text{-pz})$, $[Rh(CO)_2Cl]_2(\mu\text{-bpy})$,

TABLE 7.4 The Water Gas Shift Reaction[a]

Catalyst precursor	Products[b]			Products[b]		
	Solvent	H_2	CO_2	*Solvent*	H_2	CO_2
$RhH(PEt_3)_3$ (**1**)	Pyridine	10.0	10.0	Acetone	4.4	4.3
$RhH[P(i\text{-}Pr)_3]_3$ (**2**)	Pyridine	32.2	33.9	Acetone	27.2	29.4
$Rh_2H_2(\mu\text{-}N_2)[P(c\text{-}C_6H_{11})_3]_4$ (**3**)	Pyridine	17.8	17.2	Acetone	9.4	10.0
$RhH(N_2)[PPh(t\text{-}Bu)_2]_2$ (**4**)				Acetone		0.5
$RhH[P(t\text{-}Bu)_2]_2$ (**5**)				Acetone		0.2
$RhCl(PPh_3)_3$				Acetone		0.06
$RhCl(CO)(PPh_3)_2$				Acetone		0.9[c]

[a]*The reaction was carried out by using catalyst precursor (0.1 mmol), H_2O (2 mL) and CO (20 kg/cm^2) in the solvent specified (5 mL) at 100 °C for 18 h; for details see Experimental Section.*
[b]*Mol/mol of Rh per h.*
[c]*The reaction temperature was 150 °C.*
Taken from Table 2 of the T. Yoshida, T. Okano, Y. Ueda, S. Otsuka, J. Am. Chem. Soc. 103 (1981) 3411.

$[Rh(bPY)_3](ClO_4)_3$, $[Rh(bPY)_2(H_2O)_2](ClO_4)_3$, $[Rh(bpy)(CO)_2]ClO_4$ and Rh(bpy)(CO)(H)$(Cl)_2$ complexes.

Reaction of $[Rh(bpy)_2]Cl$ with CO at 50-95 °C in ethanol-water solution for several days produces H_2 and CO_2. Turnover frequency of 3 h^{-1} is obtained at 95 °C. Similarly, H_2 and CO_2 were produced when the solvent was *p*-dioxane or ethylene glycol, but the rate decreased substantially in *N,N*-dimethylacetamide. With the alcohol-water mixtures, rates were greater with 10-50% alcohol than in pure water and greatest in weakly acidic solution (pH −4), being negligible in both basic (pH >12) and strongly acidic (pH <1) solutions.

The yield of CO_2 was 1.0 for H_2 and was typically 0.95 in 15:10 ethanol-water. The pH of the reaction solution after a 20 h run at 90 °C was 4.0-4.7 with 15:10 ethanol-water as solvent and 5.4 in 3:22 ethanol-water and did not change with reaction time. When H_2O was replaced by D_2O, the 'hydrogen' produced was >70% D_2, indicating the source of H_2 to be water or water-exchangeable protons as from the —OH group of alcohols. The other catalysts were also investigated for WGS reaction. The turnover rate decreased more than a factor of 10 when $Rh(bpy)_2^+$ was replaced with Rh $(Phen)_2^+$ as a catalyst precursor. The $[Rh(CO)_2Cl]_2$ complex was almost inactive, but some WGS activity was observed when bpy, phen or Me_2bpy was added to alcohol-water solutions of the $[Rh(CO)_2Cl]_2$, dimer (Rh:L = 1:1), with the reactivity order L = bpy > Me_2bpy > phen.

Then in 1989 Pardey and Ford [14] investigated various Rh complexes in pyridine solutions for homogeneous WGS reaction. They prepared the *cis*-$[Rh(CO)_2L_2]PF_6$ (L = 4-picoline, pyridine, 2-picoline or 2,6-lutidine) and *trans*-$[Rhpy_4Cl_2]Cl$ rhodium complexes using the methyl pyridines (4-picoline, 3-picoline and 2-picoline) and dimethyl pyridines (2,6-lutidine and 3,5-lutidine). Table 7.5 presents the turnover frequencies of rhodium-based

TABLE 7.5 WGSR Activities by $RhCl_3·3H_2O$ in Aqueous Pyridine and Substituted Pyridines[a]

Amin	pK_a[b]	$TF(H_2)$[c]
4-Picoline	6.00	98
3-Picoline	5.52	84
Pyridine	5.27	68
3,5-Lutidine	6.23	60
2-Picoline	5.97	3
2,6-Lutidine	6.75	1

[a] *P_{CO} = 0.9 atm at 100 °C, [Rh] = 10 mM; 10 mL of solution amine/water, 8/2, v/v.*
[b] *From K. Schofield, Hetero-Aromatic Nitrogen Compound, Plenum Press, New York, (1967) 146–14.*
[c] *Mol H_2/mol $RhCl_3·3H_2O$/day. Experimental uncertainty is about ±10%.*
Taken from Table 2 of A.J. Pardey, P.C. Ford, J. Mol. Catal. 53 (1989) 241.

catalysts in various solvents for WGS reaction. The TF values for this rhodium-based catalyst proved to be strongly dependent on the solvent medium. Here the amine to water ratio is 8/2. Modest differences were noted for those pyridines methylated in the meta or para positions with aqueous 4-picoline solutions displaying the highest activity. However, much more dramatic differences were observed for the solutions of pyridines substituted in the ortho position; for example, 2,6-lutidine solutions proved to be roughly two orders of magnitude less active than those of 4-picoline. Then they investigated the effect of water content in the amine system on the WGS activity. Interestingly, no activity was observed in the presence of pure amine. The WGS activity increases with increasing water content up to 8/2 and decreases upon further increase in the water content.

Then they investigated the effect of Rh concentration, reaction temperature and CO partial pressure on the WGS activity of Rh complexes. The results are presented in Figures 7.1 and 7.2. For various concentrations of rhodium (0.005-0.050 M) and temperatures (80-120 °C), a linear dependence of turnover frequencies was observed with respect to the CO partial pressure. However, an increase in Rh concentration from 0.005 to 0.020 M resulted in a decrease in turnover frequency followed by nearly constant values at higher Rh concentration.

Based on their spectroscopic measurements, they proposed two types of mechanisms for the homogeneous WGS reaction in pyridine solutions for Rh-based catalysts. The mechanisms are presented in Scheme 7.1a and b. In Scheme 7.1a the key step is addition of CO to the Rh complex, and in Scheme 7.1b the key step is addition of another ligand (L) to the Rh complex. In Scheme 7.1a, the added CO would serve to enhance the electrophilicity of all the coordinated carbonyls, hence activate these toward attack by the relatively

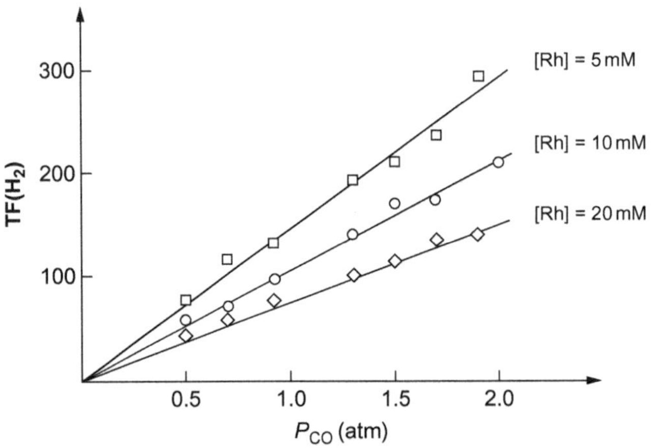

FIGURE 7.1 Plots of TF(H_2) vs. P_{CO} for different rhodium concentrations in 80% aqueous 4-picoline at 100 °C. *(Taken from Figure 4 of A.J. Pardey, P.C. Ford, J. Mol. Catal. 53 (1989) 241.)*

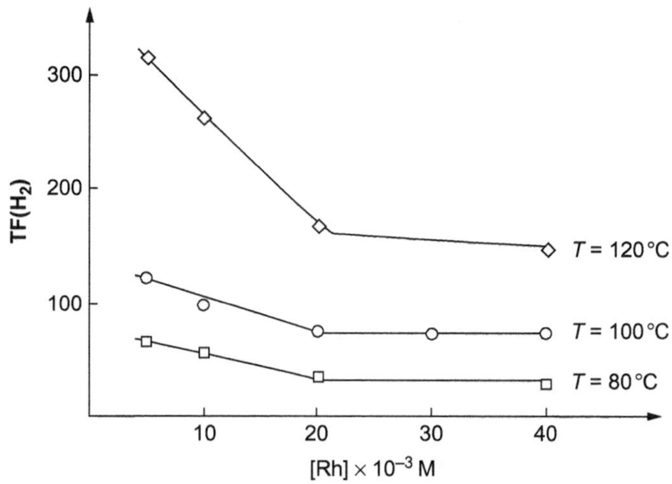

FIGURE 7.2 Plots of TF(H_2) vs. [Rh] for different temperatures in 80% aq. 4-picoline under 0.9 atm CO. *(Taken from Figure 5 of A.J. Pardey, P.C. Ford J. Mol. Catal. 53 (1989) 241.)*

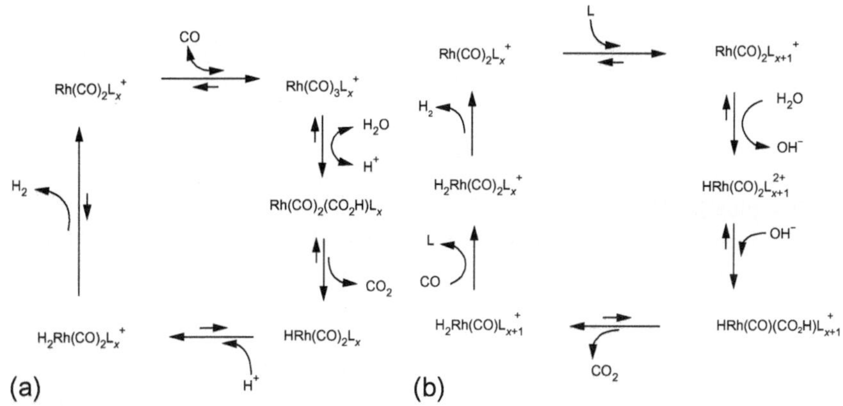

SCHEME 7.1 Proposed mechanisms for WGS catalysis by monomeric species. *(Taken from the Scheme 1a and 1b of A.J. Pardey, P.C. Ford, J. Mol. Catal. 53 (1989) 241.)*

weak nucleophile H_2O. If this were rate limiting, then the catalysis rate would be dependent on the concentration of the tricarbonyl species. Addition of another L to the coordination sphere would strongly enhance the basicity of the Rh(I) centre. Protonation of this by H_2O would give, formally, a Rh(III) hydride complex plus hydroxide in solution. Nucleophilic attack of OH^- or H_2O on a CO coordinated to this more electrophilic centre would give, sequentially, the Rh(III) hydroxycarbonyl and dihydride species. The equilibrium to form the $H_2Rh(CO)_2L_2^+$ species prior to rate-limiting reductive elimination of H_2.

After that not much research happened over Rh-based complexes because of its high cost and batch process.

7.3 Ru-BASED CATALYSTS

Prof. Ford group did most of the research with the Ru-based complexes. In 1967, Laine from Prof. Ford group [5] evaluated Ru carbonyl catalysts in alkaline solution for the homogeneous WGS reaction for the first time. They mixed $Ru_3(CO)_{12}$ with the KOH, H_2O and ethoxyethanol and heated to 100 °C under 1 atm of CO.

Analysis of the gas phase after 73 h gave a composition of 5.6% H_2, 6.5% CO_2, plus 88% CO, while analysis after 144 h gave 10.4% H_2, 8.5% CO_2, plus 77% CO. Also, the catalyst activity remains approximately constant or is perhaps slightly enhanced after flushing the system with fresh CO and restarting. When the temperature of reaction solutions is raised to 110 °C the rate of H_2 formation is approximately quadrupled over that at 100 °C and a similar increase is noted again for the 10 °C raise to 120 °C. Over a period of 30 days the total hydrogen produced by this system equalled 3×10^{-2} mol, which represents a ratio of 150 mol of H_2 per mole of $Ru_3(CO)_{12}$ initially added or 3 mol of H_2 per mole of KOH added. Thus, formation of H_2 clearly exceeds the molar quantities of KOH and $Ru_3(CO)_{12}$ added and thus cannot represent the stoichiometric reaction of base with coordinated carbon monoxide. That this system can be catalytic in base undoubtedly results from the fact that under the reaction conditions potassium bicarbonate in ethoxyethanol is unstable and decomposes to CO_2 plus KOH.

Then same group investigated [15] Ru/Fe mixed carbonyl system for the homogeneous WGS reaction. Also they investigated Ru, Fe, Ru/Fe systems in different media, i.e., acidic and basic. The catalytic activity results of various pure Ru, Fe and Ru/Fe systems are presented in Table 7.6. $Fe(CO)_5$ system exhibits very pure activity in all the media, i.e., KOH/ethoxyethanol, piperidine/ethoxyethanol, pyridine and H_2SO_4/diglyme solutions. On the other hand ruthenium carbonyl system exhibits fairly good activity in the other reaction media including piperidine/ethoxyethanol, pyridine and H_2SO_4/diglyme solutions. Notably, marked enhancement of activity is again seen for the mixed Fe/Ru catalysts in the piperidine and pyridine solutions but not in the acidic diglyme solutions. They said that the high activity of the ruthenium catalysts in acidic solution may be simply the result of shifts in pH-dependent equilibria.

Then they prepared mixed-metal cluster $H_2FeRu_3(CO)_{13}$ and evaluated for homogeneous WGS reaction [16]. The mixed-metal cluster $H_2FeRu_3(CO)_{13}$ is more active than either ruthenium carbonyl or iron carbonyl individually. This synergetic behaviour of the mixed-metal system is also observed for catalyst solutions prepared initially from $Ru_3(CO)_{12}$ plus $Fe(CO)_5$ or $Fe_3(CO)_{12}$ or $H_4Ru_4(CO)_{12}$ plus $Fe(CO)_5$ in alkaline solution. The catalytic activity of these

TABLE 7.6 **Catalysis by Ruthenium Carbonyl, by Iron Carbonyl and by Iron/Ruthenium Carbonyl Mixtures in Various Media ($T = 100$ °C, $P_{CO} = 0.9$ atm)**

Complex[a]	Reaction medium	Activity[b]
$Ru_3(CO)_{12}$	KOH/ethoxyethanol[c]	2.8 (0.9)
$Fe(CO)_5$	KOH/ethoxyethanol[c]	1.0 (1.0)
$Ru_3(CO)_{12}/Fe(CO)_5$[d]	KOH/ethoxyethanol[c]	7.4 (4.2)
$H_4Ru_4(CO)_{12}$	Piperidine/ethoxyethanol[e]	8.0 (2.0)
$Fe(CO)_5$	Piperidine/ethoxyethanol[e]	0.9 (0.9)[f]
$H_4Ru_4(CO)_{12}/Fe(CO)_5$	Piperidine/ethoxyethanol[e]	30.0 (12.0)
$H_4Ru_4(CO)_{12}$	Pyridine[g]	13.2 (3.3)
$Fe(CO)_5$	Pyridine[g]	~1-3 (~1-3)[h]
$H_4Ru_4(CO)_{12}/Fe(CO)_5$	Pyridine[g]	~12-30 (5-12)[h]
$H_4Ru_4(CO)_{12}$	H_2SO_4/ethoxyethanol[i]	~18 (4.5)
$H_4Ru_4(CO)_{12}$	H_2SO_4/diglyme[j]	31.6 (7.9)
$Ru_3(CO)_{12}$	H_2SO_4/diglyme[j]	33.3 (11.1)
$Fe(CO)_5$	H_2SO_4/diglyme[j]	0.0 (0.0)
$Ru_3(CO)_{12}/Fe(CO)_5$	H_2SO_4/diglyme[j]	14.8 (7.4)

[a]*Except where noted, 0.04 nmol of each complex listed was added to the solution.*
[b]*Moles of H_2 per day per mole of complex (moles of $Ru_3(CO)_{12}$ or $H_4Ru_4(CO)_{12}$ plus moles of Fe $(CO)_5$). Values in parentheses are normalized activities: moles of H_2 per day per gram-atom of metal added initially to the system.*
[c]*2 mmol of KOH, 0.02 mol of H_2O, 3.0 mL of ethoxyethanol.*
[d]*0.04 mmol of $Ru_3(CO)_{12}$/0.064 mmol of $Fe(CO)_5$.*
[e]*1.5 mL of piperidine, 2.8 mL of ethoxyethanol, 0.02 mol of H_2O.*
[f]*Low CO_2/H_2 ratios.*
[g]*3.0 mL of pyridine, 0.02 mol of H_2O.*
[h]*Reproducibility of this system was poor, a possible indication of heterogeneity.*
[i]*0.18 mmol of H_2SO_4, 0.02 mol of H_2O, 3.0 mL of ethoxyethanol. Activity decreased markedly after several days owing to the sublimation of $Ru_3(CO)_{12}$ out of the solution.*
[j]*1.8 mmol of H_2SO_4, 0.02 mol of H_2O, 3.0 mL of diglyme.*
Taken from Table 2 of P.C. Ford, R.G. Rinker, C. Ungermann, R.M. Laine, V. Landis, S.A. Moya, J. Am. Chem. Soc. 100 (1978) 4595.

solutions appears higher when the initial base added is $KHCO_3$ or $KHCO_3$ and K_2CO_3 rather than KOH.

In 1981, Slegeir *et al.* [17] reported the influence of reaction conditions on the WGS activity of ruthenium carbonyl system. The CO pressure has no or little effect on the WGS activity of $Ru_3(CO)_{12}$ catalyst system. On the other hand, at fixed CO pressure, little change in the CO concentration leads to the greater changes in the activity. At 0.1 mM ruthenium carbonyl concentration the catalyst exhibits very good activity and it decreases drastically with increasing Rh concentration to 1.0 mM, and further increase in the Rh concentration do not have much effect on the activity.

Solvents play an important role in the rate of hydrogen production. The ideal solvents are tetrahydrofuran, diglyme and dimethoxyethane. Small amounts of hydrocarbon added to the normal tetrahydrofuran or diglyme result in greater enhance in the WGS activity; however, larger quantities inhibit the reaction. The type of base has very important effect on the WGS activity of ruthenium carbonyl. The WGS activity of ruthenium carbonyl systems in various bases is presented in Table 7.7. Amines show much better activity compared to Bronsted bases. Among the various bases trimethyl amine shows much better activity compared to other bases.

The results of catalysis with primary and secondary amines indicate that ruthenium carbonyl forms active catalysts in aqueous ethylenediamines, diethanolamine, pyrrolidine and piperidine solutions. No detectable amounts of hydrogen are formed with aromatic and unsaturated primary and secondary amines under the present WGS conditions. Tertiary amines were found not to initiate catalyst systems as active as those produced by the best primary and secondary amines. Aliphatic tertiary amines exhibited only weak activity.

TABLE 7.7 Effect of Base on the Ruthenium Carbonyl-Catalyzed Water Gas Shift Reaction

Base	Amt Base (g)	Amt H_2O (g)	t_{H_2}
NMe_3	5	15	5740
NEt_3	5	15	860
NBu_3	5	15	540
N-Me pyrrolidine	5	15	~2400
$NHMe_2$	5	15	2200
Pyridine	4	16	~300
NH_3	6	15	420
Na_2CO_3	3	20	<50
Li_2CO_3	2	20	<50
Me_4NOH	0.2	20	<50
Bu_4NOH	0.03	20	<50
None	0.0	20	<50

Conditions: 0.05 mmol $Ru_3(CO)_{12}$, 92 mmol 1-butene, base and water diluted to 100 mL with diglyme, 750 psi CO, 100 °C, 10 h, 0, 31 L reactor.
Taken from the W.A.R. Slegeir, R.S. Sapienza, B. Easterling, ACS Symp. Ser. 152 (1981) 325.

Of the aromatic amines and diamines, pyridine, 1,10-phenanthroline and neocuproine form moderate active systems with $Ru_3(CO)_{12}$. Exudation of turnover frequencies in Table 7.7 reveals that a steric effect may instead be operative in the $Ru_3(CO)_{12}$-amine system, since the rates follow the order pyrrolidine > piperidine, $(CH_3)_2NCH_2CH_2OH > (C_2H_5)_2NCH_2CH_2OH$ and pyridine > quinoline.

In 1983, Prof. Ford group [18] reported the activity of Ru catalysts in acidic medium. They added 0.036 M of ruthenium, 0.25 M H_2SO_4 and 8 M H_2O in diglyme. The reaction was carried out at 0.9 atm of CO and 100 °C. An activity profile of a typical run is shown in Figure 7.3. Initially, there was very little WGSR activity once the solution was brought to the reaction. However, over a period of about 6-10 h (the induction period), the activity rose to a level close to the maximum turnover frequency of 30 mol Hz/g-atom Ru/24 h. The catalytic activity remained roughly constant for several days but then slowly decayed. DME, THF and methanol all proved to be suitable solvents for the WGSR, showing activities comparable to the diglyme solutions. However, these solvents also formed much less stable catalyst systems than found with diglyme as the solvent.

The phosphine-substituted complex $Ru_3(CO)_9(PPh_3)_3$ was also surveyed as a shift reaction catalyst. The catalyst activities were also relatively insensitive to the H_2SO_4 and H_2O concentrations of the system. Changes in the sulphuric acid and water concentrations over the respective ranges 0.25-1.5 and 4-12 M have small effects on the catalysis rates. However, the activity of the system was found to be

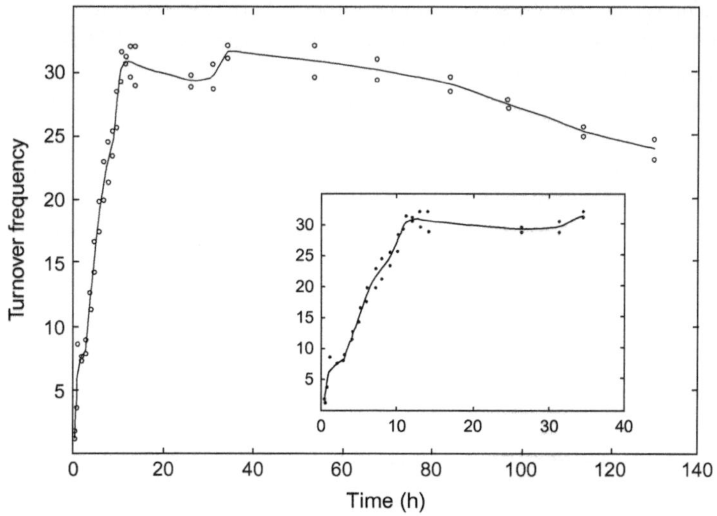

FIGURE 7.3 Plot of turnover frequency (H_2/mol Ru/day) vs. time for a typical WGSR run carried out under the following conditions: [Ru] = 0.036 M, P_{CO} = 0.9 atm, T = 100 °C, [H_2SO_4] = 0.25 M, [H_2O] = 8.0 in diglyme solution. *(Taken from Figure 1 of P. Yarrow, H. Cohen, C. Ungermann, D. Vandenberg, P.C. Ford, R.G. Rinker, J. Mol. Catal. 22 (1983) 239.)*

sensitive to the nature of the acid used, with sulphuric acid giving the highest activity among those studied. The approximate turnover frequencies were oxalic acid (3), H_2PO_4 (2), CF_3CO_2H (2), fumaric acid (2), succinic acid (1), acetic acid (1) and malonic acid (1). The stronger acids appear to give greater activities.

In the same year, Doi and Tamura [19] reported $RuCl_3$ catalysts for WGS reaction in alkaline medium. The activity results are presented in Table 7.8. The $RuCl_3$-KOH catalyst system exhibited much higher activity than the $Ru_3(CO)_{12}$-KOH system. An increase in the initial concentration of KOH resulted in the acceleration in the rate of H_2 formation for the WGSR using $RuCl_3$. However, higher initial concentrations of KOH over 1.0 M caused heterogeneity of the catalyst solution.

7.4 Fe-BASED CATALYSTS

The use of $Fe(CO)_5$ as a homogeneous catalyst for the WGSR was suggested by Reppe and Reindl as early as 1953 [1]. In 1976, Kang *et al.* [6] first reported Fe $(CO)_5$ catalyst for homogeneous WGS reaction in the open literature. However, this catalyst exhibits very less activity. In 1978, King *et al.* [20] reported the influence of CO partial pressure, reaction temperature and solvent on the WGS activity of Fe catalyst system along with Mo and W catalysts. The catalyst with initial CO pressure 14.1, reaction temperature 160 °C exhibits higher activity compared to other catalysts. It produces 200 mol of hydrogen per day per mole of metal. The effect of CO partial pressure results are presented in Table 7.9.

In 1980, King *et al.* [21] published a detailed study on the WGS activity of iron carbonyl system. In a typical experiment, $Fe(CO)_5$ dissolved in KOH and mixed with methanol-water solution (3:1) and charged with a CO-Ar mixture to the desired pressure. Inspection of these data reveals that the reaction is indeed catalytic in $Fe(CO)_5$ as well as in base that more than 300 mmol of Hr is produced by a system containing 2.23 mmol of $Fe(CO)_5$ and 39 mmol of base. The sum of partial pressures of CO and H_2 is nearly constant indicating that 1 mol of CO is consumed for every mol of H_2 produced during the course of the experiment. Figure 7.4 compares the rate of hydrogen production for a series of reactions using methanol-water mixtures of various compositions. The data reveal that a mixture containing 25% water and 75% methanol by volume is the preferred solvent system for the WGS reaction. Pure water is seen to be a very poor choice of solvent. Among the alcohol-water mixtures, it is seen that hydrogen production falls off more abruptly in the mixture containing 10% water than in the solutions richer in water. The rate of production of hydrogen increases rapidly with temperature as seen in Figure 7.5.

7.5 Ir-BASED CATALYSTS

In 1976, Kang *et al.* [6] reported Ir-based catalysts for WGS reaction along with other metal carbonyls. They ran experiment for 10 h in a 300-mL stirred

TABLE 7.8 Comparison of WGSR Activities Between $RuCl_3$-KOH and $Ru_3(CO)_{12}$-KOH Catalyst Systems[a]

			Solvent			
Run	Catalyst system	Temperature (°C)	H_2O (cm^3)	$C_2H_5OC_2H_4OH$ (cm^3)	KOH (mmol)	WGSR activity[b]
1	$RuCl_3$-KOH	90	15	0	2.5	3.0
2	$RuCl_3$-KOH	90	12	3	2.5	1.6
3[c]	$RuCl_3$-KOH	90	5	10	2.5	1.4
4[c]	$Ru_3(CO)_{12}$-KOH	90	15	0	2.5	0.3
5	$Ru_3(CO)_{12}$-KOH	90	5	10	2.5	0.2
6[d]	$Ru_3(CO)_{12}$-KOH	100	0.3	3	2.0	0.9
7	$RuCl_3$-KOH	90	15	0	5.0	2.1
8	$RuCl_3$-KOH	90	15	0	5.0	4.4
9[c]	$RuCl_3$-KOH	90	15	0	15.0	5.4
10[c]	$RuCl_3$-KOH	90	15	0	25.0	8.5

[a]*Reaction conditions: $RuCl_3$=0.1 mmol, $Ru_3(CO)_{12}$=0.03 mmol, initial CO pressure=250 Torr.*
[b]*Normalized activity=number of H_2 produced per ruthenium atom in 24 h.*
[c]*Runs 3, 4, 9 and 10 exhibited heterogenity in the catalyst solution.*
[d]*From C. Ungermann, V. Landies,S.A. Moya, H. Cohen, H. Walker, R.G. Pearson, R.G. Rinker, P.C. For J. Am. Chem. Soc. 101 (1979) 5922–5944. Reaction conditions: $Ru_3(CO)_{12}$=0.04 mmol, initial CO pressure=680 Torr.*
Taken from Table 1 of the Y. Doi, S. Tamura, Inorg. Chim. Acta 54 (1981) L235.

TABLE 7.9 WGS Activity of $Fe(CO)_5$ Catalysts (Base: NaOH, Solvent: *n*-BuoH, Metal Concentration: 0.0112, Base to Metal Ratio: 28)

Initial CO pressure	Temperature (°C)	Mol of H_2/mol of metal/day
28.2	137	16
28.2	164	52
28.2	180	72
14.1	161	200
23.2	150	0.01

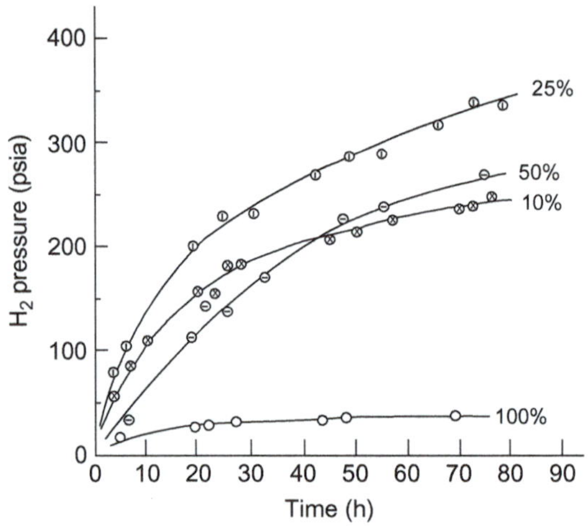

FIGURE 7.4 Partial pressure of H_2 as a function of time for various solvent mixtures. Catalyst solutions: $Fe(CO)_5$ (2.23 mmol), KOH (39 mmol), 200 mL of $H_2O—CH_3OH$ solution with composition O, 100% ⊖, 50%; Φ, 25%: ⊗, 10% water by volume. Temperature: 140 °C. Loading pressure of CO corrected to 140 °C: ~550 psia for each experiment. *(Taken from Figure 3 of D. King, Jr., R. B. King, D.B. Yang, J. Am. Chem. Soc. 102 (1980) 1928.)*

autoclave containing 0.05 mmol of catalyst, 22 mL of 25% aqueous trimethylamine, 78 mL of THF, 350 psi of CO at 125 °C. The $Ir_4(CO)_{12}$ catalyst produces 300 mol of hydrogen per mole of catalyst during the 10 h of reaction. Then, in 1981, Kaspar *et al.* reported $[Ir(COD)L_2]$ complexes with mono- (L = PPhJ, PMePh) or bidentate ligands such as $Ph_2P(CH_2)_2PPh_2$ (DPE), phenanthroline (Phen), methyl-substituted phenanthrolines (4,7-MezPhen, 3,4,7,8-M%Phen), phenyl-substituted phenanthroline (4,7-Ph_2Phen) and the sodium salt derivative 4,7-diphenyl-phenanthroline sulphonate (PhenS) for homogeneous WGS reaction. The activity results are presented in Table 7.10.

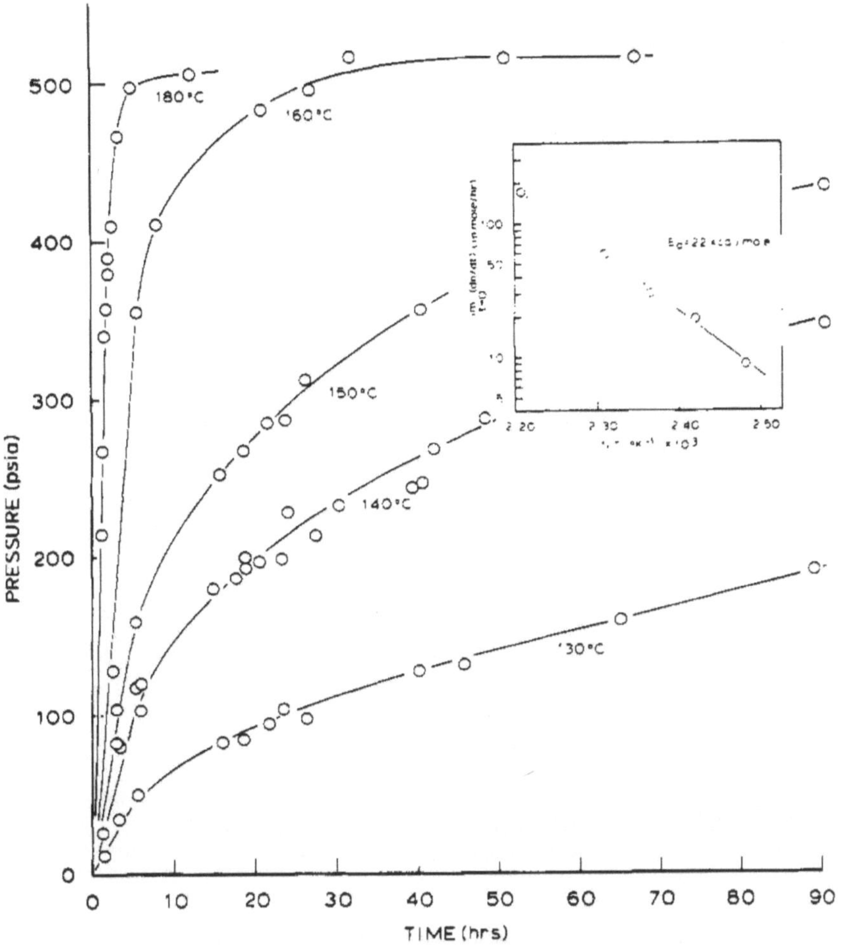

FIGURE 7.5 Partial pressure of H_2 and CO as a function of time for series of experiments differing only with respect to temperature. Catalyst solutions: $Fe(CO)_5$ (2.23 mmol), KOH (39 mmol), 200 mL (25% H_2O, 75% CH_3OH). Temperature: 140 °C. Loading pressure of CO connected to the temperature of each experiment. 130 °C, 575 psia; 140 °C, 556 psia; 150 °C, 642 psia; 160 °C, 585 psia; 180 °C, 592 psia. *(Taken from Figure 5 of A. D. King, Jr., R. B. King, D. B. Yang, J. Am. Chem. Sot., 102, 1980, 1028)*

Under the reaction conditions the starting complexes react with CO with displacement of coordinated diolefin to form iridium carbonyl compounds, which are the active catalytic species for the WGSR. The catalytic activity slightly increases when bidentate ligands are used and appears to be rather sensitive to temperature.

In 1984, Vandenberg *et al.* [22] reported iridium carbonyl catalysts for homogeneous WGS reaction. The WGSR catalysis study was typically carried out in a 100 mL glass bulb using KOH as the base and 2-ethoxyethanol-water

TABLE 7.10 WGSR Catalyzed by Ir[I] Complexes[a]

Run	Complex	Concentration ($n \times 10^5$)	Temperature (°C)	Turnover number[b]
1	$[Ir(COD)(PMePh_2)_2]^+$	10	140	1.5
2	$[Ir(COD)(PPh_3)_2]^+$	10	140	1.5
3	$[Ir(COD)DPE]^+$	10	140	7
4	$[Ir(COD)(Phen)]^+$	7.5	140	9.5
5	$[Ir(COD)(4,7\text{-}Ph_2Phen)]^+$	7.5	140	9.5
6	$[Ir(COD)(4,7\text{-}Me_2Phen)]^+$	7.5	140	11
7	$[Ir(COD)(3,4,7,8\text{-}Me_4Phen)]^+$	7.5	140	14
8	$[Ir(COD)\ (Phen\ S)]^-$	20	140	7.5
9	$[Ir(COD)\ (Phen\ S)]^-$	40	140	7
10	$[Ir(COD)\ (Phen\ S)]^-$	80	140	6.5
11	$[Ir(COD)\ (Phen\ S)]^-$	20	160	22.5
12	$[Ir(COD)\ (Phen\ S)]^-$	20	180	43.5
13	$[Ir(COD)\ (Phen\ S)]^-$	20	140	13.5[c]

[a]*Reaction conditions: A water-dioxane (1/1) mixture was used as solvent for runs 1-3. In the other cases water was used. The CO pressure varied over the range 6-10 atm depending on the working temperature. Both H_2 and CO_2 were determined by GLC analysis using a Porapak column and argon as carrier gas. A 190 mL autoclave was used in all the experiments. Reaction time 21 h. The complexes were used in runs 1-3 as PF_6^- or Cl^- in runs 4-7, and as sodium salts in runs 8-13.*
[b]*Turnover number defined as mol of substrate per mol of catalyst per hour.*
[c]*NEt_3 added, $CO/NEt_3 = 1$.*
Taken from Table 1 of the H.C. Kang, C.H. Mauldon, T. Cole, W. Slegeir, K. Cann, R. Pettit, J. Am. Chem. Soc. 9 (1977) 8323.

(4/l volume ratio) as the solvent. Figure 7.6 shows a typical catalysis run using K_2CO_3 as the initial base plotted in terms of turnover frequency (TF = mol of H, produced/gram atomic weight of iridium/day) as a function of time. In this experiment the solution was degassed and recharged to the same initial P_{CO} (0.9 atm at 100 °C) after each sampling of the gas mixture above the solution. Notably, the catalytic activity was found to be higher in the early stages but rapidly decreased to a significantly lower value over the longer run. Furthermore, there was a continued, although gradual, decrease in TF over a period of days.

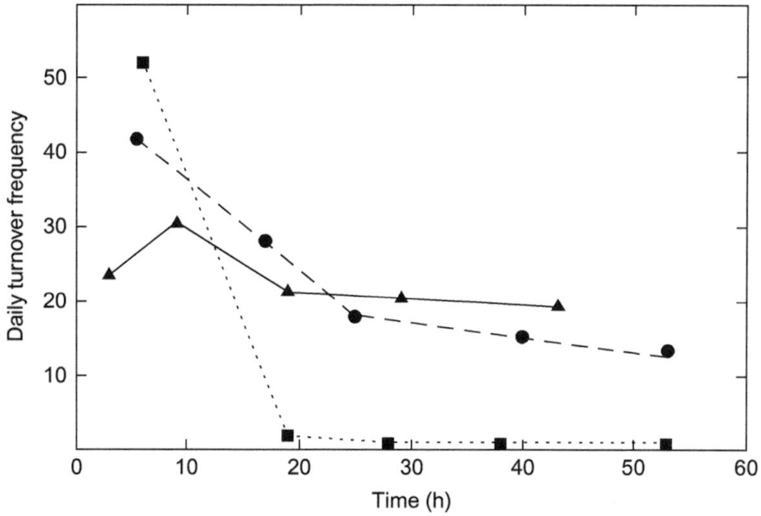

FIGURE 7.6 H_2 WGSR activities vs. time, prepared from $Ir_4(CO)_{12}$ (0.020 mmol) in 5 mL aqueous 2-ethoxyethanol (20% water by volume) with various bases at same normality ● 0.2 M K_2CO_3: ▲ 0.4 M $KHCO_3$: ■ 0.4 M KOH. Runs done at 100 °C under 0.9 atm CO (system was flushed and recharged after each reading). The turnover frequency represents the moles of H_2 produced/gram atomic weight of Ir/day. Lines drawn for illustrative purposes. *(Taken from Figure 1 of D.M. Vandenberg, T. Suzuki, P.C. Ford, J. Organomet. Chem. 272 (1984) 309.)*

7.6 Cr-, Mo- AND W-BASED CATALYSTS

In 1978, King *et al.* [20] first reported Cr, Mo and W catalysts for homogeneous WGS reaction. The catalytic activity results of various catalysts are presented in Table 7.11.

Among the various catalysts tungsten-based catalysts show better activity. In 1981, they reported detailed study of the Cr, Mo and W catalysts for homogeneous WGS reaction [23].

7.6.1 Catalytic Measurements

The WGSRs were carried out in small unlined stainless steel (type 304) autoclaves having an internal volume of 700 mL. Each autoclave was mounted in an electrical oven and the temperature of the contents was regulated with a proportional controller by using a thermocouple sensor mounted in a thermocouple well extending into the interior of the autoclave. After an initial warm-up period (−25 min) temperatures within a given autoclave could be maintained constant with this control system. The pressure of the contents was monitored throughout each experiment by using a Bourdon gauge mounted through the closure at the top of each autoclave.

TABLE 7.11 Systems Derived from the Mononuclear Metal Carbonyls as Catalysts for the Water Gas Shift Reaction[a]

Run	Metal carbonyl	Base	Solvent	Initial CO pressure (atm)	Metal concentration (M)	Base/metal mol ratio	Temperature (°C)	Mol of H_2/mol of metal per day
7	$Mo(CO)_6$	KOH	MeOH	4.3	0.00731	53	125	40
							135	115
							125	24
							125	22
7a	Filtrate of reaction mixture from run 7						125	12
8	$Mo(CO)_6$	KOH	MeOH	4.3	0.00207	87	120	5
							140	24
9	$MO(CO)_6$	KOH	MeOH	4.3	0.0022	760	115	9
							135	10
							160	40
							150	26
9a	Solids precipitating from the reaction mixture during run 9+MeOH+KOH						155	4
10	$Mo(CO)_6$	KOH	MeOH	11	0.00208	800	120	0
							130	3
							145	130

11	$Mo(CO)_6$	KOH	*n*-BuOH	11	0.0026	650	140	31
12	$Mo(CO)_6$	KOH	THF	11	0.00208	800	160	11
13	$Mo(CO)_6$	KBH_4	MeOH	4.3	0.0036	3	105	1
							120	5
							140	30
14	$Mo(CO)_6$	None	MeOH	4.3	.00255	0	160	0
15	$W(CO)_6$	KOH	MeOH	7.7	0.00132	1265	95	0
							110	13
							130	140
16	$W(CO)_6$	KOH	MeOH	7.7	0.00177	111	155	460[b]
17	$W(CO)_6$	KOH	MeOH	7.7	0.00105	1600	170	~900[d]

[a]*For runs 1–5 170 ml of n-BuOH and 30 ml of 2.08 Aq NaOH were used. For runs 6-17 100 ml of solvent was used.*
[b]*In this run over 16 hours period 2.77 mol of hydrogen was obtained for each mol of KOH.*
[d]*Data from R.M. Laine, R.G. Rinker P.C. Ford J. Am. Chem. Soc. 101 (1979) 5922–5944.*
Taken from Table 1 of R.B. King, C.C. Frazier, R.M. Hanes, A.D. King, Jr., J. Am. Chem. Soc. 100 (1978) 2925.

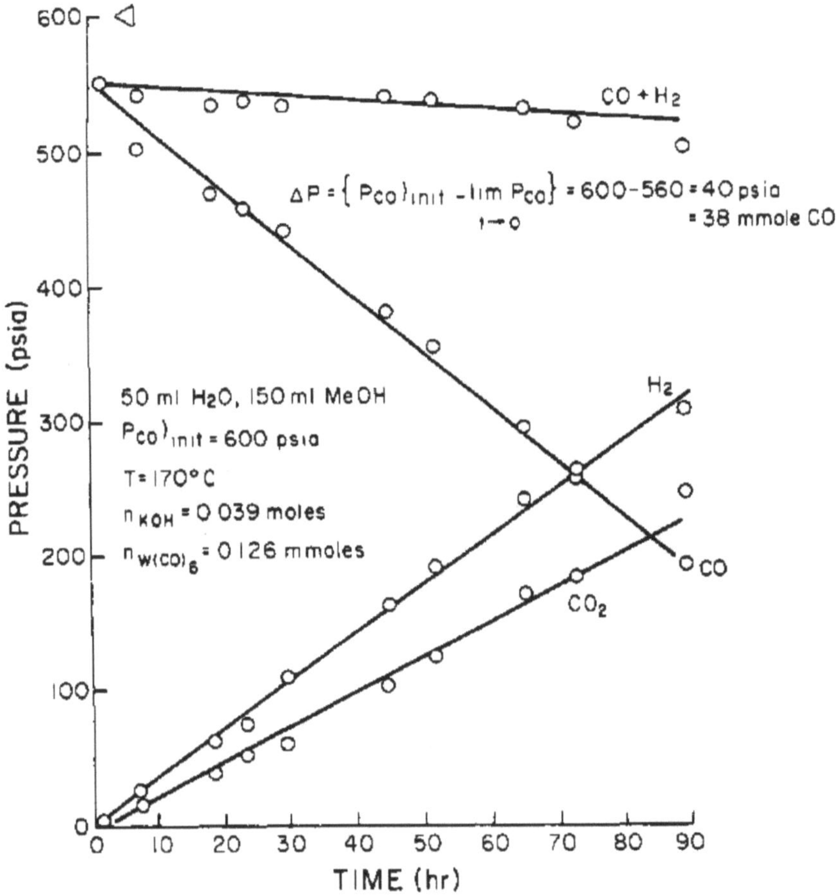

FIGURE 7.7 Partial pressure of H_2, CO_2 and CO as a function of time. Catalyst solution: $W(CO)_6$ (0.126 mmol), KOH (39 mmol), 25% H_2O-75% CH_3OH (200 mL). Temperature: 170 °C. Loading pressure of CO corrected to 170 °C: 600 psia. *(Taken from Figure 1 of A.D. King, Jr., R.B. King, D. B. Yang, J. Am. Chem. Soc. 103 (1981) 2699.)*

Figure 7.7 shows the WGS activity of $W(CO)_6$ catalyst at 170 °C for 90 h in 1:3 water-methanol solution by using KOH as base. The data in Figure 7.7 clearly illustrate that the reaction is clearly catalytic with respect to both W $(CO)_6$ and base since beyond 10 h into the run the mmol of H_2 produced exceeds the amounts of $W(CO)_6$ (0.126 mmol) and base (39 mmol) present in the system. Second, the fact that the sum of the partial pressures of CO and H_2 is nearly horizontal indicates that the mole consumption of CO exactly matches the production of H_2 as required by the stoichiometry. Thus one infers that no side reactions occur over the course of the WGS reaction.

Table 7.12 shows the influence of methanol-water ratio on the WGS activity of the Cr, Mo and W catalysts. It is seen that the less polar solvent mixtures

TABLE 7.12 The Effect of Solvent on the Rate of Production of H_2[a]

Composition	Rate (day^{-1})		
	$Cr(CO)_6$ (180 °C)	*$Mo(CO)_6$ (160 °C)*	*$W(CO)_6$ (160 °C)*
10% H_2O-90% CH_3OH	2500	265	330
25% H_2O-75% CH_3OH	1570	220	270
50% H_2O-50% CH_3OH	480	40	115

[a]*Solution: 200 mL of methanol-water mixtures of varying composition, each containing 39 mmol of KOH and 0.126 mmol of transition metal carbonyl. Loading pressure of CO: 400 psig at ~25 °C.*
Taken from Table 1 of A.D. King, Jr., R.B. King, D.B. Yang, J. Am. Chem. Soc. 103 (1981) 2699.

containing 10% and 25% water by volume are superior to that containing 50% water in the reaction medium. The solubilities of these transition metal carbonyls in water are too low to allow a comparison to be made between these alcohol solutions and pure water. They investigated the effect of CO initial pressure over the WGS activity of Cr-, Mo- and W-based catalysts. It is seen that CO inhibits the WGS reaction and does so to about the same extent with each catalyst system. Then they investigated the effect of base concentration of KOH, $NaHCO_2$ and $N(C_2H_5)_3$ for Cr-, Mo- and W-based catalysts. The rate of H_2 production is seen to increase with base concentration irrespective of the specific nature of the base. It seems that WGS activity increases with increasing reaction temperature. Among the various catalysts $M(CO)_6$ catalyst exhibits higher activity.

In 1982, Slegeir *et al.* [24] also reported tungsten-based catalysts for homogeneous WGS reaction. In 1992 Ng and Tsakiri [25] reported $Mo(CO)_6$ catalysts for homogeneous WGS reaction. They investigated methanol-water system and toluene-water systems by using KOH as a base. Table 7.13 shows the effect of $Mo(CO)_6$ and KOH concentrations on the WGS activity of $Mo(CO)_6$ for methanol-water system.

Increase in the KOH concentration increases the H_2 production up to 56.3 mmol, and further increase in the KOH concentration decreases the WGS activity.

$Mo(CO)_6$ was also used as a catalyst for the reaction in a toluene-water emulsion (toluene:water = 3:1 by volume). The amount of H_2 production increases almost linearly with increasing $Mo(CO)_6$ concentration up to 2.84 mmol (0.04 M) and then decreases with further increase in catalyst concentration for toluene-water system. Also increasing KOH concentration increases the WGS activity. The catalytic results show that for CO loading

TABLE 7.13 Methanol-Water System: Effect of $[Mo(CO)_6]$ and [KOH] on Hydrogen Production at 180 °C (Methanol:Water = 3:1 vol.; 70 mL Solution; P_{CO} = 2.1 MPa)

$Mo(CO)_6$ concentration (mmol)	KOH concentration (mmol)	H_2 mmol/mmol catalyst/day	CO_2 mmol/mmol catalyst/day
0.0	71.4	3.3	3.4
3.8	0.32	0.0	13.4
0.38	10.0	17.8	22.7
0.38	35	63	59
0.38	56.3	97	79.5
0.38	71.4	65	62.5

pressure between 1 and 3.1 MPa, H_2 production increase with increasing pressure. At CO loading pressure above 4.1 MPa, H_2 production decrease with increasing pressure. In the intermediate region of 3.1-4.1 MPa, it appears that CO loading pressure has little effect on H_2 production.

They also investigated other hydrocarbon:water solvents for the homogeneous WGS reaction. The results are presented in Table 7.14. Under the reaction conditions used, methanol is found to be the most effective solvent. Water is found to be three times less effective than methanol even though the dipole moment of water is higher than methanol. However, the toluene-water system (toluene:water = 1:1 by volume) exhibits similar activity to that of pure water and is more effective than the toluene-water system with toluene:water = 3:1 by

TABLE 7.14 Solvent Effect on the Water Gas Shift Reaction (0.38 mmol $Mo(CO)_6$, 180 °C, 2.5 h, P_{CO}: 2.1 MPa, 70 mL Solution, 71.4 mmol KOH, Hydrocarbon:Water 3:1)

Solvent	Dipole moment	H_2 produced (mmol)
Methanol	5.62	56.6
Toluene	1.4	19.7
Water	6.15	17.9
Decane	0.00	6.7
Xylene	0.26	4.3

volume. Decane and xylenes with lower dipole moment than toluene are found to be less effective solvents but still sustain the reaction at reasonable levels. The order of activity of these solvents is apparently somewhat related to their polarity.

7.7 Pt-, Pd-BASED CATALYSTS

In 1977, Yoshida *et al.* [26] reported importance of water activation for the homogeneous WGS reaction over Pt catalysts. The results of the WGSRs catalyzed by some PtL_3 complexes are summarized in Table 7.15. Typically, the catalytic solution prepared from PtL_3 (0.1 mmol) and H_2O (2 mL) in an aprotic solvent (5 mL) was placed into a stainless steel bomb (100 mL) under a N_2 atmosphere and subsequently was charged with CO (20 atm). A remarkable solvent effect is observed for the catalysis. Coordinating pyridine drastically reduces the catalytic activity of $Pt[P(i\text{-}Pr)_3]_3$. The catalytic activity of PtL decreases in the order $P(i\text{-}Pr)_3 > PEt_3 \gg PPh_3$. The complete lack of catalytic activity in $Pt(PPh_3)_3$ is apparently due to the incapability of water adduct formation. The importance of water molecule activation is obvious.

In 1978, Cheng and Eisenberg [27] reported the Pt-Sn system for homogeneous WGSR: 0.1 g of K_2PtCl_4 (0.23 mmol) and 2.5 g of $SnCl_4ċ5H_2O$ (7.1 mmol) dissolved in 40 mL of glacial acetic acid, 10 mL of concentrated HCl and 10 mL of H_2O. They made the following conclusions from Figure 7.8: (i) no reaction takes place in the absence of the platinum chloride salt; (ii) the omission of tin chlorides greatly decreases catalytic activity; (iii) large excesses of tin chlorides inhibit catalysis; (iv) if acetic acid is replaced

TABLE 7.15 The Water Gas Shift Reaction[a]

Catalyst	Solvent	Gaseous products[b]			
		Temperature (°C)	*H_2*	*CO_2*	*Turnover[c]*
$Pt[P(i\text{-}Pr)_3]_3$	Acetone	100	9.4	8.2	8.8
	Acetone	125	16.0	15.8	159
	Acetone	153	37.5	31.6	345
	THF	100	7.0	6.6	68
	Pyridine	100	0.7	1.0	9
$Pt(PEt_3)_3$	Acetone	100	1.1	1.1	11
$Pt(PPh_3)_3$	THF	100	0	0	0

[a]*The reaction conditions shown in the text.*
[b]*Millimoles.*
[c]*Moles/mole of catalyst, 18 h.*
Taken from Table 1 of T. Yoshida, Y. Leda, S. Otsuka, J. Am. Chem. Soc. 100 (1978) 3941.

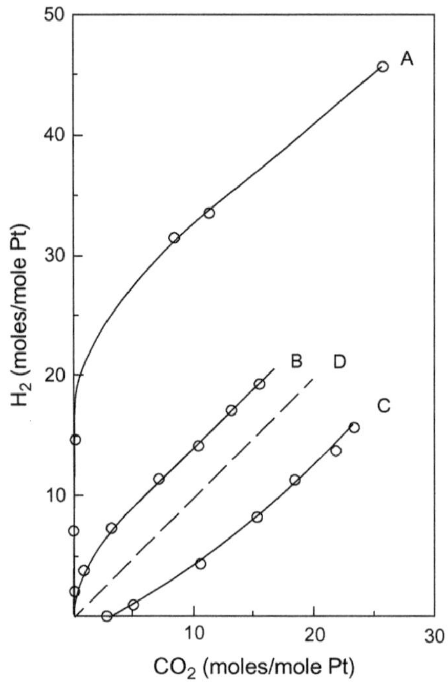

FIGURE 7.8 Product distribution for different initial [Sn(II)]:[Pt] ratios, run A corresponds to an initial [Sn(II)]:[PT] ratio of 30:1, run B to a ratio of 10:1 and run C to a ratio of 0:1. In all of these runs, the $[Sn_{total}]$:[Pt] ratio was held fixed at 30:1. The dashed line of unit slope corresponds to the product distribution for true WGS catalysis from the start of a run. *(Taken from Figure 1 of C.H. Cheng, R. Eisenberg, J. Am. Chem. Soc. 102 (1978) 5968.)*

by H_2O, only a trace of CO_2 is produced and a dark brown precipitate is obtained; and (v) no catalysis occurs in the absence of either a strong protic acid or a chloride source.

Frew *et al.* [28] also reported di platinum catalysts for homogeneous WGS reaction. The other metals investigated for homogeneous WGS reaction are Ni, Cu, Re and Os. However, very few literature reports are available on these systems.

Different research groups reported different mechanistic aspects for the homogeneous WGS reaction for various catalysts. The catalytic cycle mainly depends on the primary step, whether it is CO activation or H_2O activation, and the medium of the reaction, acidic, basic or neutral.

Before presenting individual mechanism for the specific catalyst or specific medium here we presented basic mechanism of the homogeneous WGS reaction.

7.8 BASIC MECHANISM OF THE HOMOGENEOUS WGS REACTION

Many researchers believe that the homogeneous WGS reaction mechanism cycle starts with activation of CO followed by its reaction with water (OH^-

or H_3O^+). CO activation normally begins with complexation of free CO to a coordinatively unsaturated metal. The bounded CO reactivity towards water, OH^- or H^+ depends on the metal's oxidation state, the overall charge on the complex and the extent to which the complex can stabilize the various possible intermediates that could form reaction with water, base or acid. The type of CO activation mainly depends on metal oxidation states.

7.8.1 Metallocarboxylic Acid Intermediate Formation

If the metal is neutral or negatively charged (have good electron donor ligands) it will donate electron density to the coordinated CO. Hence, only strong nucleophile such as OH^- will be capable of reacting with the activated CO and forms anionic metallocarboxylic acid.

$$MCO + OH^- \rightarrow M(CO_2H)^- \quad (7.1)$$

If the metal or overall complex is positively charged then H_2O reacts with the activated CO and forms metallocarboxylic acid.

$$MCO + H_2O \rightarrow M(CO_2H) + H^+ \quad (7.2)$$

7.8.2 Metallo-Formic Acid Intermediate Formation

The alternative route of CO activation begins with the reaction of CO with OH^- to produce formate and formic acid with excess water and does not rely on the unsaturated metal.

$$CO + OH^- \rightarrow HCO^- \quad (7.3)$$

$$HCO_2{}^- + H_2O \rightarrow HCO_2H + OH^- \quad (7.4)$$

Then the formate anion or formic acid binds to a coordinatively unsaturated metal and forms metalloformate rather than a metallocarboxylic acid.

$$M + HCO_2H \rightarrow HM(O_2CH) \quad (7.5)$$

$$M + HCO^- \rightarrow [M(O_2CH)] \quad (7.6)$$

Metal formates could also arise via rearrangement of a metallocarboxylic acid, as shown in reaction (7.7)

$$HM(CO_2H) \rightarrow M + HCO_2H \rightarrow HM(O_2CH) \quad (7.7)$$

The existence of metallocarboxylic acids and evidence for their formation upon reaction of carbonyls with OH— are now well established, beginning with the initial report by Deeming and Shaw [29] and continuing with the more recent work of Darensbourg *et al.* [32,33], Gross and Ford [35], Laine *et al.* [5] and others. The following complexes have been isolated and/or identified spectroscopically: $IrCl_2(CO_2H)(CO)(PMe_2Ph)_2$, $CpFe(CO)(PPh_3)CO_2H$,

$[Ru\text{-}(bpy)_2CO_2H]PF_6$, $CpRu(CO)(PPh_3)CO_2H$, $((Et_3P)_2PtCl)CO_2H$, *trans*-PtH $(CO_2K)[P(i\text{-}Pr_3)_3]$, $CpRe(NO)(CO)CO_2H$, $Mn(CO)_5CO_2H$, $Ru_3(CO)_{11}CO_2H$ and $Fe(CO)_4CO_2H$ (gas phase).

7.8.3 CO_2 Formation

The majority of the above-cited metallocarboxylic acids and metalloformates decompose rather readily to produce CO_2 and a metal hydride; however, the exact mechanisms can vary. Deeming and Shaw [29] were the first to observe that metallocarboxylic acids can decompose with loss of CO_2.

$$IrCl_2(CO_2H)(CO)(PMe_2Ph)_2 \rightarrow IrHCl_2(CO)(PMe_2Ph)_2 + CO_2 \tag{7.8}$$

Three other bimolecular decomposition pathways have been proposed by Darensbourg *et al.* [33]

$$MCO_2H + OH^- \rightarrow [MC(OH)_2O^-] \rightarrow MH + HCO_2 \tag{7.9}$$

$$MCO_2{}^- + H_2O \rightarrow [MC(OH)_2O^-] \rightarrow MH + HCO_3{}^- \tag{7.10}$$

$$MCO_2H + MH \rightarrow H_2 + M_2 + CO_2 \tag{7.11}$$

7.8.4 H_2 Formation

Uni-molecular reductive elimination provides the simplest possible route for generation of H_2.

$$MH_2 \rightarrow H_2 + \text{'M'} \tag{7.12}$$

A number of complexes were reported in the literature including $H_2Fe(CO)_4$ [34], $H_2RhCl(PPh_3)_3$, $H_2IrCl(CO)(PPh_3)_2$ and $H_2Co[P(OR)_3]_4$, which follows uni-molecular reductive elimination mechanism for H_2. Reports are also available on bimolecular reductive elimination mechanism for the H_2 elimination.

$$MH + MH \rightarrow M_2 + H_2 \tag{7.13}$$

$$2HCO(CO)_4 \rightarrow CO_2(CO)_8 + H_2 \tag{7.14}$$

$$2H_2Os(CO)_4 \rightarrow H_2Os_2(CO)_8 + H_2 \tag{7.15}$$

In most cases bridging hydride intermediate formation takes place in bimolecular reductive mechanism which then leads to H_2 and dimeric species.

Alternate mechanisms are available for H_2 production that does not proceed through reductive elimination. For example, metal hydrides are known to react with acids according to reaction (7.8):

$$MH + HX \rightarrow MX + H_2 \tag{7.16}$$

The main catalysts used for homogeneous WGS reaction are Rh, Ru and Ir complexes. The other catalysts used are transition metal carbonyls. The reaction can be conducted in acidic, basic or aqueous media.

7.8.4.1 Mechanism of Homogeneous WGS Reaction in Basic Conditions

7.8.4.1.1 Iron-Based Catalysts

As described earlier Kang *et al.* [6] first time reported the Fe catalysts in the basic medium. Kang *et al.* [6] reported that aqueous alcoholic solutions of Fe(CO)s made basic with alkali metal hydroxides also promoted the WGSR. They proposed the mechanism by using their IR data.

$$Fe(CO)_5 + OH^- \rightarrow Fe(CO)_4CO_2H^-$$

$$Fe(CO)_4CO_2H^- \rightarrow HFe(CO)_4{}^- + CO_2$$

$$HFe(CO)_4{}^- + H_2O \rightarrow H_2Fe(CO)_4 + OH^-$$

$$H_2Fe(CO)_4{}^- \rightarrow H_2 + Fe(CO)_4$$

$$Fe(CO)_4 + CO \rightarrow Fe(CO)_5$$

Pearson and Mauermann's studies provide evidence that loss of CO_2 proceeds via deprotonation of the metallocarboxylic acid, rather than directly from the metallocarboxylic acid. Contrary to the work of Pearson and Mauermann, and Laine *et al.*, recent kinetic studies by Gross and Ford [35], on the $Fe(CO)_5$/OH^- system in mixed THF/MeOH/H_2O solvent systems, indicate that decarboxylation proceeds via the metallocarboxylic acid, rather than the dianion.

7.8.4.1.2 Ruthenium-Based Catalysts

Gross and Ford [35] and Bricker *et al.* [37] first time proposed proper mechanism over Ru-based catalysts.

$$Ru_3(CO)_{12} + OH^- \rightarrow Ru_3(CO)_{11}(CO_2H)^-$$

$$Ru_3(CO)_{11}(CO_2H)^- \rightarrow CO_2 + HRu_3(CO)_{11}{}^-$$

$$HRu_3(CO)_{11}{}^- + CO \rightarrow [HRu_3(CO)_{12}]^-$$

$$[HRu_3(CO)_{12}]^- + H_2O \rightarrow Ru_3(CO)_{12} + H_2 + OH^-$$

$$HRu_3(CO)_{11}{}^- + H_2 \leftrightarrow H_3Ru_4(CO)_{12}{}^-$$

$$H_3Ru_4(CO)_{12}{}^- + CO \leftrightarrow HRu_4(CO)_{13}{}^- + H_2$$

$$HRu_4(CO)_{13}{}^- + 6CO \leftrightarrow 3HRu_3(CO)_{11}{}^- + Ru_3(CO)_{12}$$

Slegeir *et al.* [17] have briefly studied a number of aspects of ruthenium catalysis of the WGSR using amine as base. Their studies with the

$(CH_3)_3N/THF/H_2O$ solvent system suggest the following mechanism for amine-promoted WGSR catalysis:

$$Ru_3(CO)_{12} + 3CO \leftrightarrow 3Ru(CO)_5$$

$$Ru(CO)_5 + (CH_3)_3N \leftrightarrow (CO)_4RuC(=O)N(CH_3)_3$$

$$(CO)_4RuC(=O)N(CH_3)_3 + H_2O \leftrightarrow HRu(CO)_4^- + CO_2 + (CH_3)_3NH^+$$

$$HRu(CO)_4^- + H_2O \leftrightarrow H_2Ru(CO)_4 + OH^-$$

$$H_4Ru(CO)_4 \leftrightarrow H_2 + Ru(CO)_4$$

$$Ru(CO)_4 + CO \leftrightarrow Ru(CO)_5$$

Most recently, a new type of mononuclear/dinuclear ruthenium WGSR catalysis system was described by Shvo *et al.* [38]. The system is based on the reaction of tetraphenylcyclone (CPD) ruthenium tricarbonyl, $(\eta^4\text{-CPD})^-$ $Ru(CO)_3$ with OH^-.

$$CPDRu(CO)_3 + OH^- \rightarrow [CPDRu(CO)_2CO_2H]^-$$

$$[CPDRu(CO)_2CO_2H]^- \rightarrow CPDRu(CO)_2H^- + CO_2$$

$$CPDRu(CO)_2H^- + H_2O \rightarrow CPDRu(CO)_2H_2 + OH^-$$

$$CPDRu(CO)_2H_2 \rightarrow CPDRu(CO)_2 + H_2$$

7.8.4.1.3 Group VI Metal Complexes as Catalysts for the WGSR

As described earlier, King *et al.* [21] first time reported the mechanism over the group VI metals.

$$CO + OH^- \rightarrow HCO_2^-$$

$$M(CO)_6 \rightarrow M(CO)_5 + CO$$

$$M(CO)_5 + HCO_2^- \rightarrow M(CO)_5(O_2CH)^-$$

$$M(CO)_5(O_2CH)^- \rightarrow HM(CO)_5^- + CO_2$$

$$HM(CO)_5^- + H_2O \rightarrow H_2M(CO)_5^- + OH^-$$

$$H_2M(CO)_5 \rightarrow M(CO)_5 + H_2$$

7.8.4.1.4 Rh-Based Catalysts

Laine and Thomson [36,39] have briefly explored the WGSR catalysis chemistry of $Rh_6(CO)_{16}/KOH$ systems.

$$Rh_6(CO)_{16} + OH^- \rightarrow HRh_6(CO)_{15}^- + CO_2$$

$$HRh_6(CO)_{15^-} \rightarrow H_2 + Rh_{12}(CO)_{30}{}^{2-}$$

$$Rh_{12}(CO)_{30}{}^{2-} + 2CO + 2H_2O \rightarrow 2Rh_6(CO)_{16} + 2H_2 + 2OH^-$$

Kaneda *et al.* [11] have studied $Rh_6(CO)_{16}$/amine WGSR catalyst systems. As described earlier diamines such as ethylenediamine considerably enhance the activity of the catalyst. No detailed studies have been reported to date on the kinetics or mechanisms of the $Rh_6(CO)_{16}$/amine WGSR system, again emphasizing the need for research in the area of amine promoters.

7.8.4.1.5 Ir-Based Catalysts

Although iridium-catalyzed WGSR systems were described some time ago, detailed kinetic studies have only recently been reported by Vandenberg *et al.* [22]. Their observations lead them to propose the following catalytic cycle:

$$Ir_4(CO)_{12} + OH^- \leftrightarrow HIr_4(CO)_{11}{}^- + CO_2$$

$$HIr_4(CO)_{11}{}^- + OH^- \leftrightarrow [HIr_4(CO)_{10}(CO_2H)]^{2-}$$

$$[HIr_4(CO)_{10}(CO_2H)]^{2-} \leftrightarrow H_2Ir_4(CO)_{10}{}^{2-} + CO_2$$

$$H_2Ir_4(CO)_{10}{}^{2-} + H_2O \leftrightarrow H_3Ir_4(CO)_{10}{}^- + OH^-$$

$$H_3Ir(CO)_{10}{}^- + CO \leftrightarrow HIr_4(CO)_{11}{}^- + H_2$$

7.8.4.2 Mechanism of WGS Reaction in Acidic Conditions

7.8.4.2.1 Rh-Based Catalysts

In 1977, Cheng *et al.* [7] first time proposed the mechanism of the homogeneous WGS reaction for Rh catalysts in acetic acid/HI solution. Then Zudin *et al.* [8] reported the Pd complexes in trifluoroacetic acid. Baker *et al.* [10] described kinetic and mechanistic investigations that suggest the catalytic cycle.

$$HRh(CO)_2I_3 + HI \leftrightarrow H_2 + Rh(CO)_2I_4{}^-$$

$$Rh(CO)_2I_4{}^- + CO \leftrightarrow I^- + Rh(CO)_3I_3$$

$$Rh(CO)_3I_3 + H_2O \leftrightarrow H^+ + Rh(CO)_2I_3(CO_2H)^-$$

$$Rh(CO)_2I_3(CO_2H)^- \leftrightarrow HI + Rh(CO)_2I_2{}^- + CO_2$$

$$Rh_2(CO)_2I_2{}^- + HI \leftrightarrow HRh(CO)_2I_3{}^-$$

Furthermore, they proposed that at low temperatures the rate-limiting step is oxidation of Rh(I) by HI to Rh(III), and at high temperatures the rate-limiting step is reduction of a Rh(III) carbonyl species with concomitant release of CO_2.

Singleton and Forster have also examined the same WGSR catalyst system but under more forcing conditions [40,41]. Although they observe similar catalyst species, they find different reaction kinetics. The major difference appears to be that at higher temperatures and pressures. WGSR is nearly independent of CO pressure at high acidity but is inversely affected by changes in CO pressure at low acidities.

Mahajan *et al.* [13] have reported the studies of polypyridine rhodium complexes that provide some mechanistic explanations about the WGSR catalysis systems originally described by Marnot *et al.* In their studies of rhodium WGSR catalysis in the presence of phen, bipy, pyrazine and 4,4′-dimethylbipy, they were able to obtain spectroscopic and kinetic evidence that suggests that the active catalyst species under acidic conditions is likely to be $Rh(bipy)_2(CO)H_2^+$.

7.8.4.2.2 Pt-Based Catalysts

Likholobov *et al.* [42] have followed up on the original report by Zudin *et al.* [8] and proposed the following catalytic cycle for Ph_3P-complexed palladium WGSR catalysis system run in 20% aqueous trifluoroacetic acid.

$$P_2PdX_2CO \leftrightarrow [P_2Pd(X)CO]^+ + X^-$$

$$[P_2Pd(X)CO]^+ + H_2O \leftrightarrow [P_2Pd(H_2O)CO]^{2+} + X^-$$

$$[P_2Pd(H_2O)CO]^{2+} + H_2O \leftrightarrow [P_2Pd(OH)CO]^+ + H_3O^+$$

$$[P_2Pd(OH)CO]^+ + X^- \rightarrow P_2Pd(CO_2H)_X$$

$$P_2Pd(CO_2H)X \rightarrow P_2PdHX + CO_2$$

$$P_2PdHX + HX \rightarrow P_2PdX_2 + H_2$$

$$P = Ph_3P, \quad HX = CF_3CO_2H$$

The mechanism proposed was confirmed by the work of Cariati *et al.* [43]. In addition to their studies on rhodium WGSR catalysis in acid media, Cheng and Eisenberg have also reported [27] that mixtures of platinum chloride and tin chloride are active WGSR catalysts in an acetic acid/HCl solvent system. They proposed the following mechanism (Scheme 7.2). The catalytic cycle appears to involve the Sn(II)/Sn(IV) redox couple. The formation of H_2 coincides with the oxidation of Sn(II) to Sn(IV) and CO is oxidized to CO_2 concurrent with reduction of Sn(IV) to Sn(II).

Yarrow *et al.* [18] proposed the reaction mechanism.

$$HRu_2(CO)_{9-x}(OR_2)_x{}^+ + CO \rightarrow HRu_2(CO)_{10-x}(OR_2)_{x-1}{}^+ + OR_2$$

$$HRu_2(CO)_{10-x}(OR_2)_{x-1}{}^+ + H_2O \rightarrow HRu_2(CO)_{9-x}(CO_2H)(OR_2)_{x-1} + H^+$$

$$HRu_2(CO)_{9-x}(CO_2H)(OR_2)_{x-1} \rightarrow H_2Ru_2(CO)_{9-x}(OR_2)_{x-1} + CO_2$$

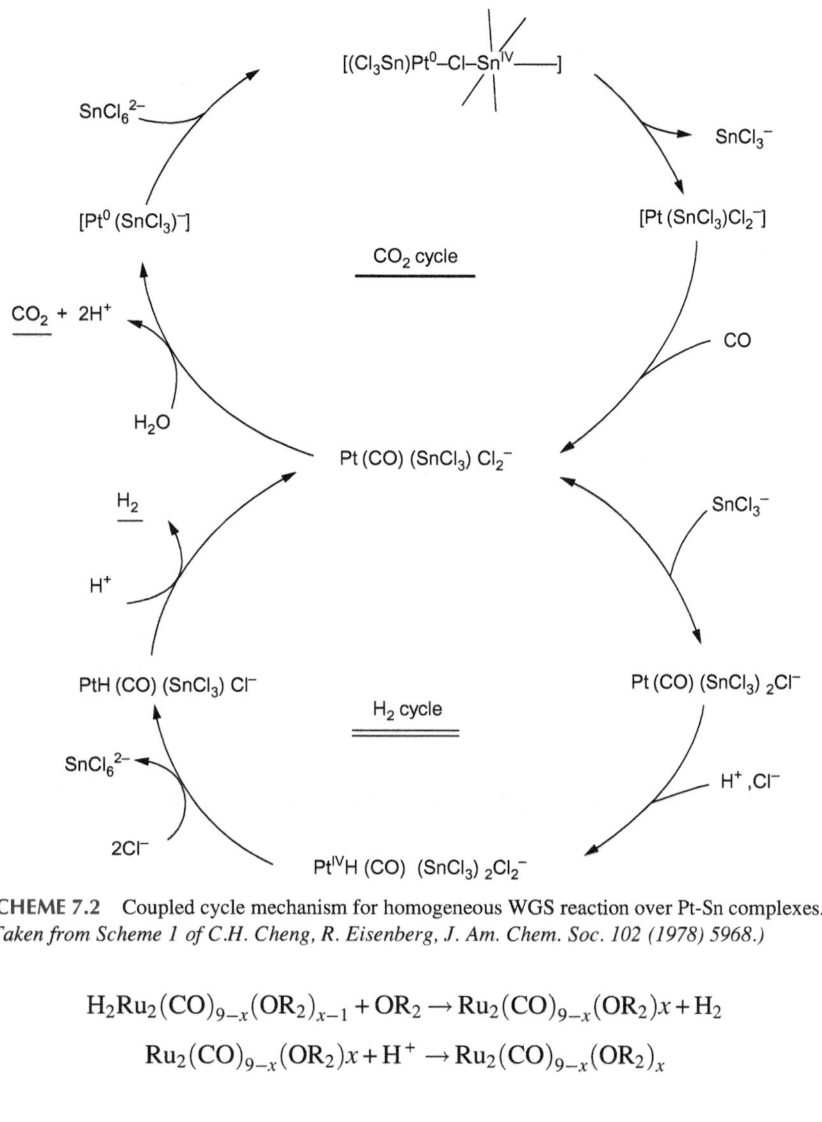

SCHEME 7.2 Coupled cycle mechanism for homogeneous WGS reaction over Pt-Sn complexes. *(Taken from Scheme 1 of C.H. Cheng, R. Eisenberg, J. Am. Chem. Soc. 102 (1978) 5968.)*

$$H_2Ru_2(CO)_{9-x}(OR_2)_{x-1} + OR_2 \rightarrow Ru_2(CO)_{9-x}(OR_2)x + H_2$$

$$Ru_2(CO)_{9-x}(OR_2)x + H^+ \rightarrow Ru_2(CO)_{9-x}(OR_2)_x$$

7.8.4.2.3 Homogeneous Catalysts in Both Acidic and Basic Medium

Only three WGSR catalyst systems have been described wherein it has not been necessary to activate and maintain the catalytic reactivity of the system through addition of either base or acid. One reaction system employs platinum phosphine catalysts, and the other two systems use rhodium phosphine catalysts. Kaspar *et al.* [44] describe an active WGSR system based on rhodium/phosphine catalysts species that functions in dioxane/water mixtures both in base and in acid. The acid-promoted catalyst systems are much more active than the base-promoted systems. The one platinum system and one of the rhodium systems were described by Yoshida *et al.* [26]. These investigators report catalyst systems that function by H_2O activation rather than CO activation, which appears to be a requirement of all of the systems described above. They proposed the following mechanism.

$$PtL_3 \leftrightarrow PtL_2 + L$$

$$PtL_2 + H_2O \leftrightarrow PtH(OH)L_2$$

$$PtH(OH)L_2 + S \leftrightarrow [PtH(S)L_2]OH$$

$$[PtH(S)L_2]OH + CO \leftrightarrow [PtH(CO)L_2]OH + S$$

$$[PtH(CO)L_2]OH \leftrightarrow [PtH(CO_2H)L_2]$$

$$[PtH(CO_2H)L_2] \leftrightarrow PtH_2L_2 + CO_2$$

$$PtH_2L_2 + L \leftrightarrow PtL_2 + H_2$$

7.8.4.2.4 Mechanism of Homogeneous WGS Reaction in Neutral Medium

In the neutral rhodium system, the complex RhHLs, where $L = P(i\text{-}Pr)_3$ or $P(c\text{-}C_6H_{11})_3$, was found to be an active WGSR catalyst in either acetone, THF or pyridine. The following reaction sequence is proposed for the catalytic cycle:

$$RhHL_3 + H_2O + 2py \leftrightarrow [RhH_2(py)_2L_2]OH + L$$

$$[RhH_2(py)_2L_2]OH + CO \leftrightarrow [RhCO(py)_2L_2]OH + H_2$$

$$[RhCO(py)_2L_2]OH + CO \leftrightarrow [Rh(CO)_2(py)L_2]OH + py$$

$$[Rh(CO)_2(py)L_2]OH \rightarrow [Rh(CO_2H)(py)L_2]$$

$$[Rh(CO_2H)(py)L_2] + CO \rightarrow RhH(CO)_2L_2 + CO_2 + py$$

$$RhH(CO)_2L_2 + H_2O \leftrightarrow [RhH_2(CO)L_2]OH$$

$$[RhH_2(CO)L_2]OH + py \leftrightarrow [Rh(Py)(CO)L_2]OH + H_2$$

A recent set of papers describe the use of dpm (bis(diphenylphosphino) methane) ligands to form cluster complexes of rhodium [9,45], iridium [46] and platinum [28,47] that all promote the WGSR under conditions that do not require the addition of acid or base. The earliest report was that of Kubiak and Eisenberg [9] who mentioned that the complex $[Rh_2(\mu\text{-}H)(\mu\text{-}CO)(CO)_2(dpm)_2]^+$ was an active WGSR catalyst under neutral conditions. In a later paper [45], they report that the catalyst system was most active in the presence of 1 equiv. of toluenesulphonic acid and 2 equiv. of a salt (e.g., LiX, X = Cl, Br). Most recently, Sutherland and Cowie [46] have extended the work of Kubiak and Eisenberg through studies of both the original rhodium system and its

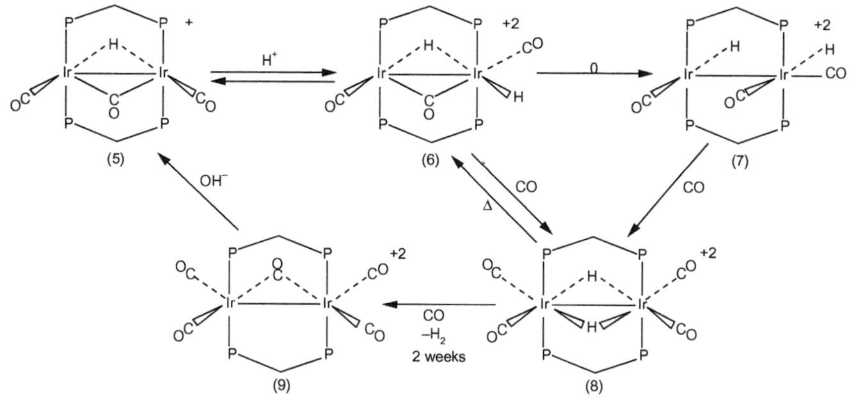

SCHEME 7.3 W. Reppe, W. Schweckendiek, K. Kroeper (BASF) 2658075 (1953) *(Taken from B. R. Sutherland, M. Cowie, Organometallics 4 (1985) 1637.)*

iridium analogue. Their work was initiated based on the hypothesis that a bridging hydroxide would be a likely intermediate in the catalytic cycle; however, they disprove this hypothesis. Sutherland and Cowie suggest a potential catalytic cycle based on their iridium studies (Scheme 7.3).

REFERENCES

[1] W. Reppe, U.S. Patent, 1953.

[2] D.M. Fenton, U.S. Patent 3,490,872, 1970.

[3] D.M. Fenton, U.S. Patent 3,539,298, 1970.

[4] D.M. Fenton, U.S. Patent 3,781,364, 1973.

[5] R. M. Laine, R. G. Rinker, P. C. Ford, Homogeneous catalysis by ruthenium carbonyl in alkaline solution: the water gas shift reaction, J. Am. Chem. Soc. 99 (1977) 252–253.

[6] H. C. Kang, C. H. Mauldon, T. Cole, W. Slegeir, K. Cann, R. Pettit, Reductions with carbon monoxide and water in place of hydrogen. 1. Hydroformylation reaction and water gas shift reaction, J. Am. Chem. Soc. 99 (1977) 8323–8325.

[7] C. H. Cheng, D. E. Hendriksen, R. Eisenberg, Homogeneous catalysis of the water gas shift reaction using rhodium carbonyl iodide, J. Am. Chem. Soc. 99 (1977) 2791–2792.

[8] V. N. Zudin, V. A. Likholobov, I. Yermakov, N. K. Yeremenko, Kinet. Katal. 18 (1977) 524.

[9] C. P. Kubiak, R. Eisenberg, Molecular A frames. Synthesis from binuclear rhodium(0) precursors and catalytic activity in the water gas shift reaction and alkyne hydrogenation, J. Am. Chem. Soc. 102 (1980) 3637–3639.

[10] E. C. Baker, D. E. Hendriksen, R. Eisenberg, Mechanistic studies of the homogeneous catalysis of the water gas shift reaction by rhodium carbonyl iodide, J. Am. Chem. Soc. 102 (1980) 1020–1027.

[11] K. Kaneda, M. Hiraki, K. Sano, T. Imanaka, S. Teranishi, An active catalyst for the water gas shift reaction using a rhodium carbonyl cluster-diamine system, J. Mol. Catal. 9 (1980) 227–230.

[12] T. Yoshida, T. Okano, Y. Ueda, S. Otsuka, Activation of water molecule. 5. Rhodium(I) hydride catalyzed water gas shift reaction. Identification of the elemental reactions comprising the catalytic cycles, J. Am. Chem. Soc. 103 (1981) 3411–3422.

[13] D. Mahajan, C. Creutz, N. Sutin, Homogeneous catalysis of the water gas shift reaction by (po1ypyridine) rhodium complexes, Inorg. Chem. 24 (1985) 2063–2067.
[14] A. J. Pardey, P. C. Ford, Homogeneous catalysis of the water-gas shift reaction by rhodium complexes in aqueous substituted pyridine solutions, J. Mol. Catal. 53 (1989) 247–263.
[15] P. C. Ford, R. G. Rinker, C. Ungermann, R. M. Laine, V. Landis, S. A. Moya, Homogeneous catalysis of the water gas shift reaction by mixed-metal (iron/ruthenium) catalysts, J. Am. Chem. Soc. 100 (1978) 4595–4597.
[16] C. Ungermann, V. Landis, S. A. Moya, H. Cohen, H. Walker, R. G. Pearson, R. G. Rinker, P. C. Ford, Homogeneous catalysis of the water gas shift reaction by ruthenium and other metal carbonyls. Studies in alkaline solutions, J. Am. Chem. Soc. 101 (1979) 5922–5929.
[17] W. A. R. Slegeir, R. S. Sapienza, B. Easterling, Mechanistic aspects of the homogeneous water gas shift reaction, ACS Symp. Ser. 152 (1981) 325–343.
[18] P. Yarrow, H. Cohen, C. Ungermann, D. Vandenberg, P. C. Ford, R. G. Rinker, Homogeneous catalysis of the water-gas shift reaction by ruthenium carbonyl complexes: studies in acidic solutions, J. Mol. Catal. 22 (1983) 239–256.
[19] Y. Doi, S. Tamura, Homogeneous catalysis of the water gas shift reaction using ruthenium chloride in aqueous alkaline solution, Inorg. Chim. Acta 54 (1981) L235–L236.
[20] R. B. King, C. C. Frazier, R. M. Hanes, A. D. King Jr., Active homogeneous catalysts for the water gas shift reaction derived from the simple mononuclear carbonyls of iron, chromium, molybdenum, and tungsten, J. Am. Chem. Soc. 100 (1978) 2925–2927.
[21] A. D. King Jr., R. B. King, D. B. Yang, Homogeneous catalysis of the water gas shift reaction using iron pentacarbonyl, J. Am. Chem. Soc. 102 (1980) 1028–1032.
[22] D. M. Vandenberg, T. Suzuki, P. C. Ford, Homogeneous catalysis of the water gas shift reaction by iridium carbonyl in alkaline solution, J. Organomet. Chem. 272 (1984) 309–320.
[23] A. D. King Jr., R. B. King, D. B. Yang, Homogeneous catalysis of the water gas shift reaction with the use of Group 6 transition metal carbonyls: hexacarbonylchromium, hexacarbonylmolybdenum, and hexacarbonyltungsten, J. Am. Chem. Soc. 103 (1981) 2699–2704.
[24] W. A. R. Slegeir, R. S. Saplenra, R. Rayford, L. Lam, Role of homogeneous formate complexes in the water gas shift reaction catalyzed by the Group VI metal carbonyls, Organometallics 1 (1982) 1728–1730.
[25] F. T. T. Ng, S. K. Tsakiri, Homogeneous catalysis of water-gas shift reaction in emulsions, Fuel 72 (1993) 211–215.
[26] T. Yoshida, Y. Leda, S. Otsuka, Activation of water molecule. 1. Intermediates bearing on the water gas shift reaction catalyzed by platinum(0) complexes, J. Am. Chem. Soc. 100 (1978) 3941–3942.
[27] C. H. Cheng, R. Eisenberg, Homogeneous catalysis of the water gas shift reaction using a platinum chloride-tin chloride system, J. Am. Chem. Soc. 100 (1978) 5968–5970.
[28] A. A. Frew, R. H. Hill, L. ManojloviC-Muir, K. W. Muir, R. J. Puddephatt, Dimerisation of a diplatinum to a tetraplatinum complex during catalysis of the water gas shift reaction: the X-ray crystal structure of $[Pt_4(\mu_2\text{-}CO)_2(\mu_2\text{-}h_2PCH_2PPh_2)_3\{Ph_2PCH_2P(:O)Ph_2\}]$, J. Chem. Soc. Chem. Commun. (1982) 198–199.
[29] A. J. Deeming, B. L. Shaw, Oxidative addition of alkyl or aryl chloroformates to the complexes $[IrCl(CO)L_2]$(L = dimethylphenyl-phoshine or -arsine, or trimethylphosphine). Iridium–carboxy- complexes, J. Chem. Soc. A. (1969) 443–448.
[30] J. R. Sweet, W. A. G. Graham, Hydrido and hydroxycarbonyl compounds of the carbonyl (.eta.-clopentadienyl)nitrosylrhenium group, Organometallics 1 (1982) 982–986.

[31] R. A. Grey, G. P. Pez, A. Wallo, Anionic metal hydride catalysts. 2. Application to the hydrogenation of ketones, aldehydes, carboxylic acid esters, and nitriles, J. Am. Chem. Soc. 103 (1981) 7536–7542.

[32] D. J. Darensbourg, C. S. Day, M. B. Fischer, Crystal and molecular structure of dicarbonyl(.eta.5-cyclopentadienyl)(formato)iron(II), Inorg. Chem. 20 (1981) 3577–3579.

[33] D. J. Darensbourg, A. Rokicki, M. Darensbourg, Facile reduction of carbon dioxide by anionic Group 6b metal hydrides. Chemistry relevant to catalysis of the water-gas shift reaction, J. Am. Chem. Soc. 103 (1981) 3223–3224.

[34] R. Eisenberg, D. E. Hendriksen, The binding and activation of carbon monoxide, carbon dioxide, and nitric oxide and their homogeneously catalyzed reactions, Adv. Catal. 28 (1979) 79–172.

[35] D. C. Gross, P. C. Ford, Nucleophilic activation of coordinated carbon monoxide. 3. Hydroxide and methoxide reactions with the trinuclear clusters $M_3(CO)_{12}$ (M = Fe, Ru, or Os). Implications with regard to catalysis of the water gas shift reaction, J. Am. Chem. Soc. 107 (1985) 585–586.

[36] W. J. Thomson, R. M. Laine, Homogeneous catalytic reduction of benzaldehyde with carbon monoxide and water applications of the water gas shift reaction, ACS Symp. Ser. 152 (1980) 133–145.

[37] J. C. Bricker, C. C. Nagel, S. G. Schore, Reactivities of ruthenium cluster anions: implications for catalysis of the water-gas shift reaction, J. Am. Chem. Soc. 104 (1982) 1444–1445.

[38] Y. Shvo, Y. Rahamim, D. Czerki, in: 12th International Conference on Organometallic Chemistry, Vienna, September 1985, Paper 0339.

[39] R. M. Laine, Applications of the water-gas shift reaction. Hydroformylation and hydrohydroxymethylation with carbon monoxide and water, J. Am. Chem. Soc. 100 (1978) 6451–6454.

[40] T.C. Singleton, U.S. Patent 4,151,107, 1979.

[41] D. Forster, T. C. Singleton, Homogeneous catalytic reactions of methanol with carbon monoxide, J. Mol. Catal. 17 (1982) 299–314.

[42] Q. A. Likholobov, V. N. Zudin, I. Yermakov, in: Proceedings of the Fifth Japan-Soviet Seminar on Catalysis, Osaka, Japan, 1980, pp. 9–18.

[43] F. Cariati, R. Ugo, F. Bonati, Reactions of inorganic acids with zerovalent platinum, palladium, and nickel compounds having triphenylphosphine or 1,2-bis(diphenylphosphino)ethane as ligands, Inorg. Chem. 5 (1966) 1128–1132.

[44] R. Kaspar, A. Spogliarch, M. G. Cernogoraz, Homogeneous catalytic reduction of benzalacetone with carbon monoxide and water, J. Organomet. Chem. 255 (1983) 371–376.

[45] C. P. Kubiak, C. Woodcock, R. Eisenberg, Rhodium A-frame complexes with small-molecular bridgeheads. Synthesis and structure of $[Rh_2(.mu.-H)(.mu.-CO)(CO)_2(PPh_2CH_2PPh_2)_2]$ $(p-CH_3C_6H_4SO_3)$.2THF and catalytic properties, Inorg. Chem. 21 (1982) 2119–2129.

[46] B. R. Sutherland, M. Cowie, Hydroxy- and hydrido-bridged binuclear complexes of iridium: synthesis, characterization, and attempts to model binuclear water-gas shift catalysts. Structure of $[Ir_2(CO)_2(.mu.-H.Cl)(Ph_2PCH_2PPh_2)_2]$, Organometallics 4 (1985) 1637–1648.

[47] J. R. Fisher, A. J. Mills, S. Sumner, M. P. Brown, M. A. Thomson, R. J. Puddephatt, A. A. Frew, L. Manojovic-Muir, K. W. Muir, Reversible displacement of dihydrogen by carbon monoxide in binuclear platinum complexes. Characterization of binuclear carbonyl complexes of platinum(I), Organometallics 1 (1982) 1421–1429.

Chapter 8

Photo-Catalytic Water-Gas Shift Reaction

8.1 PHOTOCHEMICAL REACTION

Photochemical reaction is a chemical reaction initiated by the absorption of energy in the form of light. The consequence of molecules' absorbing light is the creation of transient excited states whose chemical and physical properties differ greatly from the original molecules. Excited states are stronger acids and stronger reductants than the original ground states. Photochemical reactions and the properties of excited states are also critical in many commercial processes and devices. Both photography and xerography are based upon photochemical processes, while the manufacture of semiconductor chips or the preparation of masks for printing newspapers relies on UV light to destroy molecules in selected regions of polymer masks.

8.2 PHOTO-CATALYTIC REACTION

Photo-catalysis is the acceleration of a photochemical reaction in the presence of a catalyst. In photo-catalysis, light is absorbed by the catalyst. In photo-generated catalysis, the photo-catalytic activity depends on the ability of the catalyst to create electron-hole pairs, which generate free radicals (e.g. hydroxyl radicals: •OH) able to undergo secondary reactions. Its practical application was made possible by the discovery of water electrolysis by means of titanium dioxide.

8.3 PHOTO-WATER-GAS SHIFT REACTION

Production of hydrogen in the presence of UV or visible light by using water-gas shift (WGS) reaction is known as photo-WGS reaction. In the photo-WGS reaction we use light as an energy source instead of heat.

8.4 Pt/TiO_2

In 1980, Sato and White [1] reported the production of hydrogen via photo-WGS reaction. They began to investigate Pt/TiO_2 catalysts for various photo-catalytic applications. They found that UV illuminated Pt/TiO_2 catalyst

Water Gas Shift Reaction. http://dx.doi.org/10.1016/B978-0-12-420154-5.00008-5

decomposes liquid water. However, pure titania did not exhibit any activity. Then they investigated these catalysts for reactions of gaseous H_2O with hydrocarbons, activated carbon, lignite and CO to produce H_2 and CO_2. Photo-WGS reaction is thermodynamically unfavourable on Pt/TiO_2 and the reaction involves photo-decomposition of water. They observed that Pt/TiO_2 exhibited three times higher hydrogen yield compared to titania alone. ^{18}O isotopic labelled studies revealed that a part of the oxygen in the product CO_2 comes from the H_2O. They studied the effect of CO pressure on Pt/TiO_2 catalyst; rate of the photo-assisted WGS reaction is almost constant until the pressure of CO falls to about 0.3 Torr. The reaction accelerates slightly as the CO pressure drops below 0.3 Torr. A slight increase in the initial rate (10-20%) was observed when the H_2O pressure was reduced from 24 Torr to about 5 Torr, while the initial CO pressure was fixed at about 0.6 Torr. They tested the wavelength dependence by using three filters. The 415-nm cutoff filter completely retarded the reaction, while the Plexiglass and Pyrex filters suppressed the rate to 4% and 67%, respectively. The Pt/TiO_2 prepared from un-doped TiO_2 (no H_2 treatment) showed much lower (10%) photo-catalytic activity than the Pt/H_2-doped TiO_2 for the WGS and the water decomposition reaction.

In the photo-assisted WGS reaction up on irradiation and excitation of electron into the conduction band of TiO_2, charge separation will compete with hole-electron recombination. Holes will tend to migrate to the Pt-TiO_2, interface, and the electrons will migrate away from it. When a photo-stationary state is reached, the bending of the valence and conduction bands in the semiconductor will be reduced and the electrons and holes can be characterized by quasi-Fermi levels'. Photo-catalyzed process is described in Figure 8.1. In this mechanism, holes come to the TiO_2-gas interface where they react with adsorbed water to form an adsorbed OH radical and a proton. Then CO reacts with the OH radical to give CO_2 and a hydrogen atom which recombines with other hydrogen atoms at Pt sites.

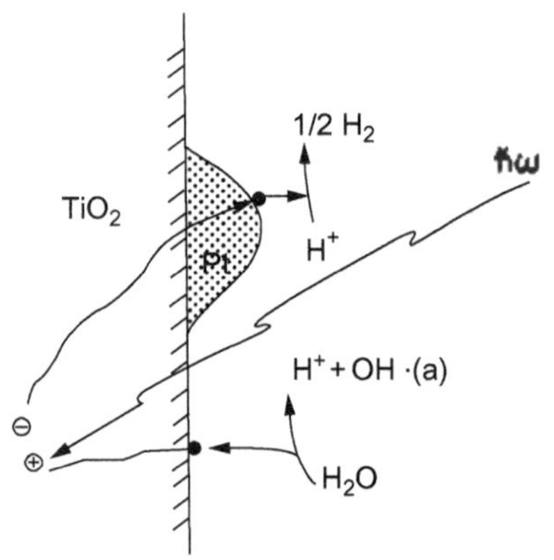

FIGURE 8.1 Pictorial model of the photo-induced redox processes occurring at the Pt/TiO_2 surface. *(Taken from Figure 8 of S. Sato, J.M. White, J. Am. Chem. Soc. 102 (1980) 7206.)*

Based on these observations they proposed the reaction mechanism

$$p^+ + H_2O \rightarrow OH_{(a)} + H^+ \tag{8.1}$$

$$p^+ + OH_{(a)} \rightarrow O_{(a)} + H^+ \tag{8.2}$$

$$CO_{(g)} \rightarrow CO_{(a)} \tag{8.3}$$

$$CO_{(a)} + OH_{(a)} \rightarrow CO_{2(g)} + H_{(a)} \tag{8.4}$$

$$CO_{(a)} + O_{(a)} \rightarrow CO_{2(g)} \tag{8.5}$$

$$H^+ + e^- \rightarrow H_{(a)} \tag{8.6}$$

$$2H_{(a)} \rightarrow H_{2(g)} \tag{8.7}$$

Then, they prepared NaOH-coated Pt/TiO_2 [2] catalysts for the photo-WGS reaction. When Pt/TiO_2 was coated with NaOH, the photo-catalytic activity for WGS reaction increased significantly. 4% coated NaOH increases the rate of reaction by the factor of 2.5. At the beginning of the reaction, CO_2 formation is very slow because of the reaction of CO_2 with NaOH. As the reaction proceeds, CO_2 formation becomes faster. When the CO pressure falls down to about 0.05 Torr, the consumption of CO, as well as the formation of H_2, becomes slower presumably because the diffusion of CO through the water layer on the catalyst becomes important. The photo-catalytic WGS activity increases with increasing NaOH loading up to 10 wt%. At the beginning of the reaction the H_2 formation was very fast; however, it declined with time.

Then Fang *et al.* [3] reported the effect of synthesis method on the WGS activity of Pt/TiO_2 catalysts. They synthesized 2 and 10 wt% Pt/TiO_2 catalysts by photo-decomposition and impregnation methods. Then they impregnated NaOH over Pt/TiO_2 catalyst. WGS activity of various Pt/TiO_2 catalysts is presented in Table 8.1. Influence of H_2 reduction on various Pt/TiO_2 catalysts is presented in Figure 8.2. Table 8.1 shows that addition of NaOH to the Pt/TiO_2 catalyst increases the WGS activity by about 2 orders of magnitude in the absence of H_2 reduction (catalysts 2 and 4). However, when the catalysts are reduced in the presence of H_2 at 875 °C the H_2 production increased from 0.04 to 9 μmol/g (catalysts 1 and 3). Increase of H_2 reduction temperature from 700 to 875 °C did not alter the rate that much (catalysts 3 and 7). Significant amount of rutile titania was observed when catalyst reduced at 875 °C and only anatase was observed when catalyst reduced at 700 °C. These results suggest that titania phase did not influence the WGS activity. By comparing the photo-decomposition method with the impregnation method with and without NaOH WGS activity measurements suggest that H_2 production rates are independent of synthesis method. Table 8.1 shows that increase in the Pt wt% from 2 to 10 has little effect on the WGS rate. The rate of WGS reaction increases from 0.5 to 8 to 19 μmol/(h g) with increasing NaOH loading from 0% to 5.3%. Fang *et al.* also reported that the rate of WGS reaction is also dependent on the light

TABLE 8.1 Catalyst Preparation Parameters and Performance for Photo-assisted WGS Reaction

No.	Catalyst	Pt (wt%)	NaOH (wt%)	Prep. method	Doping temp. (°C)	H_2 production[a] (μmol/(g h))
1	NaOH/ Pt/TiO_2	2	3.5	I[b]	875	9
2	NaOH/ Pt/TiO_2	2	3.5	I		3
3	Pt/TiO_2	2	0	I	875	0.2
4	Pt/TiO_2	2	0	I		0.04
5	NaOH/ Pt/TiO_2	2	3.5	I	700	9
6	NaOH/ Pt/TiO_2	2	3.5	P[c]	700	9
7	Pt/TiO_2	2	0	I	700	0.4
8	Pt/TiO_2	2	0	P	700	0.6
9	NaOH/ Pt/TiO_2	10	3.5	P	700	9
10	NaOH/ Pt/TiO_2	2	3.5		700	8
11	NaOH/ Pt/TiO_2	Tr[d]	3.5		700	8[e]
12	NaOH/ TiO_2	0	3.5		700	0.4
13	NaOH/ Pt/TiO_2	2	5.3	P	700	19
14	Pt/TiO_2	2	0	P	700	0.5

[a] *The flux of photons with energy greater than the band gap (3.1 eV) of anatase is about 10^{10} cm^{-2} g^{-1}. The rate quoted was measured 50 and 75 min by using data shown in Figure 1 of S.M. Fang, B.H. Chen, J.M. White, J. Phys. Chem. 86 (1982) 3126.*
[b] *Impregnation method.*
[c] *Photo-decomposition method.*
[d] *A small amount of residual Pt from the previous experiment remained after cleaning the reactor ultrasonically with water.*
[e] *Initial rate.*
Taken from Table 1 of S.M. Fang, B.H. Chen, J.M. White, J. Phys. Chem. 86 (1982) 3126.

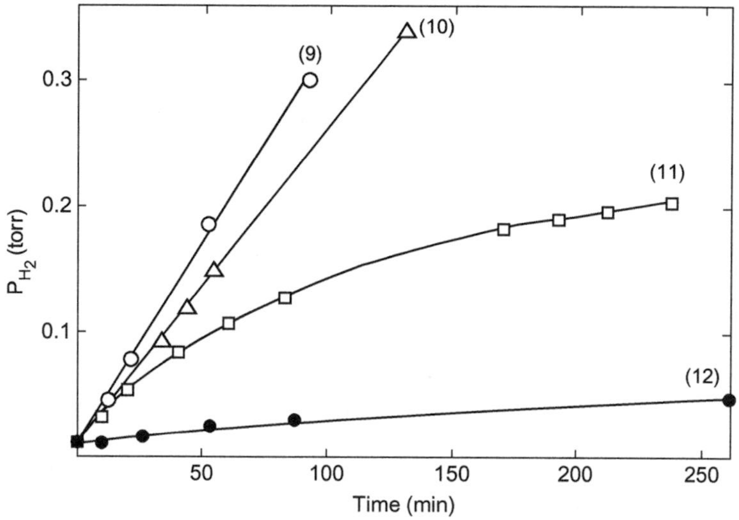

FIGURE 8.2 Effect of Pt loading on 3.5 wt% NaOH/(700 °C reduced) TiO_2: (○) 10 wt%, (△) 2 wt%, (□) trace of Pt, (●) Pt free. *(Taken from Figure 3 of S.M. Fang, B.H. Chen, J.M. White, J. Phys. Chem. 86 (1982) 3126.)*

intensity. The WGS rate is 0.5 μmol/(h g) in the presence of dark and it is 19 μmol/(h g) under illumination of full arc.

The reduction of Pt/TiO_2 with hydrogen removes lattice oxygen from leaving defects and Ti^{3+} ions. This process increases the conductivity of pellets by making n-type conductor. The authors reported that strong metal support interaction has no influence on the WGS activity. However, the addition of 3.5 wt% NaOH increases the WGS rate by 16 times.

In the same year, Yixuan *et al.* [4] also reported Pt/TiO_2 catalysts for photo-assisted WGS reaction. They synthesized 1.5% Pt/TiO_2 by photo-deposition method and reduced the catalysts in different environments and evaluated for photo-WGS reaction in gas phase and liquid phase. They prepared three types of catalysts and the details are presented in Table 8.2.

They evaluated B54 catalyst for the gas phase photo-WGS reaction. Increasing reaction temperature from 30 to 50 °C increases the H_2 evolution from 115 to 540 μL/h. Also increasing hydrogen pre-treatment temperature from 500 to 700 °C increases the H_2 evolution from 70 to 115 μL/h. The reactions in liquid phase also support this observation. Also, the catalytic activity improves in the presence of 100% illuminated light compared to the dark conditions.

B54 and B501 catalysts treated at temperatures higher than 200 °C also exhibit very good activity in the absence of light. These results suggest that light illumination is not the important factor for the WGS activity. They also evaluated B54 and TC 26 catalysts for liquid phase WGS reaction under illumination as well as in dark. As expected the catalysts exhibit higher activity under

TABLE 8.2 Series of Pt-TiO_2 Catalyst Samples

Sample series	Pre-treatment of support TiO_2	Post-treatment in H_2 of catalyst Pt-TiO_2
B501	Air, 500 °C, 2 h	No treatment, 200, 500, 700 °C
TC26	H_2, 500 °C, 2 h	No treatment, 500 °C
B54	H_2, 700 °C, 2 h	No treatment, 150, 200, 400, 500, 600, 700 °C

Taken from Table 1 of C. Yixuan, W. Zhaobin, C. Yanxin, L. Huaxin, H. Zupei, L. Huiqing, D. Yonglei, Y. Chunying, L. Wenzhao, J. Mol. Catal. 21 (1983) 275.

illumination compared to dark. Also, increasing hydrogen pre-treatment temperature increases the activity. On the whole, catalytic activity results show that high temperature hydrogen pre-treatment is necessary for the enhancement of photo-catalytic activity of Pt/TiO_2. They also proposed mechanism based on the photo-decomposition of water.

$$\begin{aligned}
&\mathrm{TiO_2} \xrightarrow{h\nu} \oplus + \ominus \\
&\mathrm{2H_2O} + 2\ominus \rightarrow \mathrm{2H_{(a)}} + \mathrm{2OH^-_{(a)}} \\
&\mathrm{2H_{(a)}} \rightarrow \mathrm{H_2} \\
&\mathrm{CO} + \mathrm{2OH^-_{(a)}} + 2\oplus \rightarrow \mathrm{CO_2} + \mathrm{H_2O}
\end{aligned}$$

Then Tsai *et al.* [5] reported photo-WGS reaction over Pt/TiO_2 (100) crystal surface. They used unreduced and reduced TiO_2 surfaces to investigate the effect of reduction environment. Pt was deposited on the TiO_2 surface using a resistively heated tungsten filament evaporator. The experimental setup is shown in Figure 8.3. Figure 8.4 shows the hydrogen evolution over reduced Pt/TiO_2 (100) catalyst number average of 1 × 1015 atoms at 363 K in the presence of 18 Torr of water vapour and 0.3 Torr of CO with a UV input power of 250 W. Hydrogen is produced with an initial rate of 3.6×10^{15} and then the reaction rate decreases rapidly with time and a constant value of 0.9×10^{15} is attained after 40-50 min of reaction. The hydrogen production rate increases with increasing Pt loading and attains a maximum at a total Pt coverage of 6×10^{14} and declines thereafter. Water and CO pressures have no effect on the hydrogen reaction rate.

Based on the catalytic results and literature reports they proposed mechanism for the photo-catalytic reaction. During reduction the TiO_2 reduced to Ti_2O_3 and forms an oxygen vacancy.

$$\begin{aligned}
&\mathrm{2TiO_{2(s)}} \rightarrow \mathrm{2Ti_2O_{3(s)}} + \frac{1}{2}\mathrm{O_{2(g)}} \\
&\Delta G^{\mathrm{o}} = 61.6\mathrm{kcal}
\end{aligned}$$

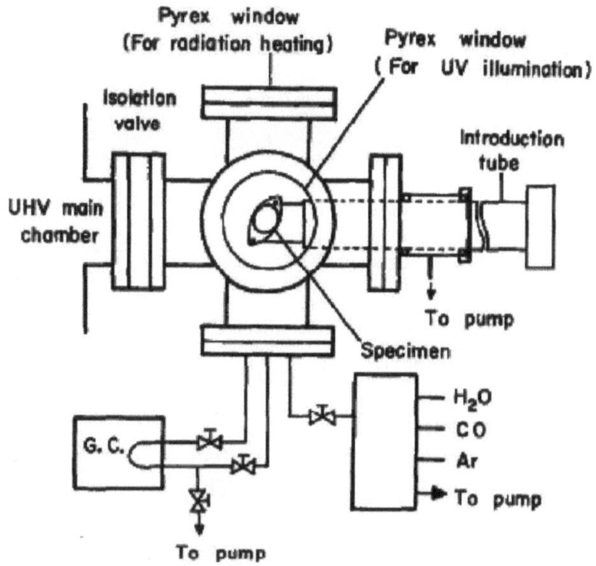

FIGURE 8.3 A schematic diagram of the experimental setup for performing photo and thermal WGS reactions. *(Taken from Figure 1 of S.C. Tsai, C.C. Kao, Y.W. Chung, J. Catal. 79 (1983) 451.)*

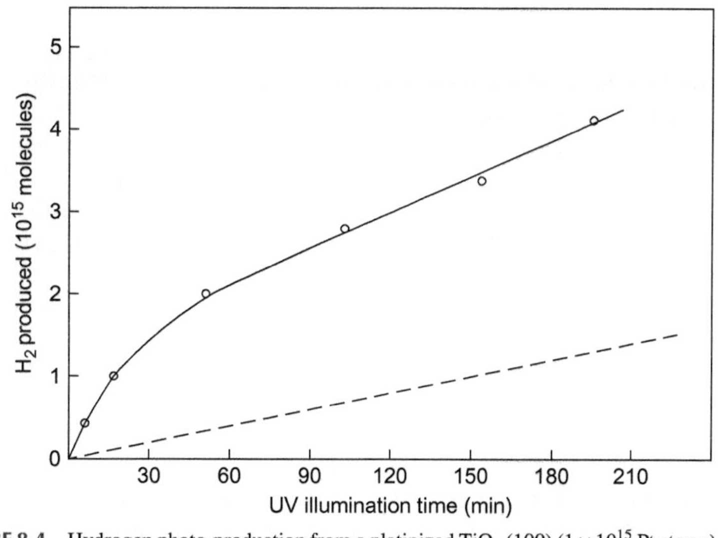

FIGURE 8.4 Hydrogen photo-production from a platinized TiO_2 (100) (1×10^{15} Pt atoms) crystal under UV illumination at 363 K in the presence of 18 Torr of water vapour and 0.3 Torr of CO. Background contribution (dashed line) is also included. *(Taken from Figure 1 of S.C. Tsai, C.C. Kao, Y.W. Chung, J. Catal. 79 (1983) 451.)*

The deposition of Pr over the surface of TiO_2 enhances the reduction of TiO_2. Based on these facts the reaction mechanism would be

$$hv \rightarrow h + e \tag{8.8}$$

$$Pt/Ti^{4+} - O^{2-} + 2h + CO_{ads} \xrightarrow{k} Pt/Ti^{4+} - \Box s + CO_{2(g)} \tag{8.9}$$

$$Pt/Ti^{4+} - \Box s + H_2O_{ads} + 2e \rightarrow Pt/Ti^{4+} - O^{2-} + H_{2(g)} \tag{8.10}$$

Here □ denotes oxygen vacancy on the surface, and Pt/Ti^{4+}-O^{2-} represents the Pt-TiO_2 periphery site.

And the rate of hydrogen production was given by

$$r = k\left[Pt/Ti^{4+} - O^{2-}\right]p\left[Co_{ads}\right]$$

Here p is the hole concentration.

Then Tsai and Chung [6] reported the effect of TiO_2 particle size on the photo-catalytic activity of Pt/TiO_2 catalysts. They prepared the TiO_2 powders with two different sizes 800 and 150 Å. They deposited the Pt by using the photo-deposition method. Before the Pt deposition the catalysts were reduced at 600 °C in the presence of hydrogen. The scanning electron microscopy measurements show that there is no significant sintering after reduction treatment. The catalyst with 800 Å particle size exhibits a photo-reaction rate of 1240×10^{15} and the catalysts with 150 Å particle size exhibits photo-reaction rate of 50×10^{15}.

After that not much research happened on the photo-catalytic WGS reaction because the amount of hydrogen produced is very less compared to the other processes like steam reforming and traditional WGS reaction. Also, the process is not a continuous process.

However, in 2002 Millard and Bowker [7] reported Pd/TiO_2 catalysts for photo-WGS reaction. They prepared 0.01, 0.1, 0.5, 1 and 2 wt% Pd/TiO_2 catalysts and evaluated for photo-WGS reaction. Initially they investigated the effect of CO concentration of after gas shift activity of 0.5 wt% Pd/TiO_2 catalyst. The rate of WGS reaction increases with increasing CO concentration from 0.1 to 1 mL and further increases in CO concentration results in decrease in the activity and reaches 0 at 10 mL of CO.

Then they investigated the effect of Pd loading on WGS activity of Pd/TiO_2 catalysts by using 1 mL CO. The results are presented in Figure 8.5.

The rate of photo-WGS reaction increases with increasing Pd loading up to 0.5 wt% and further increase in the Pd loading decreases the WGS activity. Effectively there is no reaction for the 2 wt% Pd catalyst. This reactivity pattern can be related to the boundary of palladium, as reactivity is postulated by the proposed mechanism to take place at the titania-palladium interface. Boundary length would increase with particle size, which would increase with weight loading. It would reach a maximum rate when the palladium particles are big enough to almost touch, where a maximum perimeter is reached. Any greater

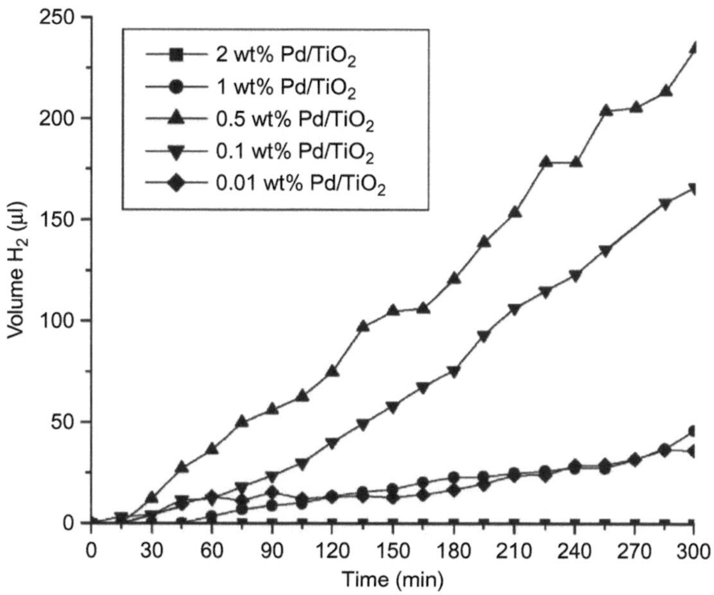

FIGURE 8.5 The effect of palladium loading on the CO reaction. *(Taken from L. Millard, M. Bowker, J. Photochem. Photobiol. A Chem. 148 (2002) 91.)*

loading would decrease the rate of hydrogen evolution as the perimeter decreases due to agglomeration of particles.

Then they investigated the effect of oxygen addition on the WGS activity of 0.5% Pd/TiO_2 catalyst. As shown in Figure 8.6 the WGS activity decreases when they injected 0.2 mL of O_2 into the reaction chamber.

In 2013 Sastre *et al.* [8] reported gas phase photo-catalytic WGS reaction over various catalysts for the first time. They used TiO_2 and CeO_2 as supports and Pt, Pd, Cu, Au and Ag as metals. The Au/TiO_2 and Au/CeO_2 catalysts were prepared by deposition-precipitation method. Pt-, Pd- and Au-supported catalysts were synthesized by impregnation method. The BET surface area, nano-particle size and metal wt% of various catalysts.

Table 8.3 shows photo-catalytic activity for hydrogen evolution and product distribution after 22 h irradiation with simulated sunlight (100 mW/cm^2). No activity is observed in the absence of light. As shown in Table 8.3, although both TiO_2 and CeO_2 supports show a low activity for photo-catalytic version of WGS the presence of noble metals considerably increases their photo-WGS activity. In particular, Pd and Pt NPs at 1% loading increase H_2 formation by one order of magnitude with respect to the photo-catalytic activity of P25. Similarly the presence of Au enhances considerably the photo-catalytic activity of TiO_2. In the case of CeO_2 we found that the optimal loading of supported Au was about 0.5 wt%. As can be seen in Table 8.3, the most active photo-catalyst tested was Au/TiO_2 which achieves CO conversion of about 40% in 4 h under

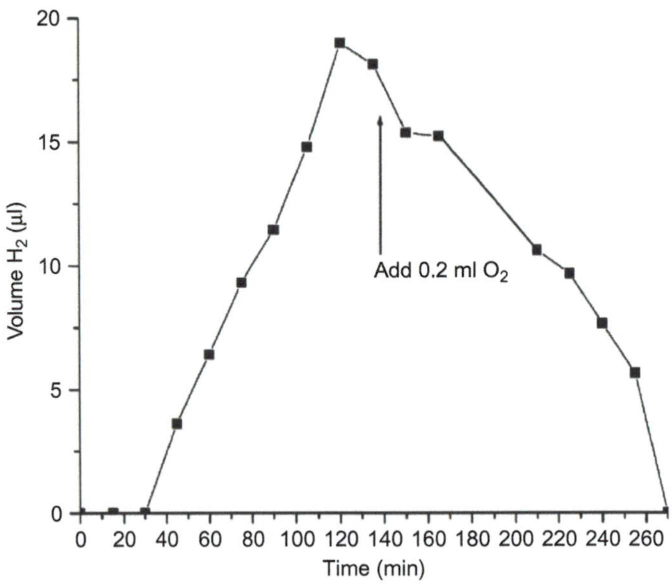

FIGURE 8.6 The effect of injecting oxygen into the carbon monoxide reaction. *(Taken from L. Millard, M. Bowker, J. Photochem. Photobiol. A Chem. 148 (2002) 91.)*

TABLE 8.3 Photo-catalytic Activity for Hydrogen Evolution and Product Distribution after 22 h Irradiation with Concentrated Sunlight

Photo-catalyst	CO conversion (%)	Molar balance	Production (μmol/g catalyst)		
			H_2	CO_2	CH_4
TiO_2	3	99.99	264	649	4
Pd/TiO_2 (1%)	24	99.11	4178	4385	1
Cu/TiO_2 (1%)	4	99.99	591	951	7
Pd/TiO_2 (1%)	25	99.27	4279	4763	64
Au/TiO_2 (1%)	71	95.25	10,506	13,447	7
Ag/TiO_2 (1%)	10	99.11	1650	1833	8
CeO_2	4	99.98	350	607	5
Au/CeO_2 (0.2%)	8	98.37	1380	1526	15
Au/CeO_2 (0.5%)	28	92.14	4880	4762	2
Au/CeO_2 (1%)	31	96.36	1358	1341	<1

Irradiation Condition: Mixture of Moisture Saturated N_2 (80%) and CO (20% 4.9 mmol), Irradiation Time: 22 h, Photo-catalyst Mass: 250 mg, Irradiation Source: Solar Simulator (1000 mW/cm^2).
Taken from F. Sastre, M. Oteri, A. Corma, H. Garcia, Energy Environ. Sci. 6 (2013) 2211.

our reaction conditions. Longer irradiation times lead to higher conversions of up to 71% in 22 h and lower light fluencies lead to lower conversions.

Reports are also available on homogeneous photo-catalytic WGS reaction. In 1980 Hamilton [9] first ever reported the photo-catalyst. The complex [RuCl (CO)(bipy)$_2$]CI, a product of the long-term reaction of $RuC1_3 \cdot 3H_2O$ with 2 mol of bipyridyl (bipy) in refluxing dimethylformamide, produces H_2, CO and CO_2 on photolysis in water; under CO the photochemical WGS reaction is catalytic. Irradiation of aqueous solutions of [RuCl(CO)(bipy)$_2$]Cl with visible light at 100 °C in the presence or absence of air produces hydrogen, but no oxygen. Since CO and CO are also products, this is clearly a photochemical WGS reaction, rather than a straightforward photo-dissociation of water. The final ruthenium product is [RuCl(H_2O) (bipy)$_2$]$^+$. Under carbon monoxide (1 atmos), the reaction is catalytic and, although slow, appears to proceed without loss of activity for several days. He also proposed mechanism based on the catalytic results. The mechanism is presented in Scheme 8.1.

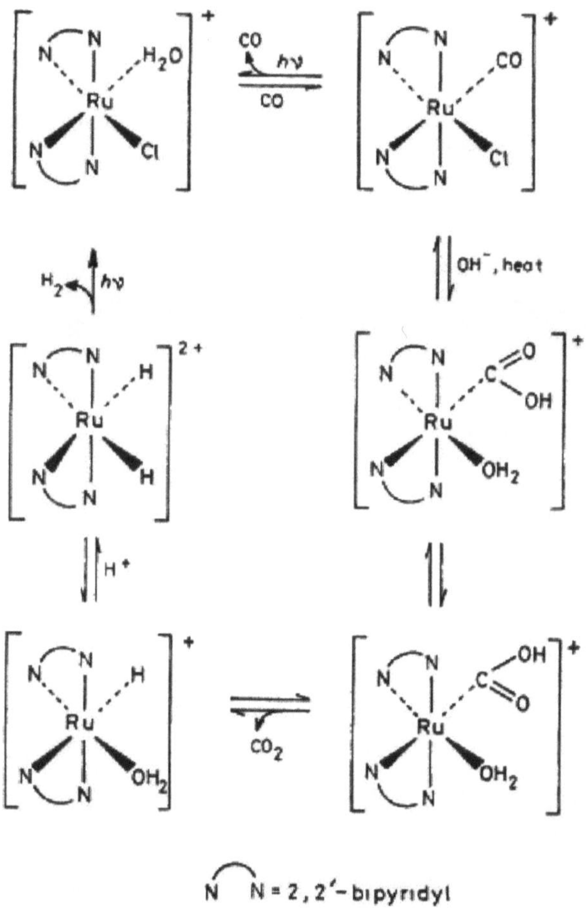

SCHEME 8.1 The mechanism of WGS reaction over Ru complexes. ***(Taken from D.J.C. Hamilton, J. Chem. Soc. Chem. Comm. (1980) 1213.)***

8.5 MECHANISM

In 1993 Ziessel [10] reported detailed mechanistic studies of homogeneous photochemical WGS reaction catalyzed by iridium(111) complexes. He prepared several Ir cationic complexes using standard Schlenk or vacuum-line techniques and are shown in Scheme 8.2.

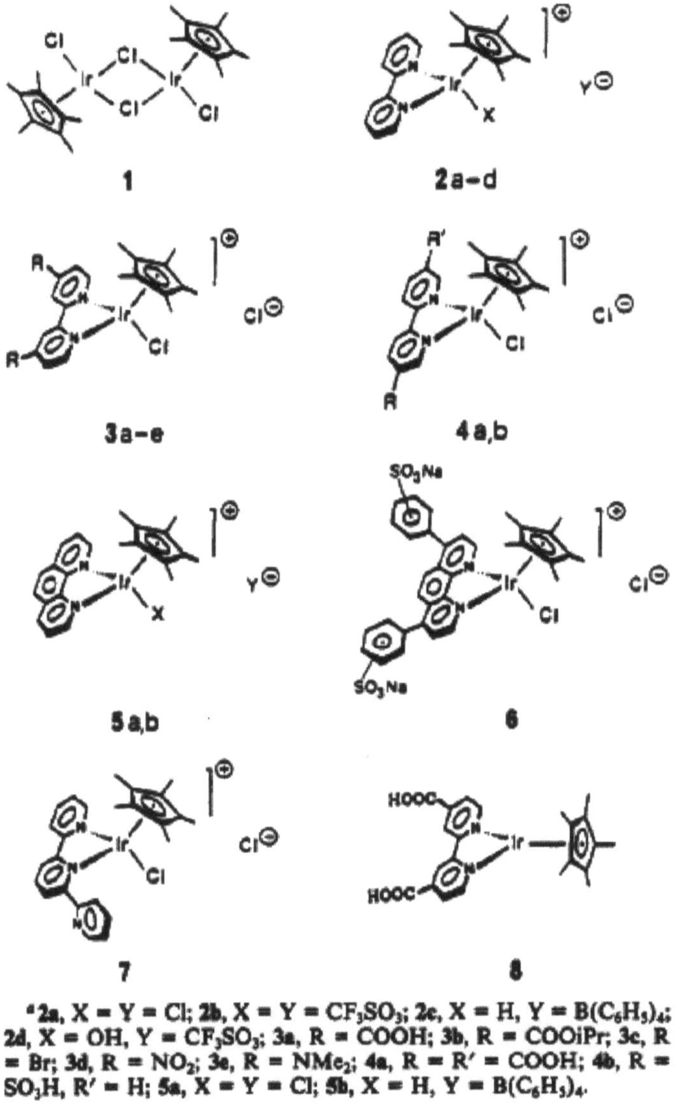

SCHEME 8.2 (2a) X=Y=Cl; (2b) X=Y=CF_3SO_3; (2c) X=H, Y=$B(C_6H_5)_4$; (2d), X=OH, Y=CH_3SO_3; (3a) R=COOH; (3b) R=COOiPr; (3c) R=Br; (3d) R=NO_2; (3e) R=NMe_2; (4a) R=R′=COOH; (4b) R=SO_3H, R′=H; (5a) X=Y=Cl; (5b) X=H, Y=$B(C_6H_5)_4$. *(Taken from R. Ziessel, J. Am. Chem. Soc. 115 (1993) 118.)*

General Procedure for the WGS Reaction Studies

The WGS reaction was carried out either in a 30-mL thermostated Schlenk flask containing 10 mL of a 0.01 M phosphate buffer (pH 2-13) or in a 50-mL nonthermostated Schlenk flask (room temperature) containing 15 mL of a 0.01 M phosphate buffer and complexes **1-8**. Irradiation was conducted with a 250-W halogen lamp (slide projector) focused on the solution.

The results obtained for some typical experiments using complexes **1-8** are given in Table 8.4. No system produced significant amounts of H_2 in the dark or in the absence of CO. All of the complexes show some catalytic activity, analyzed as the amount of H_2 and CO_2 produced, with varying efficiencies, when irradiated in the presence of CO. The catalytic activity of all complexes was strongly dependent on pH with an optimum activity lying between pH **5 and 7**.

The rate of H_2 formation increases with increasing catalyst concentration. Between **0.045** and **0.2** mM concentration, the reaction showed a first-order rate dependence. Above **0.2** mM a saturation effect was observed, due to a

TABLE 8.4 Photochemical WGSR catalyzed by Ir complexes

Expt no.[a]	Complex, conditions	pH[b]	V_{H_2} (mL)	V_{CO_2}[c] (mL)	Overall turnover[d]
1	**1**	7.0			
2	**2a, 2c, 3a**, dark[e]	7.0			
3	**2a**	5.0	0.6	0.6	3
		6.0	0.9	0.3	5
		7.0	1.1	0.2 (1.1)	6
		8.0	0.9	Traces	5
		9.0	0.6	Traces	3
		10.0	0.2		1
4	**2a**, 25 equiv bpy[f]	7.0	1.4	0.3 (1.4)	8
		8.0	1.2	Traces	6
		10.0	0.3		1
3	**2b**	7.0	1.1	0.2	6
6	**2c**[g]	7.0	0.4	0.1	3
7	**2d**	7.0	1.1	0.2	6

Continued

TABLE 8.4 **Photochemical WGSR catalyzed by Ir complexes—Cont'd**

Expt no.	Complex, conditions	pH	V_{H_2} (mL)	V_{CO_2} (mL)	Overall turnover
8	**3a**	5.0	3.3	3.2	6
		6.0	7.2	2.0 (7.1)	38
		7.0	12.3	2.5 (12.2)	58
		9.0	11.0	Traces	52
		10.0	9.1		43
		11.0	7.6		36
9	**3b**	2.0	1.5	1.5	8
		5.0	4.2	1.5 (4.1)	22
		7.0			
10	**3c**	2.6	0.3	0.2	1
		5.0	1.2	1.1	6
		7.0	0.6	0.1	3
11	**3d**[h]	5.0-9.0	Traces		
12	**3e**	5.0			
		7.0	0.5	0.1	2
		9.0	1.2	Traces	6
13	**4a**	5.0	3.9	3.8	19
		7.0	10.1	2.0 (9.8)	48
		9.0	12.9	Traces	61
14	**4b**	5.0	4.3	4.2	20
		7.0	11.1	2.2 (10.6)	52
		9.0	11.3	Traces	53
15	**5a**[i]	7.0	1.7	0.3 (1.6)	10
	5a, 15 equiv phen[i]	7.0	2.2	0.4 (2.1)	13
16	**5b**[g]	7.0	0.5	0.1	4
17	**6**	2.0	0.8	0.8	6
		5.0	2.5	2.4	20
		7.0	3.6	0.7 (3.4)	29
		9.0	2.2	Traces	15
		11.0	0.7		5

TABLE 8.4 Photochemical WGSR catalyzed by Ir complexes—Cont'd

Expt no.	Complex, conditions	pH	V_{H_2} (mL)	V_{CO_2} (mL)	Overall turnover
18	**7**	2.0-7.0	Traces		
19	**8**	7.0	7.9	1.6 (7.8)	37

[a]*At room temperature and various pH. 2-h irradiation with white light (λ > 400 nm) of 15-mL solution containing 5-5.5 × 10^{-4} M of complex. Each experiment has been performed twice and showed good reproducibility with an error of ±5%.*
[b]*Adjusted using a 0.01 M phosphate buffer.*
[c]*Because of the* $CO_2 \cdot H_2O/HCO_3^-$, and HCO_3^-/CO_3^{2-} equilibrium the stoichiometric amount of CO_2 was only obtained with a pH value lower or equal to 3. The values given in brackets were obtained by prior acidification of the solution with HCl (1 M).
[d]*Calculated by dividing the number of mol of H_2 obtained by the number of mol of the starting complex.*
[e]*Dark reaction with a 15 h contact time.*
[f]*Compared to the starting Ir^{III} complex and added as a CH_3CN (0.5 mL) solution.*
[g]*7.5 mL of 0.01 M phosphate buffer and 7.5 mL of CH_3CN.*
[h]*Decoloration of the solution after a few minutes irradiation.*
[i]*A yellow product deposited on the wall of the flask during the irradiation.*
Taken from R. Ziessel, J. Am. Chem. Soc. 115 (1993) 118.

photofiltering effect of the complex on the photoactivation process. The photochemical WGSR has also been studied as a function of time using complex **3a**. The activity studies showed H_2 generation to be nonlinear with respect to time when CO pressure was not adjusted every hour.

They also investigated the effect of ligand substitution on the WGS activity of Ir complexes. The effects of changing, (i) the monodentate ligand (active catalytic site), (ii) the chelate ligand (photosensitive effect), (iii) the nature of substituent groups of the bidentate ligand (electronic effect), (iv) the substitution position of the electron withdrawing groups on the 2,2′-bipyridine ligand, were studied. Replacement of the chloro anion by a more labile triflate or hydroxo ligand had no effect on the rate of reaction. The use of various polypyridine chelates showed that 1,10-phenanthroline ligands increase the efficiency compared to the 2,2′-bipyridine analogues. A remarkable increase in catalytic activity (by one order of magnitude) was observed when the nonsubstituted 2,2′-bipyridine ligand in complex **2a** was substituted with electron withdrawing groups such as –COOH, –COOiPr, or $-SO_3H$. Conversely, a decrease in activity was observed when an electron donating substituent such as $-NMe_2$ was used. Thus, the order of catalytic activity of the various complexes with respect to CO oxidation is bpy-4-4^1 $(COOH)_2$ ~ bpy-5-SO_3H > bpy-5-5^1 $(COOH)_2$ > bpy-4-4^1 $(COOiPr)_2$ > Phen C_6H_4 –$(SO_3Na)_2$ > phen > bpy ≫ bpy-4-4^1 $(NMe_2)_2$.

They also proposed reaction mechanism based on their catalytic studies (Scheme 8.3).

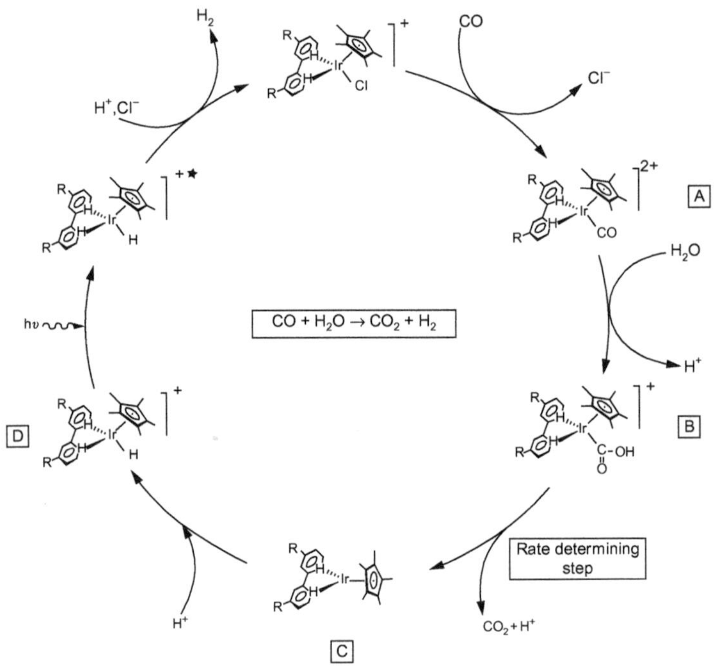

SCHEME 8.3 Proposed reaction mechanism for the homogeneous catalysis of the photochemical WGSR. *(Taken from R. Ziessel, J. Am. Chem. Soc. 115 (1993) 118.)*

REFERENCES

[1] S. Sato, J. M. White, Photoassisted water-gas shift reaction over platinized titanium dioxide catalysts, J. Am. Chem. Soc. 102 (1980) 7206–7210.

[2] S. Sato, J. M. White, Photocatalytic water decomposition and water-gas shift reactions over NaOH-coated, platinized TiO_2, J. Catal. 69 (1981) 128–139.

[3] S. M. Fang, B. H. Chen, J. M. White, Photoassisted water-gas shift reaction on platinized titania. The influence of preparation parameters, J. Phys. Chem. 86 (1982) 3126–3130.

[4] C. Yixuan, W. Zhaobin, C. Y. Anxin, L. Huaxin, H. Zupei, L. Huiqing, D. Yonglei, Y. Chunying, L. Wenzhao, Metal-semiconductor catalyst: photocatalytic and electrochemical behavior of Pt-TiO_2 for the water-gas shift reaction, J. Mol. Catal. 21 (1983) 275–289.

[5] S. C. Tsai, C. C. Kao, Y. W. Chung, Metal-semiconductor catalyst: photocatalytic and electrochemical behavior of Pt-TiO_2 for the water-gas shift reaction, J. Catal. 79 (1983) 451–461.

[6] S. C. Tsai, Y. W. Chung, Effects of particle size on photoassisted water-gas shift reaction over Pt-TiO_2, J. Catal. 86 (1984) 231–234.

[7] L. Millard, M. Bowker, Photocatalytic water-gas shift reaction at ambient temperature, J. Photochem. Photobiol. A Chem. 148 (2002) 91–95.

[8] F. Sastre, M. Oteri, A. Corma, H. Garcia, Photocatalytic water gas shift using visible or simulated solar light for the efficient, room-temperature hydrogen generation, Energy Environ. Sci. 6 (2013) 2211–2215.

[9] D. J. C. Hamilton, Photocatalysis of the homogeneous water-gas shift reaction, J. Chem. Soc. Chem. Commun. (1980) 1213–1215.

[10] R. Ziessel, Photocatalysis. Mechanistic studies of homogeneous photochemical water gas shift reaction catalyzed under mild conditions by novel cationic iridium(III) complexes, J. Am. Chem. Soc. 115 (1993) 118–127.

Chapter 9

Mechanism and Kinetics of the WGS Reaction

9.1 MECHANISM

The mechanism of water gas shift reaction (WGSR) has been proposed long time back; however, there is still controversy about the reaction mechanism of the WGSR. Mainly two mechanisms have been proposed for the WGSR: one is regenerative or associative and other one is dissociative or redox mechanism. There is a general agreement on the regenerative mechanism or redox mechanism over the high-temperature shift catalysts; however, there is considerable uncertainty about the operative mechanism at low temperatures over the Cu catalysts and precious metal-precious metal reducible metal oxide catalysts. In 1975 Newsome and in 1996 Kochloffl reviewed the mechanism of the WGSR [1a,b]. All the reports proposed two rate-determining steps: (a) those that involve a rate-determining step in which a molecule of H_2O or CO, from the gas phase, reacts with a surface species (oxygen vacancy or a surface oxygen atom) and (b) those that involve a rate-determining step in which the reaction takes place between two adsorbed species (the Langmuir-Hinshelwood mechanism). The mechanism initially involves dissociation of H_2O to liberate H_2 and removal of oxygen atom. Then CO takes the oxygen and forms CO_2. The CO can react with the H_2O, OH or O either from the gas phase, the adsorbed state or the surface lattice. The redox mechanism was first proposed by Temkin and Kulkova [2a] in 1949 and later supported by the results of Boreskov *et al.* [3]. The multi-step formate mechanism was first proposed by the Oki *et al.* in 1973 [4].

9.1.1 Mechanism of the High-Temperature WGSR

In 1949 Kulkova and Temkin [2a] proposed oxidation reduction mechanism for the first time. They proposed regenerative mechanism of the Eley-Rideal type in which water is dissociated on the surface of the catalyst to produce hydrogen. The cycle is completed by the reduction of the surface of the catalyst by the CO that produces CO_2.

The mechanism proposed by Temkin *et al.* is

Water Gas Shift Reaction. http://dx.doi.org/10.1016/B978-0-12-420154-5.00009-7

$$H_2O + (\cdot) \leftrightarrow H_2 + (O) \tag{9.I}$$

$$CO + (O) \leftrightarrow CO_2 \tag{9.II}$$

Here $(\cdot)$ is the vacant site on the oxide surface caused by the removal of oxygen atom and (O) is the oxygen atom of the surface oxide.

The redox mechanism implied in Equations (9.I) and (9.II) is essentially the same as the Mars-Van Krevelen mechanism, with the difference that the oxygen used to oxidize the catalyst comes from the water rather than the gas-phase oxygen.

Several kinetic expressions have been derived by considering this mechanism. By considering Equation (9.II) as a rate-limiting step, Kulkova and Temkin [2a] determine the following kinetic expression:

$$r = k_+ P_{CO}\left(\frac{P_{H_2O}}{P_{H_2}}\right)^{1/2} - k_- P_{CO_2}\left(\frac{P_{H_2}}{P_{H_2O}}\right)^{1/2} \tag{9.1}$$

r is the rate of reaction, k_+, k_- is the rate constant for the forward and backward reaction and P is the partial pressure of the particular species. Shchibrya *et al.* [2b] considered a heterogeneous surface with low concentrations of oxygen vacancies with no rate-limiting step and proposed the following kinetic expression:

$$r = k\,\frac{P_{H_2O}P_{CO} - k^{-1}P_{CO_2}P_{H_2}}{AP_{H_2O} + P_{CO_2}} \tag{9.2}$$

Several experimental studies have been performed to validate this model. This mechanism was mainly validated by Boreskov *et al.* [3]. They directly measured the oxidation and reduction of an iron oxide-based catalyst with CO and H_2O. They concluded that the catalyst surface is reduced and oxidized by the CO and H_2O and the rate of reactions is in good agreement with the rate conversion of CO in the WGSR.

Oki and associates [4] did the most fundamental and extensive mechanistic studies on the iron catalyzed WGSR. They used the stoichiometric number method developed by the Horiuti. They simultaneously measured the exchange rates of various isotopes between reactants and products and free energies of the reaction. Initially, deuterium tracer experiments were performed by the Kaneko and Oki and they found the stoichiometric number to be between 1 and 2. However, when they performed ^{14}C tracer experiments they found that the reaction is away from equilibrium and the stoichiometric number is nearly 2. They interpreted this data based on the mechanism by considering a single rate-determining step. However, in later papers based on additional experimental evidence they concluded that their initial interpretation was wrong. They conducted ^{18}O tracer experiments and found that ^{18}O exchange between CO_2 and H_2O was very fast and the stoichiometric number approached infinity. However, ^{18}O exchange between CO and CO_2 was not so fast and stoichiometric number was nearly 2. Based on these results they proposed several reaction mechanisms, presented in Table 9.1.

TABLE 9.1 Proposed Reaction Mechanisms for High-Temperature WGSR

Possible mechanisms for water gas shift reaction and observable stoichiometric number[a]						
Mechanism	*Elementary step*		υ	*Apparent stoichiometric no.*		
				D	^{14}C	^{18}O
I	$CO \rightarrow CO(a)$	(I)	1	∞	1	1, ∞
	$H_2O \rightarrow 2H(a) + O(a)$	(II)	1	1	∞	1, ∞
	$CO(a) + O(a) \rightarrow CO_2(a)$	(III)	1	∞	1	1
	$CO_2(a) \rightarrow CO_2$	(IV)	1	∞	1	1
	$2H(a) \rightarrow H_2$	(V)	1	1	∞	∞
II	$CO \rightarrow CO(a)$	(I)	1	∞	1	1, ∞
	$H_2O \rightarrow OH(a) + H(a)$	(II)	1	1	∞	1, ∞
	$CO(a) + OH(a) \rightarrow COOH(a)$	(IIIa)	1	1, ∞	1	1
	$COOH(a) \rightarrow CO_2(a) + H(a)$	(IIIb)	1	1, ∞	1	1
	$CO_2(a) \rightarrow CO_2$	(IV)	1	∞	1	1
	$2H(a) \rightarrow H_2$	(V)	1	1	∞	∞
III	$H_2O \rightarrow 2H(a) + O(a)$	(I)	1	1	∞	1, ∞
	$CO + O(a) \rightarrow CO_2$	(II)	1	∞	1	1, ∞
	$2H(a) \rightarrow H_2$	(III)	1	1	∞	∞
IV	$H_2O \rightarrow H_2O(a)$	(I)	1	1	∞	1, ∞
	$CO \rightarrow CO(a)$	(II)	1	∞	1	1, ∞
	$CO(a) + H_2O(a) \rightarrow CO_2 + H_2$	(III)	1	1	1	1
V	$CO \rightarrow CO(a)$	(I)	1	∞	1	1, ∞
	$H_2O + CO(a) \rightarrow CO_2 + H_2$	(II)	1	1	1	1

[a]*Taken from Ref. [61].*

In Table 9.1 mechanism III is the oxidation reduction mechanism proposed by Temkin *et al.* [2a].

In Table 9.1 mechanism II is also called as associative mechanism. Several authors indicated that the reaction mechanism over iron oxide catalysts is dependent on the temperature. Armstrong and Hilditch suggested that the activity of magnetite at lower temperatures may be limited by the dissociation of stream and this indicates that the associative mechanism is dominant at those conditions.

Rethwisch and Dumesic [5] studied the adsorption of CO/CO_2 and $CO/CO/H_2O$ gas mixtures over iron oxide catalysts and concluded regarding the regenerative mechanism in the magnetite catalysts. In 1982 Lund and Dumesic [6] concluded that catalysts that are not active for adsorption of CO and CO_2 are not active for water gas shift (WGS). Tinkle and Dumesic [7] performed isotopic label experiments to investigate the adsorption/desorption and the inter-conversion of CO and CO_2 on chromia-promoted magnetite to support this observation. They concluded that interconversion of CO and CO_2 is rapid relative to the adsorption and desorption of either species and that a common adsorbed species results from the adsorption of CO and CO_2.

In recent studies about new formulations of the high-temperature catalysts, diffuse reflectance infrared Fourier transform spectroscopy was used to elucidate the reaction mechanism. The results showed that the catalyst surface can provide oxygen to CO even in the absence of water. When the surface comes in contact with water, the surface of the catalyst recovers oxygen. The presence of intermediate species on the catalyst surface was not detected. This suggested that the WGSR over Fe based catalysts proceeds by redox mechanism.

9.1.2 Mechanism of the High-Temperature WGSR Over Iron Oxide Catalysts Using Characterization Techniques

9.1.2.1 *Ferrites*

Ferrites may be defined as magnetic material composed of oxides containing ferric ions as main constituents. Iron can exist in the form of three oxides, namely hematite (Fe_2O_3), magnetite (Fe_3O_4) and wustite (FeO). The latter is unstable below 570 °C, where it decomposes to α-Fe and Fe_3O_4. In the temperature-programmed reduction (TPR) of pristine hematite (Fe_2O_3), the first reduction peak appears at 356 °C, corresponding to the reductive transition of Fe_2O_3 to Fe_3O_4 [8, 9]. The peak at 565 °C corresponds to the transformation of Fe_3O_4 to FeO. Under the experimental conditions of the present study, complete reduction of FeO to metallic Fe was not observed even up to 800 °C, and only partial reduction took place.

9.1.2.2 *Hematite*

Hematite crystallizes in the rhombohedral lattice system, and it has the same crystal structure as ilmenite and corundum with closely packed oxygen lattice with Fe^{3+} cations occupying octahedral sites. It exhibits a complex magnetic behaviour, being anti-ferromagnetic at low temperatures ($T < 80$ K), and undergoes a Morin transition at $T_M > 260$ K involving a spin flip leading to a weak ferrimagnetic state, before becoming paramagnetic (Curie temperature) at high temperatures ($T_C \cong 955$ K). Its structure can be visualized as being made up of Fe-O_3-Fe units (triplets) of closely packed oxygen atoms with Fe(III) on either side. The Fe(III) atoms in each of these Fe-O_3-Fe units have opposite spins, being anti-ferromagnetically coupled as a result of superexchange interaction through the triad of oxygen atoms. It was shown that the system is essentially anti-ferromagnetic, and that the low symmetry of the cation sites allows spin-orbit coupling to cause canting of the moments when they are in the plane perpendicular to the *c*-axis. The disappearance of the moment with a decrease in

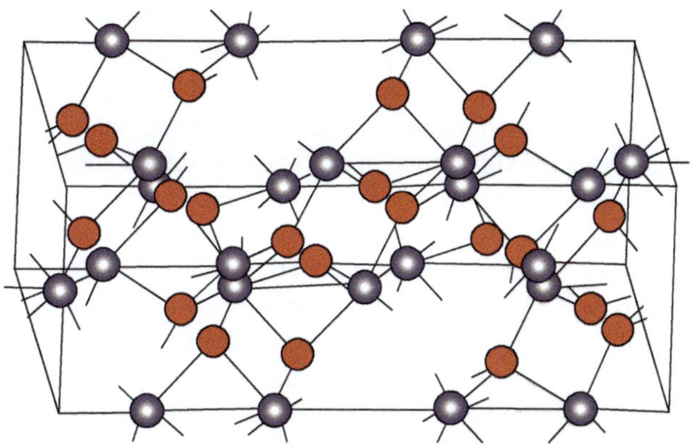

FIGURE 9.1 Crystal structure of hematite.

temperature at 260 K is caused by a change in the anisotropy which causes the moments to align along the *c*-axis. In this configuration, spin canting does not reduce the energy. The magnetic properties of bulk hematite differ from their nanoscale counterparts. For example, the Morin transition temperature of hematite decreases with a decrease in the particle size. The suppression of this transition has also been observed in some of the hematite nanoparticles, and the presence of impurities, water molecules and defects in the crystals were attributed to the absence of a Morin transition. The crystal structure of hematite is shown in Figure 9.1.

9.1.2.3 Magnetite

Magnetite is an example of naturally occurring ferrite. It has spinel structure. The unit cell of spinel ferrite consists of 32 oxygen, 16 trivalent iron and 8 divalent iron atoms. The most important feature of unit cell is that its array of oxygen ions leaves two open kinds of interstices which can be filled by metal ions. These interstices are referred to as tetrahedral or A sites and octahedral or B sites.

The ideal spinel structure consists of a cubic closely packed array of anions, with one-eighth of the tetrahedral and one-half of the octahedral interstices occupied by cations, so that the cation to anion ratio is 3:4. In a binary oxidic spinel, AB_2O_4, two extreme distributions of cations among the available sites are possible: the 'normal' distribution, $A(B_2)O_4$, and the 'inverse' distribution, $B(AB)O_4$; in each case the ions in parentheses occupy octahedral sites. Intermediate distributions also are known, in particular the 'random' or 'statistical' distribution $A_{1/3}B_{2/3}(A_{2/3}B_{4/3})O_4$. It is convenient to characterize the cation distribution by specifying the fraction of tetrahedral sites, x, occupied by B cations, i.e., $A_{1-x}B_x(A_xB_{2-x})O_4$. The parameter x equals 0, 2/3 and 1 for normal, random and inverse spinel, respectively. In the substitution spinels some of the ferric iron ions are replaced by trivalent ions of other metal. The effect of magnetization depends on the site preferred by the substituent.

Attempted lattice energy calculations, as well as the experimental observation of different cation distributions in spinels containing ions of similar charge and size, suggest that the difference in electrostatic lattice energy between normal and inverse spinel is small in common 2-3 and 2-4 spinels. Other factors, such as the 'site preference' of individual cations, will then determine the cation distribution. The crystal structure of magnetite is shown in Figure 9.2.

Magnetite possesses an inverse spinel structure with oxygen ions forming a face-centred cubic closely packed structure. The formula for describing Fe occupancy is (Fe^{3+}) $[Fe^{2+}, Fe^{3+}]O_4$ where the parentheses (·) stand for cations at tetrahedral sites while brackets [·] denote cations at octahedral lattice sites. Stoichiometric magnetite has all available substitutional sites occupied by Fe^{2+} and Fe^{3+} ions. Non-stoichiometric magnetites also exist, with various numbers of available sites being either vacant or occupied by impurity ions.

In 1973 Topsoe *et al.* [10] identified that there is a rapid electron exchange between Fe^{3+} and Fe^{2+} ion and this exchange is responsible for the higher WGS activity. The high electrical conductivity of magnetite, $\mu = 200$ ohm^{-1} cm^{-1}, indicates that a redox couple is present and is highly facile.

Topsoe and Boudart conducted several Mössbauer experiments on Cr- and Cr-Pb-promoted iron oxide catalysts. The Fe_2O_3 exhibits single six-line

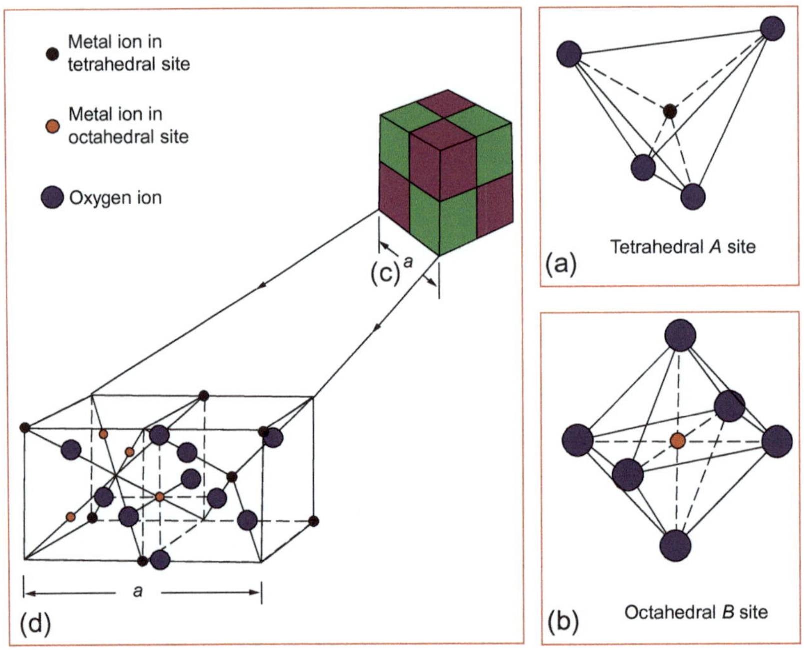

FIGURE 9.2 Crystal structure of magnetite.

spectrum in the Mössbauer spectra and magnetite exhibits two six-line spectrum in the Mössbauer spectra one due to octahedral sites and the other due to tetrahedral sites at room temperature. There is not much difference in the Mössbauer spectra of pure magnetite and Cr-promoted magnetite. For Cr catalyst the ratio between octahedral to tetrahedral sites is much lesser than two and this ratio decreases further with increasing exposure to air at room temperature and can be explained by partial oxidation to Fe_2O_3. Since the ratio is less than 2, they finally suggested the structure of Cr-Fe_3O_4 as

$$\left(Fe^{3+}\right)\left[Fe^{3+}_{1-x}Fe^{2+}Cr^{3+}_{x}\right]O_4$$

The CrPb-Fe_3O_4 catalyst behaved quite different from that of the Cr-Fe_3O_4 catalyst. The value of octahedral to tetrahedral sites was much larger than 2 and did not change with time upon exposure to air at room temperature. The first result rules out partial oxidation to Fe_2O_3, and the second finding shows that the catalyst structure is stable at room temperature, which is consistent with the fact that the CrPb-Fe_3O_4 catalyst was not, oxidized as easily at high temperature as the Cr-Fe_3O_4 catalyst.

Reithwisch and Dumesic [11] studied a number of spinel structures (normal, mixed and inverse) and concluded that only inverse and mixed spinel structures can readily undergo rapid electron exchange between 2+ and 3+ ions, thereby catalyzing the WGSR. The significance of $2+ \Leftrightarrow 3+$ redox couple was also demonstrated by Boreskov [12]. If one can facilitate tetrahedral sites to expand and octahedral sites to contract, an improvement in covalency of the system could result in promoting better electron hopping between these sites ($Fe^{2+} \Leftrightarrow Fe^{3+}$).

We also performed Mössbauer spectra measurements over various metal-promoted iron oxide catalysts [13]. We introduced selected transition/inner transition metal ions (M = Cr, Mn, Co, Ni, Cu, Zn and Ce) into the iron oxide spinel lattice that were screened for effectiveness for the WGSR. ^{57}Fe Mössbauer spectra of pristine Fe_2O_3 sample exhibits a six-line Zeeman spectrum as a result of the complete structural and magnetic transformation.

Incorporation of metal cations into the hematite (α-Fe_2O_3) crystal structure also modified the magnetic field and influences other properties of hematite particles. These effects strongly depend on the nature of the incorporated metal cation. Different substituent ions (M^{n+}) possess different ionic radii compared to Fe^{3+}. This difference will produce distortions in the crystal structure as the substituent ions (M^{n+}) diffuse into the lattice. This asymmetry is directly reflected in EFG (electric field gradient) of the Fe octahedra. The differences in the EFG in turn produce slight differences in the value of Δ. The Mössbauer spectra of our metal-doped iron oxide samples show a superposition of a Zeeman sextet, corresponding to large-sized ferromagnetic particles, with a super-paramagnetic doublet corresponding to smaller particles. The difference between the relaxation behaviour in two different scenarios can be understood as follows: in the sample of non-interacting nanoparticles, the relaxation is governed by an anisotropy energy barrier, and the relaxation time (τ) for magnetization reversals is given by Neel's expression [14],

$$\tau = \tau_0 \exp(KV/kT)$$

where K is the magnetic anisotropy constant, V is the particle volume, k is Boltzmann's constant and T is the temperature. With a typical value of τ_0 of the order of 10^{-11} s and a typical particle size distribution, only a small fraction of the particles will have relaxation times of the order of a nanosecond, which gives rise to line broadening. Therefore, the spectra consist of sextets and doublets with relatively sharp lines. If inter-particle interactions dominate in a sample, the interaction field due to interactions with neighbouring particles dominates the magnetic energy. In this case, the sub-lattice magnetization direction may fluctuate rapidly in directions around that of the interaction field and this leads to a reduced hyperfine field, thereby causing variations in interaction fields, thus leading to broadening of the lines of the sextet. Generally, the incorporation of metal cations with charges other than +3 into the crystal structure of α-Fe_2O_3 will change electrical neutrality in a crystal lattice. Counter charge will be required to restore electrical neutrality. In the case of the formation of solid solutions of M^{n+} ($n \neq 3$) ions and α-Fe_2O_3 this charge balance is most likely achieved by the creation/formation of oxygen vacancies. We observed a variation in the quadrupole-splitting values which is a direct indication of high site distortion caused by oxygen vacancies. In conclusion, the Mössbauer spectra (hyperfine magnetic parameters) were influenced by the size and substructure of α-Fe_2O_3 particles.

On activation, the hematite- like $Fe_{1.82}M_{0.18}O_3$ phase transformed into either an inverse or mixed $Fe_{2.73}M_{0.27}O_4$ magnetite-like spinel phase. In the case of the Fe_2O_3-Cr_2O_3 catalyst, reduction of Fe_2O_3 to Fe_3O_4 was observed at T_{max} 350 °C, whereas further reduction to FeO occurred at higher temperatures. Adding chromium to Fe_2O_3 did not improve the reducibility of hematite to magnetite. Based on X-ray diffraction (XRD), TPR and Mössbauer studies of the Fe_2O_3-Cr_2O_3 spinel and earlier information, we proposed that iron-chromia forms an inverse spinel structure and that Cr^{3+} replaces equal amounts of Fe^{2+} and Fe^{3+} from the octahedral sites, with the displaced Fe^{2+} consequently located in tetrahedral sites. In case of Fe/Mn catalyst there is a of the iron oxide TPR peaks to lower temperatures (by ~50 °C) as manganese is introduced into the hematite host lattice. Thus, this shift indicates that the presence of Mn facilitates the reduction of Fe^{3+} species [Fe_2O_3 to Fe_3O_4]. The 613 °C T_{max} peak corresponds to the reduction of magnetite to wustite, beyond which total reduction to metallic iron is expected to occur. The addition of cobalt to iron oxide causes the temperature of the maximum rate of hydrogen uptake [for $Fe_2O_3 \rightarrow Fe_3O_4$] to be lower by 20 °C (from 350 to 330 °C) indicating that the presence of cobalt renders the hematite easier to reduce. XRD studies indicate that the cobalt gets completely incorporated into the Fe_2O_3 crystal structure resulting in the formation of Fe-Co-O spinel. An interesting observation from the present study is that the addition of Cu to Fe_2O_3 considerably brought down the reduction temperature of hematite to magnetite to 190 °C as compared to 348 °C, which is observed in pristine hematite sample. It is well known that the presence of even 1 atomic % Cu shifts the peak reduction temperature by >100 °C [15]. Interestingly, the incorporation of Cu did not affect the stability of Fe_3O_4 phase, which was found to be stable up to 600 °C, after which transformation of the magnetite to wustite was observed in the present study. The reduction profiles of Fe/Ni and Fe/Zn are similar to pure Fe_2O_3 sample. On the other hand, promoting iron oxide with cerium causes the T_{max} of the transitions of both hematite to magnetite and magnetite to wustite to shift to lower temperatures. Also, the ceria surface shell reduction in the case of Fe/Ce system occurs at 380 °C, instead of 485 °C as in the pristine ceria sample; however, the ceria bulk reduction is not affected by the presence of iron.

Then we [16] investigated the effect of Cu on the structure and WGS activity of these spinels. Remarkably, our shift activity measurements reveal that Cu acts as a promoter for all the M-modified ferrites with M=Cr, Ni, Co, Mn and Zn except M=Ce. For the latter case, Cu acts as an inhibitor for the high-temperature WGSR. After co-doping with Cu all the M-modified ferrites exhibit three types of peaks. The reduction process of Cu was observed along with the reduction process of iron oxide and other metal oxide. In Cu reduction process the reduction of $Cu^{2+} \rightarrow Cu^{1+}$ occurred at around 180 °C and further reduction from Cu^{1+} to metallic Cu occurred at around 225 °C in all the Cu co-doped modified ferrites. Remarkably, co-doping of Cu into M-modified ferrites selectively promotes the reduction of hematite to magnetite but not either magnetite to wustite or wustite to metallic iron transformations except for the case of Fe/Ce. Interestingly, Cu also did not promote the reduction of the other metal oxide present in the M-modified ferrites except for the case of Fe/Ce. On the other hand, co-doping of Cu into the Fe/Ce catalyst promoted the reduction of hematite to magnetite as well as magnetite to wustite, besides ceria surface reduction. However, the ceria bulk reduction was unaffected.

We performed Mössbauer spectroscopy both at room and liquid nitrogen temperature over the activated catalysts after pre-reduction. Our TPR measurements show that all the catalysts transformed from hematite to magnetite after the pre reduction. Stoichiometric magnetite has all available substitutional sites occupied by Fe^{2+} and Fe^{3+} ions. Non-stoichiometric magnetites also exist, with various numbers of available sites either vacant or occupied by impurity ions. The magnetic properties of magnetite have been broadly explained on the assumption that spins on octahedral and tetrahedral sites form two antiferromagnetically coupled sub-lattices [17]. In addition, stoichiometric magnetite undergoes an order-disorder phase transition at 120 K, which is attributed to fast electron hopping among the Fe^{2+} and Fe^{3+} octahedral cations above 120 K [18]. Associated with this transition is an anomaly in both the specific heat and electrical conductivity of magnetite. On such a condition, one might expect three different Fe sites in magnetite: one from Fe^{3+} ions at tetrahedral sites, one from the Fe^{3+} ions at octahedral sites and another from the Fe^{2+} ions in the octahedral sites. In fact, in stoichiometric magnetite there are only two six-line hyperfine structures [19, 20]. Above 120 K these features are generally identified as one with Fe^{3+} ions in the tetrahedral sites, and the other to a set of Fe^{3+} and Fe^{2+} ions at octahedral sites, i.e. one six-line hyperfine structure is due to all tetrahedral cations, while the other is due to octahedral cations; below 120 K these are identified as one with Fe^{3+} ions in both sites, and the other due to Fe^{2+} ions. The explanation for the spectral change is attributed to the electron exchange between ions at the octahedral sites, which is rapid above 120 K but slows down below that temperature. Mössbauer spectra of Cu co-doped modified ferrites at 300 K are presented in Figure 9.3. All the Cu co-doped and M-modified ferrites show similar Mössbauer spectra except the cases of Fe/Ce and Fe/Ce/Cu. The typical spectra exhibit two well-developed six-line magnetic hyperfine structures typical of bulk materials. The high-field component is due to Fe^{3+} ions at T_d sites, while the low-field one arises from Fe^{2+} and Fe^{3+} ions present at O_h sites [21]. There is not much change in the isomer shift δ values for Fe at T_d sites after Cu co-doping. On the other hand, δ values for O_h (low-field component) in the Cu co-doped M-modified ferrite catalysts are noticeably higher than those in M-modified ferrite catalysts at 300 K. The O_h/T_d ratio is generally close to 2 in pure activated Fe_2O_3 catalyst. However, in the Cu co-doped ferrite catalysts the fraction decreases compared to the one in M-modified ferrite catalysts. These results show that the Cu guest enters the O_h sites of magnetite during the activation of the catalysts and alters the local structure. The Mössbauer line shape of both Fe/Ce and Fe/Ce/Cu ferrite differ from other modified ferrites; in the case of Fe/Ce catalyst, in addition to the two six-line patterns, a broad singlet is observed at near 0 mm/s. This singlet is due to the super paramagnetic behaviour of the catalysts,

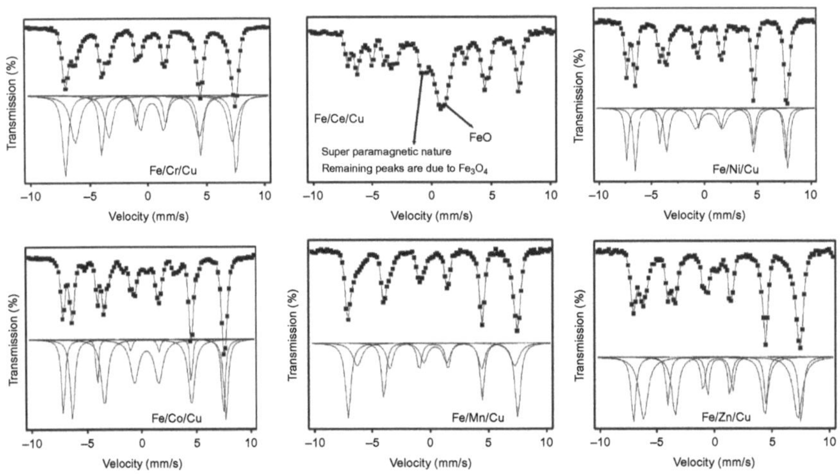

FIGURE 9.3 Mössbauer spectra of Cu co-doped modified ferrites at room temperature. *(Taken from Figure 8 of J. Phys. Chem. C 116 (2012) 11019-11031.)*

resulting from a collapse of the H-field in the nanosized particles. The Mössbauer spectrum of Fe/Ce/Cu exhibit two additional peaks at a velocity of 1.5 mm/s along with the two six-line magnetic hyperfine structures and the super paramagnetic nature. These peaks primarily belong to the wustite phase.

Mössbauer spectra of Cu co-doped modified ferrites were also measured at 78 K. The spectra show two well-developed six-line magnetic hyperfine structures typical of bulk materials with no trace of super paramagnetic behaviour except Fe/Ce/Cu. The lower field spectrum is due to Fe^{2+} ions at O_h sites, while the high-field spectrum arises from Fe^{2+} ion at both O_h and T_d sites. All Cu co-doped M-modified ferrites exhibited higher Fe^{3+}/Fe^{2+} fractions compared to the M-modified ferrites. Sorescu *et al.* [22] reported that the increasing Mn content in the Magnetite relative area of the Fe^{2+} ions decreases due to the Mn^{2+} ions substituting Fe^{2+} ions at octahedral sites and the relative area of the Fe^{3+} ions increases accordingly. These results suggest that Cu replaces the Fe^{2+} ions at octahedral sites except for the case of Fe/Ce/Cu. On the other hand, the formation of FeO was continued in Fe/Ce/Cu at liquid nitrogen temperature. On the whole, Mössbauer spectroscopic results show that Cu enters at octahedral sites of the magnetite during the activation and replaces the Fe^{2+} ions from the octahedral sites and promotes the WGS activity except for the Fe/Ce/Cu. On the other hand, the addition of Cu to the Fe/Ce catalyst leads to the formation of wustite phase during the activation and decreases the WGS activity. On the whole, our WGS activity and structural characterization measurements suggest that Fe/Ce catalyst is a better catalyst for the WGSR.

9.1.3 Mechanism of the Low-Temperature WGSR

Mechanism of the low-temperature WGSR is much more complicated than the high-temperature catalysts. The mechanism is completely different on different catalysts like Cu, Pt, Au, etc. Interestingly, the mechanism also changes if we change the support from reducible oxide to non-reducible oxide. Some of the features of the low-temperature WGSR are different from the high-temperature WGSR i.e. (1) The dissociation of H_2O to H_2 and O happens over iron oxide Fe_3O_4, and this type of dissociation is not reported in the literature on the Cu catalysts; (2) The amount of CO adsorbed on metallic copper, even at low temperatures, is less than that on the platinum group of metals at similar temperatures; and (3) The WGS rate is proportional to the CO partial pressure to the first order over the Cu-ZnO compared to the zero order observed over the Pt-based catalysts. The last feature implies that reducing the CO concentration from 1.0% [a typical value in the low-temperature shift (LTS) reactor] to 0.5% requires twice as much catalyst as reducing it from 2% to 1%, thus leading to large second-stage LTS reactors. Efforts have been made in literature to synthesize cost-effective catalysts. Nobel metal catalysts seem to be suitable catalysts.

The largely accepted mechanism for the low-temperature WGSR is associative mechanism. In 1920, Armstrong and Hilditch [23] proposed the associative mechanism of Langmuir-Hinshelwood type. In this mechanism water is dissociatively adsorbed on the surface of the catalyst to form reactive hydroxyl groups. When these hydroxyl groups are combined with the adsorbed CO, a surface intermediate structure is produced that is decomposed to CO_2 and H_2.

9.1.3.1 Formate Species Mechanism

As explained above, Oki *et al.* [5] did stoichiometric number studies and supported the associative mechanism. Extensive research efforts have been reported in the literature using various characterization techniques to prove the existence of associative mechanism over Cu and other metallic catalysts. After the proposal of Armstrong and Hilditch in 1969 Yureva *et al.* [3] described an associative mechanism for $CuCr_2O_3$. Boreskov *et al.* [24] had carried out pioneering IR studies over a wide range of Cu catalysts including Cu-ZnO catalysts to support the formate formation during the WGSR. In 1980, Van Herwijnen and de Jong [25] suggested that the step of decomposition of the formate species adsorbed on the catalysts was more likely to be the rate-determining step. Grenoble *et al.* [26] proposed a bifunctional associative mechanism for Cu/Al_2O_3 catalysts in which adsorption of CO occurs at the metal centre and hydroxyl one of the OH bonds to form the intermediate species. These authors also suggested that a shift of the intermediate formate species from the support to the metal centre is necessary in order to obtain decomposition of the intermediate. Shido and Iwasawa [27] conducted several *in situ* Fourier transform infrared spectroscopic measurements over various metal oxides and they concluded that surface OH groups react with CO and form bridged formats. Then these bridged formats were converted to bidentate formates above 170 °C. Then these formats decomposed into H_2 and CO. Jacobs *et al.* conducted isotopic studies and confirmed the formation of bridged carbonated on Pt/CeO_2. They also suggested that the rate-limiting step for low-temperature WGSR is cleavage C—H bond of the formate. On the whole, formate mechanism includes formation of bidentate formate from CO and surface OH groups which act as intermediate. This bidentate formate decomposes to H_2 and surface unidentate carbonate which further decomposes to CO_2. The main roles of Pt are catalyzation of the reduction of ceria leading to the formation of surface terminal OH groups on ceria and decomposition of the formate to H_2 and CO_2.

As explained above, the cleavage of C—H bond of the formate is the rate-determining step. Hence, addition of bases accelerates the decomposition of C—H bond and increases the WGS activity. Pigos *et al.* [28] showed that incorporation of Na into the Pt-ZrO_2 enhances the WGS activity. Their diffuse reflectance infrared Fourier transform (DRIFT) studies indicate that incorporation of Na into Pt-ZrO_2 modifies the electronic properties of the surface formate species and weakens the C—H bond. The faster decomposition of the surface formate over Pt-ZrO_2 is shown in Figure 9.4. Mojet *et al.* found that increasing the K content in the Pt-SiO_2 and Pt-K-L zeolites increased the concentration of the bridged carbonyls.

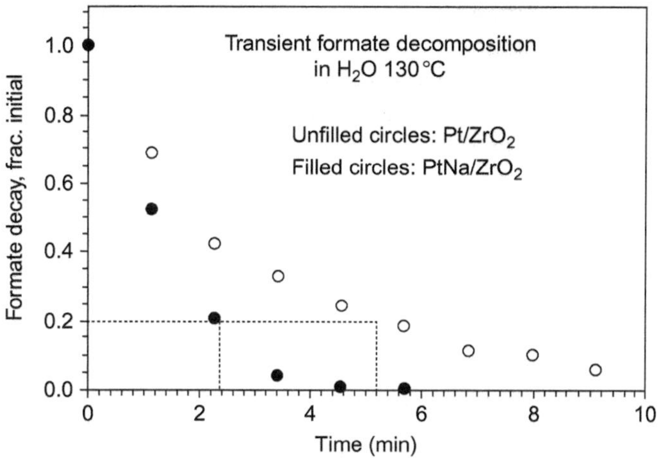

FIGURE 9.4 Formate area response to steaming at 130 °C for (unfilled circles) Pt/ZrO_2 and (filled circles) $PtNa/ZrO_2$. The 0.20-life of formate is indicated for each case, demonstrating a faster rate for $PtNa/ZrO_2$. Forward decomposition. *(Taken from Figure 9 of Appl. Catal. A Gen. 319 (2007) 47-57.)*

In 1981, Sato and White [29] doped Pt-TiO_2 with NaOH and found an improvement in the photocatalyzed WGS rate. Klier [30] also highlighted the promoting influence of alkali dopants, with their relative efficiency being decreasing in the order Cs, Rb, K, Na and Li. Klier also suggested that the alkali should be present at concentrations less than a monolayer. Campbell *et al.* [31] observed a promotion of the WGS activity of Cu(1 1 0) by Cs ions. Honda Research Inc. has also claimed [32] a remarkable improvement in the WGS activity of Pt-ZrO_2 catalysts for fuel processors for use in fuel cell applications by doping the catalysts with alkali. Among the promising compositions discovered was an important improvement when Pt-ZrO_2 was doped with Na alone or in combination with vanadium.

However, with high alkalinity a trade-off was observed such that the stability of adsorbed carbonate species was found to increase. The increase in the concentration of alkali metal ion also leads to a decrease in the catalyst surface area, blocking of active Pt sites and inhibition of surface ceria reduction. Also with increase in the atomic number of over a series of alkali metal catalysts the stability of the carbonate species (another surface species formed during the WGSR) was also found to increase. This result indicates that alkali doping is an optimization problem, that is, while improving the decomposition rates of formate species, the carbonate intermediate stability also increases, making it difficult to liberate the CO_2. An optimal amount of basicity sufficient to decompose the formate but not enough to stabilize too much of the carbonate is needed. Infrared spectroscopy results of CO adsorbed on Pt and ceria suggested that the alkali dopant is located on, and electronically modifies, both the Pt and ceria components. Alkali doping may, thus, provide a path forward for improving the WGS rate by means other than resorting to higher noble metal loadings.

Meunier *et al.* [33] investigated the mechanism of the WGSR over Au-Ce (La)O_2 catalysts using simultaneous DRIFT and kinetic analysis. The analysis of the formate exchange curves between 150 and 220 °C suggested the presence of two types of surface formates: (1) slow formates in which the rate constant is 10- to 20-fold slower than that of CO_2 and (2) fast formates that exchanged on a time scale similar to that of CO_2. The rate of formate decomposition was calculated from the diffuse reflectance infrared Fourier transform spectroscopy (DRIFTS) studies. The measurements show that the rate of CO_2 formation is much higher than that of formate decomposition, which suggests that formates are not the only intermediates in the low-temperature WGSR.

On the whole, all the above mechanistic studies suggest that not only are formate species present during the WGSR but also other intermediates like carbonates/carboxylates. These results also suggest completely different mechanism like redox mechanism can also be occur during the WGSR.

9.1.3.2 *The Carboxylate Mechanism*

For the first time, in 2008, Gokhale *et al.* [34] proposed the concept of carboxylate mechanism. They did several density functional theory (DFT) calculations to investigate the mechanism of WGSR over Cu(1 1 1). The typical mechanism involves the oxidation of CO by OH to form carboxyl (COOH) species. The carboxyl so formed may then yield CO_2. This mechanism is very similar to the redox mechanism. The typical reaction mechanism is as follows:

$$CO \rightarrow CO^* \quad (9.III)$$

$$H_2O \rightarrow H_2O^* \quad (9.IV)$$

$$H_2O + {}^* \rightarrow H^* + OH^* \quad (9.V)$$

$$CO^* + OH^* \rightarrow COOH^* \quad (9.VI)$$

$$COOH^* \rightarrow CO_2{}^* + H^* \quad (9.VII)$$

$$COOH^* + OH^* \rightarrow CO_2{}^* + H_2O^* \quad (9.VIII)$$

$$CO_2{}^* \rightarrow CO_2 \quad (9.IX)$$

$$2H^* \rightarrow H_2 \quad (9.X)$$

They showed that while it is possible to form COOH from CO and OH in a single elementary reaction step, the same is not possible with HCOO formation. That is because OH binds to the surface through its O atom, CO through its C atom, whereas formate binds through its two O atoms, not its C atom. They calculated the thermal energies or activation energies of the each step like water activation, OH dissociation, CO oxidation by O, CO oxidation by OH, carboxyl dissociation etc. The energetics of competing reaction paths point to the predominance of the COOH-mediated WGS path, whereby CO is directly oxidized

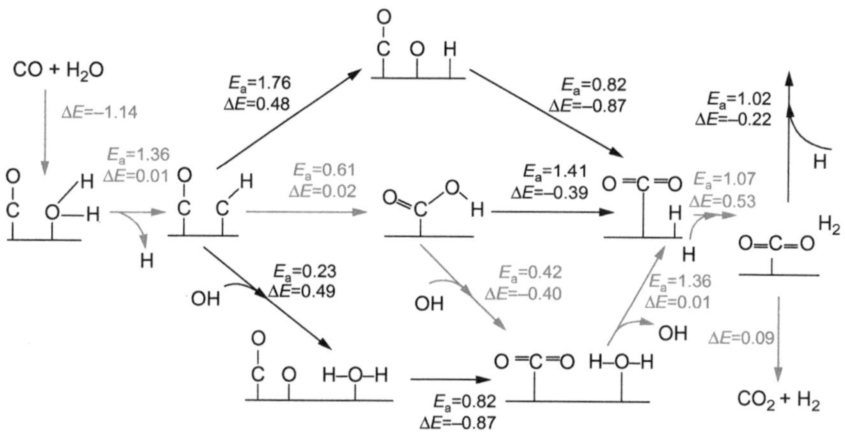

FIGURE 9.5 Reaction network for WGSR. A reaction scheme including both the surface redox mechanism and the carboxyl mechanism is outlined. The thermochemistry and the kinetic barriers for all the elementary steps are given in electronvolts. For reactions involving bond making, the activation barriers are reported with respect to the adsorbed reactants at infinite separation from each other. The minimum energy pathway for the WGSR is highlighted in green. *(Taken from Figure 7 of J. Catal. 247 (2007) 277-287.)*

by the surface OH rather than the atomic O. Based on the calculations they proposed the reaction pathway in Figure 9.5.

Finally, they concluded that (1) H abstraction from H_2O appears to be the rate-controlling step for the entire WGSR network; (2) carboxyl (COOH) is a very reactive intermediate, short-lived and likely difficult to be identified spectroscopically; and (3) formate (HCOO), formed from CO_2+H, is a spectator species which tends to block active sites and can reach substantial surface coverages, particularly at higher pressures. Based on the good 'fit' between the calculated and observed data, they suggested that Cu(1 1 1) may be a dominant active site for the WGSR on realistic industrial catalysts.

Gokhale *et al.* [35] investigated the mechanism over Pt(1 1 1). They obtained elementary steps, reaction intermediates, energetics and reaction mechanism for WGS on Pt-based catalysts through a combination of DFT calculations, micro-kinetic modelling and experimental studies. They concluded that formate cannot be formed from OH and CO in a single elementary step and does not act as a reaction intermediate but rather as a spectator species, which is formed from reacting CO_2 with H. The reaction of CO and OH results in a carboxyl (COOH) intermediate which subsequently decomposes into CO_2 and H. Their DFT calculations suggest that the step $COOH + OH \rightarrow CO_2 + H_2O$ would be the favoured CO_2 formation path, mainly because of its low activation energy barrier; the micro kinetic model showed that this reaction is limited by the small OH coverage on the surface. Instead, the direct decomposition step $COOH \rightarrow CO_2 + H$ accounts for 75-95% of the total CO_2 production under the investigated reaction conditions. Additional H_2O in the feed increases the

OH coverage and makes the OH-mediated COOH + OH low-energy decomposition path more kinetically accessible.

9.1.3.3 *Redox Mechanism*

The redox mechanism was mainly reported on Pt-CeO_2 catalysts [36]. According to the mechanism, CO adsorbs on transition metal sites and reacts with oxygen from the ceria, which, in turn, is reoxidized by H_2O. In other words, it involves the reaction of the reactants, CO and H_2O, with the surface: CO with the oxide ion of the ceria (to yield CO_2) and H_2O with the anion vacancies on ceria (generating OH groups and, eventually, H). An important role of the metal is to adsorb/activate CO and create oxygen vacancies at the metal ceria interface. It has also been established that reduced ceria can be oxidized by CO_2. While the redox mechanism is well established at high temperatures in the case of the iron oxide-chromium oxide catalysts, its applicability to LTS over Cu-ZnO and Pt-ceria catalysts is uncertain and depends on confirmation of the ability of H_2O to reoxidize the partially reduced support oxide at temperatures below 250 °C, especially in the presence of significant amounts of hydrogen, as is the case for fuel cell applications.

9.1.4 Mechanism of the WGSR Over Cu and Au Nanoparticles

The mechanism of the WGSR over Cu and Au nanoparticles is an important issue. Since the mechanism of nanoparticles is completely different from bulk oxides, the presence of support completely modifies the mechanism. Then the serious questions are what the intrinsic reactivity of these nanoparticles is? Can nanoparticles of Cu and Au catalyze the WGSR on their own without the aid of an oxide support (such as ceria or ZnO)? Rodriguez *et al.* [37] synthesized bulk Cu, Au catalysts and also catalysts supported on CeO_2, ZnO. The ZnO-supported catalysts exhibit much higher catalytic activity than bulk Au and Cu catalysts. Interestingly, ceria-supported catalysts exhibit higher activity than bulk and ZnO-supported catalysts. They investigated the WGS mechanism over Cu_{29} and Au_{29} clusters (representative of the metal nanoparticles formed on deposition on ceria or ZnO supports) and on Cu(1 0 0) and Au(1 0 0) surfaces (representative of the surface of the bulk metals) using DFT calculations. The mechanism is as follows

$$CO(g) \rightarrow CO^* \quad (9.XI)$$

$$H_2O(g) \rightarrow H_2O^* \quad (9.XII)$$

$$H_2O^* \rightarrow H^* + OH^* \quad (9.XIII)$$

$$CO^* + OH^* \rightarrow OCOH^* \quad (9.XIV)$$

$$OCOH^* \rightarrow CO_2{}^* + H^* \quad (9.XV)$$

$$2H^* \rightarrow H_2 \quad (9.XVI)$$

The adsorption of CO or H_2O on the Cu particles is exothermic. The first and the most important energy barrier is the dissociation of water into adsorbed OH and H. Then, the reaction of OH and CO produces an OCOH, carboxyl species. The final important energy barrier is for the decomposition of this OCOH carboxyl intermediate into CO_2 gas and adsorbed H, which eventually yields the H_2 gas. The DFT results indicated that a free, metallic nanoparticle of copper can catalyze the WGSR easily. A comparison with the corresponding results on the Cu(1 0 0) surface of bulk copper shows that the dissociation of H_2O on the surface of bulk copper has a larger activation energy barrier (1.13 eV vs. 0.94 eV on the nanoparticle) and that no stable OCOH, carboxyl intermediate, is formed, as a redox mechanism operates. The presence of corner or edge atoms in Cu_{29} favours the dissociation of H_2O. The Au nanoparticles and the bulk Au(1 1 0) surface could not catalyze the WGS action. Neither surface was able to adsorb and dissociate water molecules. A perfect CeO_2(1 1 1) surface does not dissociate water at low or, even, high temperatures. When O vacancies are present, however, the H_2O molecules dissociate on the partially reduced ceria surface. Au and Cu particles facilitate the reduction of the ceria surface by the CO/H_2O mixture and, thereby, facilitate the most difficult step in the WGSR, namely, the dissociation of H_2O.

9.1.5 Mechanism of WGSR Over Non-Reducible Supported Metal Catalysts

Olympiou *et al.* [38] studied the mechanism of the WGSR over alumina-supported Pt, Pd and Rh catalysts using steady-state isotopic transient kinetic analysis techniques coupled with mass spectrometry. In particular, the concentrations (mmol g^{-1}) of active intermediate species found in the carbon path from CO to the CO_2 product (using ^{13}CO) and in the hydrogen path from H_2O to H_2 (using D_2O) were determined. It was found that by increasing the reaction temperature from 350 to 500 °C, the concentrations of the active species in both the carbon and hydrogen paths increased significantly. Based on (a) the large concentration of the active species present in the hydrogen path (OH/H located on the alumina support), which was larger than six equivalent monolayers (based on the exposed platinum metal surface area), (b) the small concentration of OH groups along the periphery of the metal-support interface and (c) the significantly smaller concentration (mmol g^{-1}) of active species present in the carbon path (adsorbed CO on the noble metal and formate species on the alumina support and/or at the metal-support interface), the authors suggested that the diffusion of OH/H species on the alumina support towards catalytic sites present in the hydrogen pathway may be the slow step in the reaction mechanism. The OH/H species were considered to be formed by the dissociation of H_2O on the alumina support. The role of the noble metal was (a) to activate the CO molecule by chemisorption and (b) to promote formate decomposition into CO_2 and H_2 products. There was also a correlation between catalytic activity and the surface concentration and binding energy of CO on the noble metals. Among the alumina-supported noble metals, the order of activity was found to be Pt > Rh > Pd.

9.1.6 Universal Mechanism

In 2006 Burch [39] published a universal mechanism for the low-temperature WGSR. The mechanism is presented in Figure 9.6. As depicted in Figure 9.6b, the active sites in the WGSR are the oxygen vacancies which dissociate water into OH adsorbed on the support. Metal sites also adsorb and activate CO. In the next step, the OH groups, located on the support oxide, react with the CO adsorbed on the metal particles to form the surface intermediates, the formates and carbonates. The latter decompose to CO_2. It must, however, be emphasized that the above mechanism involving long-lived surface intermediates, like formates or carbonates/carboxylates, is valid only below, say, 350 °C. At higher temperatures, over HTS catalysts, like Fe_2O_3-Cr_2O_3, the direct oxidation and reduction of the catalyst by H_2O and CO, respectively, by Eley-Rideal type

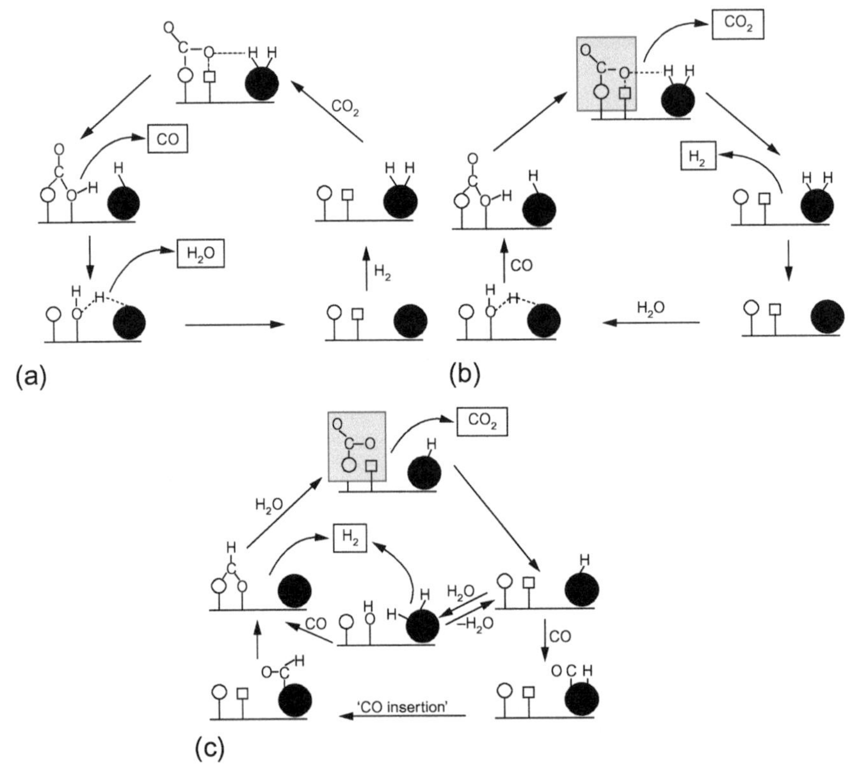

FIGURE 9.6 (a) 'Carbonate/carboxylate' mechanism for the reverse WGS reaction. (b) 'Carbonate/carboxylate' mechanism for the WGS reaction. (c) 'Universal' mechanism for the WGS reaction (116). *(Taken from Figure 51 of Catal. Rev. Sci. Eng. 51 (2009) 325-440.)*

of reactions are well known, and, hence, the redox mechanism will probably be the dominant mechanism. These Eley-Rideal processes are less favoured at low temperatures and, hence, the rates of formation or decomposition of surface intermediates, like formates and carbonates/carboxylates, assume critical importance.

Based on above studies we can make following conclusions:

(1) At high temperatures, where desorption and/or decomposition of intermediates like formate and carbonate species will be very fast, the redox processes would be expected to be important in determining the rate of the reaction. This is particularly valid in the presence of a high concentration of H_2O when the surface is covered, to a significant extent, by OH groups.
(2) At low temperatures, and, especially in the presence of a substantial amount of CO_2, the final carbonate decomposition step in the mechanism will be the slow, rate-determining step.
(3) At intermediate temperatures, especially in the presence of a large concentration of water, and a low concentration of CO_2, the formate decomposition step in the mechanism would be slow and rate determining.

9.2 KINETICS OF THE WATER GAS SHIFT REACTION

The kinetics of the WGSR has been studied for a long time and many equations have been proposed for both low- and high-temperature WGSR. The main difference among them are due to presence of impurities in the reactant gases and because the kinetic data were obtained under diffusion restrictions in many cases. Most commonly used kinetic models are presented in Table 9.2. In general the reaction rate of WGSR is fitted to Langmuir-Hinshelwood model or power model.

The Langmuir-Hinshelwood model assumes that the reactants adsorbed on the catalyst surface and the reaction takes place with these adsorbed species. The Langmuir-Hinshelwood model fits the data better when adsorption plays an important role or the reaction proceeds through an intermediate.

The power law does not consider any mechanistic aspects of adsorption and surface reaction but the fit to experimental data is quite good. The general equation of power law is as follows:

$$\begin{aligned} r &= kP^{a}_{CO}P^{b}_{H_2O}P^{c}_{CO_2}P^{d}_{H_2}(1-\beta) \\ \beta &= P_{CO_2}P_{H_2}/KP_{CO}P_{H_2O} \end{aligned} \tag{9.3}$$

where a, b, c and d are the reaction orders of CO, H_2O, CO_2 and H_2

9.2.1 Kinetics of the High-Temperature WGSR

One of the first expressions for high-temperature WGSR was proposed by Kulkova and Temkin [2a] and it was based on the regenerative mechanism

$$r = k_{+}P_{CO}(P_{H_2O}/P_{H_2})^{1/2} - k_{-}P_{CO_2}(P_{H_2}/P_{H_2O})^{1/2} \tag{9.4}$$

TABLE 9.2 Kinetic Expressions for WGSR

Model	Kinetic expression
High-temperature shift [Ref. Twigg (1989)] [82]	
Kodama *et al.*	$r=\frac{k\left[[CO][H_2O]-\frac{[CO_2][H_2]}{K}\right]}{(1+K_{CO}[CO]+K_{H_2O}[H_2O]+K_{CO_2}[CO_2]+K_{H_2}[H_2])}$
Hulburt-Vasan	$r=\frac{k[H_2O]}{1+K[H_2O]/[H_2]}$
Langmuir-Hinshelwood model	$r=\frac{kK_{CO}K_{H_2O}\left[[CO][H_2O]-\frac{[CO_2][H_2]}{K}\right]}{(1+K_{CO}[CO]+K_{H_2O}[H_2O]+K_{CO_2}[CO_2]+K_{H_2}[H_2])^2}$
Oxidation reduction model	$r=\frac{k_1k_2\{[CO][H_2O]-[CO_2][H_2]/K\}}{k_1[CO]+k_2[H_2O]+k_{-1}[CO_2]+k_{-2}[H_2]}$
Bohlboro *et al.*	$r=kP^{a}_{CO}P^{b}_{H_2O}P^{c}_{CO_2}P^{d}_{H_2}$
Low-temperature shift [Ref. Van Herwijnen and de Jong] [25]	
Campbell *et al.*	$r=k\frac{P_{CO}P_{H_2O}(1-\beta)}{(1+K_{CO}P_{CO}+K_{H_2O}P_{H_2O}+K_{CO_2}P_{CO_2}+K_{H_2}P_{H_2})^2}$
Shchibrya *et al.*	$r=\frac{kP_{H_2O}P_{CO}(1-\beta)}{AP_{H_2O}+P_{CO_2}}$
Moe	$r=kP_{CO}P_{H_2O}(1-\beta)$
Kulkova and Temkin	$r=kP_{CO}\left(\frac{P_{H_2O}}{P_{H_2}}\right)^{0.5}(1-\beta)$
Goodgidge and Quazi	$r=kP^{a}_{CO}P^{b}_{H_2O}P^{c}_{CO_2}P^{d}_{H_2}$

k is the rate constant $k=A\exp(-E_a/RT)$.
A is the pre-exponential factor.
E_a is the activation energy.
R is the universal gas constant.
T is the temperature.
r is the rate constant.
K is the equilibrium constant.
X_i is the mole fraction of the species.

where r is the reaction rate, and k_+, k_-, the rate constants of the forward and backward reactions. P_i is the partial pressure of the component i. The activation energy was 151 kJ/mol, indicating that there are no diffusion limitations.

Shchibrya *et al.* [2b] used a similar approach and proposed the following rate expression:

$$r=k\frac{P_{H_2O}P_{CO}-k^{-1}P_{CO_2}P_{H_2}}{AP_{H_2O}+P_{CO_2}} \tag{9.5}$$

TABLE 9.3 Reaction Orders and Activation Energies of Various High-Temperature WGS Catalysts

	Reaction orders					
Catalyst	*CO*	H_2O	H_2	CO_2	*E* (kJ/mol)	Reference
Fe-Cr	0.9	0.25	0	−0.6	114	[40]
Fe-Cr	1.1	0.53	0	0	95	[41]
Fe-Cr-Cu	1	0	−0.09	−0.36	111	[42]
Fe-Cr-Cu	0.9	0.31	−0.05	−0.156	88	[42]
Rh/Fe-Cr	1.45	0.44	−0.22	−0.12	63	[43]

Several authors have used the power law to fit the reaction kinetics. Table 9.3 summarizes the parameter obtained in such investigations regarding the studies about Langmuir-Hinshelwood model, Chinchen *et al.* [44] proposed an equation like

$$r = \frac{k[\mathrm{CO}][\mathrm{H_2O}][1 - \dfrac{[\mathrm{CO_2}][\mathrm{H_2}]}{K[\mathrm{CO}][\mathrm{H_2O}]}}{(1 + C_1[\mathrm{CO}] + C_2[\mathrm{H_2O}] + \cdots)^2} \tag{9.6}$$

Activation energy of 129 kJ/mol was derived from this model. The CO reaction order in not one and the reaction rate varied with pressure.

Bohlbro [45] investigated the kinetics of commercial Fe-Cr catalyst by varying the concentration of one component and keeping the other component's concentration constant. They used the power law expression. Bohlbro found that the power law expression provided fairly good accuracy for the shift reaction on a Fe-Cr catalyst in the temperature region 330-500 °C and the rate equation is

$$r = K(\mathrm{CO})^{0.9}(\mathrm{H_2O})^{0.25}(\mathrm{CO_2})^{-0.6}(\mathrm{H_2})^{0}(1-\beta) \tag{9.7}$$

Activation energy is 114 kJ/mol.

Then they investigated [46] the same catalyst at higher pressure up to 2 MPa and concluded that the same expression with the same reaction orders except for the CO order which is 1. They also found that ln *K* increases linearly with ln *P*. Atroshchenko and Loboiko [47] performed the experiment at 4 MPa and suggested zero order for H_2 and CO_2, first order for CO and an order of −0.56 and −0.3 for CO_2 depending on the temperature.

Bohlbro *et al.* [48] also investigated the kinetics of commercial Fe-Cr catalyst in the presence of H_2S. They investigated two H_2S levels, namely 50-100 and 2000 ppm. They found the catalytic activity is considerably low by the addition of 2000 ppm. The reaction orders are shown in Table 9.4.

TABLE 9.4 Reaction Orders of Fe-Cr Catalysts in the Presence of H_2S

H_2S (ppm)	CO order	H_2O order	CO_2 order	H_2 order
0	1	0.25	−0.6	0
50	0.8	0.5	−0.1	−0.15
2000	0.85	0.4	−0.1	−0.1

Bohlbro found that the rate constant is inversely proportional to H_2S concentration in the range 70-1250 ppm and at 380-500 °C reaction temperature.

Bohlbro [48] also studied the effects of alkali on some laboratory synthesized iron-based shift catalysts. The behaviour of the catalyst greatly depends on pre-treatment and CO_2 level in the feed. Also kinetics can be adequately described by the power law equation.

9.2.2 Kinetics of the Low-Temperature WGSR

Most of the studies show that some of the kinetic expressions valid for high-temperature WGSR are also valid for low-temperature WGSR. Shchibrya *et al.* found a kinetic expression similar to high-temperature kinetic equation based on $CuO/ZnO/Cr_2O_3$ [2b]. Later Uchida *et al.* [49] confirmed that the kinetic expression derived from redox mechanism accurately correlated the kinetic data obtained with the Cu-Zn catalysts. Cherednik *et al.* [50] proposed other equation for a commercial catalyst $CuO/ZnO/Cr_2O_3$

$$r = \frac{kP_{H_2O}P_{CO}}{AP_{H_2O} + P_{CO_2}} \tag{9.8}$$

Amadeo and Laborde [51] analyzed five kinetic expressions for low-temperature catalysts: two representing redox mechanism and three representing Langmuir-Hinshelwood model. Model I is proposed by Shchibrya *et al.* [2b]. Model II was a redox type model. Models III and IV were Langmuir-Hinshelwood type models that considered adsorption of four species (CO, H_2O, CO_2 and H_2) and the final model only considered the adsorption of CO and CO_2. The results of these authors indicated that only model III described the reaction behaviour in the conditions investigated.

$$\text{Model I} \quad r = \frac{kP_{H_2O}(1-\beta)}{AP_{H_2O} + P_{CO_2}} \tag{9.9}$$

$$\text{Model II} \quad r = \frac{k_1 k_2 P_{CO} P_{H_2O}(1-\beta)}{k_1 P_{CO} + k_2 P_{H_2O} + k_3 P_{CO_2}} \tag{9.10}$$

$$\text{Model III} \quad r = \frac{kP_{H_2O}P_{CO}(1-\beta)}{\left(1+\sum k_i p_i\right)^2} \tag{9.11}$$

$$\text{Model IV} \quad r = \frac{kP_{CO}(1-\beta)}{1+K_{H_2CO_2}P_{H_2}P_{CO_2}P_{H_2O}^{-1}+K_{H_2O}P_{H_2O}+K_{CO_2}P_{CO_2}} \tag{9.12}$$

$$\text{Model IV} \quad r = \frac{kP_{CO}P_{H_2O}(1-\beta)}{1+K_{H_2}P_{H_2}+K_{CO_2}P_{CO_2}} \tag{9.13}$$

The study of Choi and Stenger [52] tested two associative, two regenerative and one empirical model, obtaining good results for all of them. The best fit was provided by the above models I and II.

Power law model also provides good kinetic fit for the low-temperature WGS catalysts. Ovensen *et al.* [53] proposed microkinetic model based on surface redox mechanism and also evaluated the macroscopic power law kinetic model which was found to be an excellent representation of the kinetic data. Koryabkina *et al.* [54] determined the kinetic parameters for power law expression using catalysts based on copper over different supports. These authors suggested that there was a strong inhibition on the reaction rate by the products. They also proposed that the kinetics could be explained by a redox mechanism. The kinetic parameters obtained from different works are summarized in Table 9.5.

TABLE 9.5 Kinetic Parameters Obtained from the Power Law of the Low-Temperature WGS Catalysts

	Reaction orders					
Catalyst	*CO*	H_2O	H_2	CO_2	*E* (kJ/mol)	Reference
Cu-Zn/ Al_2O_3	1	1.5	−0.7	−0.7	78	[53]
Cu/Al_2O_3	1	1.9	−0.9	−1.4	59	[53]
8%Cu/ Al_2O_3	0.9	0.8	−0.8	−0.7	62	[54]
8%Cu-15%CeO_2/ Al_2O_3	0.7	0.6	−0.6	−0.6	32	[54]
8%Cu-CeO_2	0.9	0.4	−0.6	−0.6	56	[54]
40%Cu-Zn/Al_2O_3	0.8	0.8	−0.9	−0.9	79	[54]

In case of noble metals, Sun *et al.* studied [55] a Langmuir-Hinshelwood model that better described the experimental data than the power law model. The kinetic expression proposed by these authors was the following:

$$r = A\exp\left(\frac{-Ea}{RT}\right)\frac{P_{H_2O}P_{CO}(1-\beta)}{\left[1+A_1\exp\left(\frac{-\Delta H_1}{RT}\right)P_{CO}\right]\left[1+A_1\exp\left(\frac{-\Delta H_2}{RT}\right)P_{H_2}\right]} \tag{9.14}$$

where A, A_1 and A_2 are pre-exponential factors and H_1 and H_2 are the heat of adsorption of CO and H_2.

Ding and Chan [56] proposed a two-dimensional unsteady kinetic model for catalysts based on Pt. This model showed a good agreement with experimental data. The power law was also studied for catalysts based on noble metals, showing a reasonable fit to the experimental kinetic measurements. The kinetic parameters obtained in these investigations are summarized in Table 9.6. CO reaction orders for Au catalysts are higher than those of Pt catalysts. This could be explained by the different strength of CO adsorption on these materials. CO adsorption on Au particles is very weak and an increase of CO partial pressure favours concentration of this reactant at the metal surface, enhancing the reaction rate. On the contrary, CO adsorption over platinum metal surface is strong, producing CO coverage close to saturation. For this reason an increase of CO partial pressure has a lower effect in the reaction rate. Inhibition effect by CO_2 is higher for Au-based catalysts than in the case Pt. H_2 also inhibits the reaction rate over both types of catalysts. This can be explained by the high-adsorption capacity of H_2 on Pt and Au that hinders water adsorption.

Smith *et al.* [64] reviewed kinetics of the WGSR and proposed micro- and macro-kinetic models. The micro-kinetic method is based on the knowledge about the elementary steps that are involved in the reaction and its energetics. This method explores the detailed chemistry of the reaction. Using this method it is possible to estimate the surface coverage, reaction order and activation enthalpy. This method provides the accurate pathway and prediction of the reaction, but is computationally intensive. On the other hand, the empirical models are based on the experimental results and are typically expressed in the Arrhenius model and provide an easy and computationally lighter way to predict the rate of reaction.

Smith *et al.* [64] compiled and consolidated the detailed kinetic expressions for the high-temperature, low-temperature and noble metal catalysts. The equations are presented in Tables 9.7a, 9.7b, 9.7c, and 9.8.

TABLE 9.6 Kinetic Parameters Obtained from the Power Law of the Noble Metal Catalysts

Catalyst	Reaction orders				E (kJ/mol)	Reference
	CO	H_2O	H_2	CO_2		
Pd/CeO_2	0	0.5	−1	−0.5	38	[36]
2%Pt-1%Re/CeO_2-ZrO_2	0.07	0.67	−0.57	−0.16	71	[57]
2%Pt/CeO_2-ZrO_2	−0.05	0.85	−0.32	−0.05	71	[57]
2%Au/$CeZrO_4$	0.7	0.7	−0.9	−0.3		[39]
2.6%Pt/CeO_2	0.5	0.5	−0.5	−0.5	40	[58]
4.5%Au/CeO_2	1	1	−0.7	−0.5		[59]
1%Pt/Al_2O_3	0.1	1.1	−0.44	−0.07	84	[60]
1.66%Pt/Al_2O_3	0.11	0.82	−0.49	−0.06	81	[60]
1%Pt/CeO_2	−0.03	0.44	−0.38	−0.09	75	[60]
Pt-Re/TiO_2	0.4	0.7	−0.7	0	30	[61]
0.5%Pt/TiO_2	0.5	1	−07	0	45.2	[62]
1%Pt/CeO_2	0.14	0.66	−0.54	−0.08	91	[63]
1%Pt/TiO_2	0.30	0.85	−0.67	0	59	[63]

TABLE 9.7A Power Law Parameters of Various High Temperature WGS Catalysts [26, 41, 52, 54, 60, 63, 65, 66–72]

		Order of reaction		Arrhenius parameters	
Catalyst	Reaction conditions	*CO*	H_2O	K_o	E_a
Fe_3O_4/Cr_2O_3 8wt%Cr_2O_3 (Rhodes *et al.*) [65]	1 atm, 350-440 °C	-	-	ln K_0=11.5	112 (kJ/mol)
Fe_3O_4/Cr_2O_3 180-250 μm (Rhodes *et al.*) [65]	1 bar, 380-450 °C	-	-	ln K_0=10.1 ±0.2	118±1 (kJ/mol)
Fe_3O_4/Cr_2O_3 180-250 μm (Rhodes *et al.*) [65]	6 bar, 380-450 °C	-	-	ln K_0=12.0±0.2	124±1 (kJ/mol)
Fe_3O_4/Cr_2O_3 180-250 μm (Rhodes *et al.*) [65]	27 bar, 380-450 °C	-	-	ln K_0=7.4±0.1	111±1 (kJ/mol)
Fe_3O_4/Cr_2O_3 (Keiski *et al.*) [41]	3-5 bar, 573-633 °C	1.1	0.53	2.16×10^{11} (s^{-1}) ln K_0=26.1	95 (kJ/mol)
89%Fe_2O_3, 9%Cr_2O_3 (Keiski *et al.*) [66]	575-675 K	0.74	0.47	ln K_0=14.78	E/R=9598
80-90%Fe_2O_3, 8-13%Cr_2O_3, 1-2%CuO (San *et al.*) [79]	1 atm, 450 °C 6 mm × 6 mm	1	0	$10^{2.845}$ (mol/g_{cat} s)	111 (kJ/mol)
80-95%Fe_2O_3, 5-10%Cr_2O_3, 1-5%CuO (San *et al.*) [79]	1 atm, 450 °C 6 mm × 6 mm	0.9	0.31	$10^{0.659}$ (mol/g_{cat} s)	88 (kJ/mol)
CuO/Fe_3O_4/Cr_2O_3 180-250 μm (Rhodes *et al.*) [65]	1 bar, 380-450 °C	-	-	ln K_0=2.0±0.1	751 (kJ/mol)
CuO/Fe_3O_4/Cr_2O_3 180-250 μm (Keiski *et al.*) [66]	6 bar, 380-450 °C	-	-	ln K_0=5.5±0.1	85±2 (kJ/mol)
CuO/Fe_3O_4/Cr_2O_3 180-250 μm (Keiski *et al.*) [66]	27 bar, 380-450 °C	-	-	ln K_0=4.0±0.1	85±1 (kJ/mol)
Power gas – pilot plant data (Rase *et al.*) [68]	1/4″ × 3/8″, 2.20 g/cm^3	-	-	9.4×10^7 (1/s)	21.4 (kcal/g mol)
Girdler – pilot plant data (Rase *et al.*) [68]	1/4″ × 1/4″, 1.25 g/cm^3	-	-	1.47×10^8 (1/s)	23.8 (kcal/g mol)
ICI pilot plant data (Rase *et al.*) [68]	11.3″ × 8.5 mm, 1.36 g/cm^3	-	-	4.5×10^9 (1/s)	27.0 (kcal/g mol)

Adapted from [Smith author] *et al.*, Ref [64]

TABLE 9.7B Power Law Parameters of Various Low Temperature WGS Catalysts [26, 41, 52, 54, 60, 63, 65, 66–72]

Catalyst	Reaction conditions	Order of reaction		Arrhenius parameters	
		CO	H_2O	K_0	E_a
ICI-CuO-ZnO-Al_2O_3 (Choi and Stenger) [52]	-	1	1	3.99×10^6 (s^{-1}) (ln K_0=15.2)	52.8 (kJ/mol)
ICI 52-1 (copper-based catalyst) density=5.83 g/cm^3 (Salmi and Hakkarainen) [69]	1 atm, 200 °C	0.45	0.07	$K=5.37 \times 10^{-7}$ (mol/m^{2}s)/atm^{1+m}	
CuO-ZnO-Al_2O_3 (Choi and Stenger) [52]	-	0	1	-	41.8 (kJ/mol)
Cu-ZnO-Al_2O_3 (EX-2248) Sud Chemie (Choi and Stenger) [52]	200-250 μm size, 120-250 °C, 1:2 CO-H_2O	1	1	2.96×10^3 (s^{-1}) (ln K_0=12.6)	47.4 (kJ/mol)
42%CuO-ZnO-Al_2O_3 (G-66A) Sud Chemie (Henrik *et al.*) [80]	396-448 K CO/H_2O=1/3	-	-	4.9×10^6 (s^{-1})	71 (kJ/mol)
Cu-ZnO-Al_2O_3 (Koryabkina *et al.*) [54]	1 atm, 200 °C	0.2	0.6	-	67 (kJ/mol)
40%CuO-ZnO-Al_2O_3 (united catalyst) (Koryabkina *et al.*) [54]	1 atm, 190 °C	0.8	0.8	-	79 (kJ/mol)
Cu-ZnO-Al_2O_3 (Koryabkina *et al.*) [54]	1 atm, 180-200 °C	1	1.4	-	86 (kJ/mol)
Cu/Al_2O_3 (Choi and Stenger) [52]	-	1	1.9	-	69.3 (kJ/mol)
10%Cu-Al_2O_3 (Koryabkina *et al.*) [54]	1 atm, 130 °C	0.3	0.38	-	55 (kJ/mol)
8%CuO-Al_2O_3 (Koryabkina *et al.*) [54]	1 atm, 200 °C	0.9	0.8	-	62 (kJ/mol)
CuO/MnO_2 (Choi and Stenger) [52]	-	1	1	-	55 (kJ/mol)

Continued

TABLE 9.7B Power Law Parameters of Various Low Temperature WGS Catalysts [26, 41, 52, 54, 60, 63, 65, 66–72]—Cont'd

Catalyst	Reaction conditions	Order of reaction		Arrhenius parameters	
		CO	H_2O	K_0	E_a
8%CuO-CeO_2 (Koryabkina *et al.*) [54]	1 atm, 240 °C	0.9	0.4	-	56 (kJ/mol)
8%CuO-15%CeO_2-Al2O_3 (Koryabkina *et al.*) [54]	1 atm, 200 °C	0.7	0.6	-	32 (kJ/mol)
Cu(1 1 1) (Koryabkina *et al.*) [54]	1 atm, 340 °C	0	0.5-1	-	71 (kJ/mol)
Cu(1 1 0) (Koryabkina *et al.*) [54]	1 atm, 340 °C	0	1	-	42 (kJ/mol)
Ni (Wheeler *et al.*) [71]	0.008-0.05 s contact time	-	-	8.0×10^{7} (s^{-1})	85 (kJ/mol)
Ni/ceria (Wheeler *et al.*) [71]	Coated on alumina support	-	-	1.7×10^{8} (s^{-1})	85 (kJ/mol)
$Cu_{0.2}Ce_{0.8}O_{2-y}$ (Henrik *et al.*) [80]	473-623 K CO/H_2O=1/3	-	-	1.8×10^{3} (s^{-1})	61 (kJ/mol)
$Cu_{0.1}Ce_{0.9}O_{2-y}$ (Henrik *et al.*) [80]	573-623 K CO/H_2O=1/3	-	-	4.0×10^{3} (s^{-1})	78 (kJ/mol)
Ce(La)Ox (Li *et al.*) [72]	375-475 °C	-	-	-	58.5 (kJ/mol)
5at%Cu-Ce (10%La)Ox (Li *et al.*) [72]	175-300 °C CO/H_2O=1.5	0	1	-	30.4 (kJ/mol)
5at%Ni-Ce (10%La)Ox (Li *et al.*) [72]	275-300 °C CO/H_2O=1.5	0	1	-	38.2 (kJ/mol)

Adapted from [Smith author] *et al.*, Ref [64]

TABLE 9.7C **Power Law Parameters of Various Noble Metal WGS Catalysts**

Catalyst	Reaction conditions	Order of reaction		Arrhenius parameters	
		CO	H_2O	K_0	E_a
Ru (Wheeler *et al.*) [71]	300-1000 °C	-	-	1.6×10^7 (s^{-1})	80 (kJ/mol)
Ru/ceria (Wheeler *et al.*) [71]	0.008-0.05 s contact time	-	-	5.0×10^7 (s^{-1})	80 (kJ/mol)
Rh (Wheeler *et al.*) [71]	Coated on alumina support	-	-	3.0×10^9 (s^{-1})	130 (kJ/mol)
Rh/ceria (Wheeler *et al.*) [71]	5 wt% loading	-	-	1.5×10^{10} (s^{-1})	130 (kJ/mol)
Rh/Al_2O_3 (Wheeler *et al.*) [71]	330 °C	−0.10	0.44	5.10×10^6 (molecules/s/site)	23.0±1.3 (kcal/mol)
Rh/SiO_2 (Grenoble *et al.*) [26]	350 °C	−0.24	0.53	3.23×10^3 (molecules/s/site)	22.8±2.5 (kcal/mol)
Pd (Wheeler *et al.*) [71]	300-1000 °C	-	-	4.0×10^6 (s^{-1})	100 (kJ/mol)
Pd/ceria (Wheeler *et al.*) [71]	0.008-0.05 s contact time	-	-	4.0×10^7 (s^{-1})	100 (kJ/mol)
Pt (Wheeler *et al.*) [71]	Coated on alumina support 5 wt% loading	-	-	1.0×10^6 (s^{-1})	80 (kJ/mol)
Pt/ceria (Wheeler *et al.*) [71]	Coated on alumina support 5 wt% loading	-	-	2.5×10^7 (s^{-1})	80 (kJ/mol)
1%Pt/Al_2O_3 (Phatak *et al.*) [60]	1 atm, 225-285 °C	0.1	1.1	-	68 (kJ/mol)
1%Pt/Al_2O_3 (Phatak *et al.*) [60]	1 atm, 285-345 °C	0.06	1	-	84 (kJ/mol)
1.66%Pt/Al_2O_3 (Phatak *et al.*) [60]	1 atm, 285 °C	0.11	0.82	-	81 (kJ/mol)
1.66%Pt/Al_2O_3 (Phatak *et al.*) [60]	1 atm, 300 °C	0.1	0.77	-	81 (kJ/mol)

Continued

TABLE 9.7C Power Law Parameters of Various Noble Metal WGS Catalysts—Cont'd

Catalyst	Reaction conditions	Order of reaction		Arrhenius parameters	
		CO	H_2O	K_0	E_a
2%Pt/Al_2O_3 (Phatak *et al.*) [60]	1 atm, 270 °C	−0.21	0.75	-	82 (kJ/mol)
0.9%Pt/Al_2O_3 (Phatak *et al.*) [60]	1 atm, 100 °C	0.02	0.55	-	-
0.4%Pt/Al_2O_3 (Phatak *et al.*) [60]	1 atm, 544 °C	0.45	0.37	-	39 (kJ/mol)
Pt/Al_2O_3 (Grenoble *et al.*) [26]	270 °C	−0.21	0.75	1.9×10^6 (molecules/s/site	19.6±1.3 (kcal/mol)
1%Pt/CeO_2 (Phatak *et al.*) [60]	1 atm, 200 °C	−0.03	0.44	-	75 (kJ/mol)
1%Pt/CeO_2 (Phatak *et al.*) [60]	1 atm, 240 °C	0	1	-	46 (kJ/mol)
1%Pt/CeO_2 (Thinon *et al.*) [63]	-	0.14	0.66	-	91 (kJ/mol)
Pt/SiO_2 (Grenoble *et al.*) [26]	340 °C	−0.08	0.69	1.18×10^3 (molecules/s/site	19.1±0.8 (kcal/mol)
Pt/C (Grenoble *et al.*) [26]	340 °C	0.13	0.35	3.84×10^6 (molecules/s/site	25.5±1.4 (kcal/mol)
1%Pt/TiO_2 (Thinon *et al.*) [63]	-	0.3	0.85	-	59 (kJ/mol)
1.4%Pt-8.3% CeO_2/Al_2O_3 (Phatak *et al.*) [60]	1 atm, 260 °C	0.13	0.49	-	86 (kJ/mol)
2%Pt-1%Re/CeO_2-ZrO_2 (Phatak *et al.*) [60]	1 atm, 210-260 °C	−0.05	−0.32	-	71 (kJ/mol)

Adapted from [Smith author] *et al.*, Ref [64]

TABLE 9.8 Macro Kinetic Rate Expressions for WGSR [25, 51, 52, 56, 73–78]

Sample	Reaction conditions	Rate equation
High temperature catalysts		
Girdler (G3-b) (Elnashale and Elshishini) [73]	Particle size 0.62 cm equivalent diameter	$r = k\psi \frac{X_A X_B - \frac{X_C - X_D}{K}}{379\rho_b}, \quad k = \exp(15.95 - 4900/T)$ $\psi = 0.816 + 0.184P \quad \text{for} P \leq 11.8$ $\psi = 1.53 + 0.123P \quad \text{for} 11.8 < P \leq 20$ $\psi = 4.0 \quad \text{for} P > 20$ r is the rate of reaction
72%Fe_2O_3-8%Cr_2O_3 (Singh and sarqf) [74]	Density 4.561 g/cm^3	$r = E_{ff} \times 2.32 \times 10^{13}(X_{CO} - X_{CO}^*)\exp(-27{,}760/RT) \times R_a \times Ag_f \times P_f \times f_s$ $R_a = \exp(-8.91 + (5.553 \times 10^4/T))$ $\text{Log}\, Ag_f = (14.66 \times 10^{-4} - 2 \times 10^{-6})\tau$ $P_f = P^{(0.5-P/250)}$ $f_s = -0.276\log([H_2S] + 2.78) + 1.127$ $X_{CO}^* = X_{H_2} X_{CO_2} / X_{H_2O} k_{eq}$
Sud Chemie SHT-4 (Wei *et al.* (2008)) [81]	Density=7633.65 kg/m^3	$r = k(P_{CO}P_{H_2O} - P_{CO_2}P_{H_2}/k_{eq})$ $k = 1.78 \times 10^{22}(1 + 0.0097\delta - 1.1364\delta^2)T^{-8}\exp(-70/RT)$

Continued

TABLE 9.8 Macro Kinetic Rate Expressions for WGSR [25, 51, 52, 56, 73–78]—Cont'd

Sample	Reaction conditions	Rate equation
Low temperature WGS catalysts		
Langmuir-Hinshelwood model (Criscuoli *et al.*) [76]		$r=\frac{kK_{CO}K_{H_2O}\left[P_{CO}P_{H_2O}-\frac{P_{CO_2}P_{H_2}}{K_{eq}}\right]}{(1+K_{CO}P_{CO}+K_{H_2O}P_{H_2O}+K_{CO_2}P_{CO_2})^2}\times\frac{\rho_{cat}}{60}$ $K_{eq}=\exp\left(\frac{4577.8}{T}-4.33\right)$ $k=\exp\left(-\frac{29,364}{1.987\times T}+\frac{40.32}{1.987}\right)\,\mathrm{mol/g_{cat}\,min}$ $K_{CO}=\exp\left(-\frac{3064}{1.987\times T}-\frac{6.74}{1.987}\right)\,\mathrm{mol/g_{cat}\,min}$ $K_{H_2O}=\exp\left(-\frac{6216}{1.987\times T}+\frac{12.77}{1.987}\right)\,\mathrm{mol/g_{cat}\,min}$ $K_{CO_2}=\exp\left(-\frac{29,364}{1.97\times T}+\frac{40.32}{1.987}\right)\,\mathrm{mol/g_{cat}\,min}$
Empirical model based on industrial data Moe kinetic model (Seo *et al.*) [77]		$r=kP_{CO}P_{H_2O}\left[1-(P_{CO_2}P_{H_2}/K_{eq}P_{CO}P_{H_2O})\right]$, $K_{eq}=\exp\left(\frac{4577.8}{T}-4.33\right)$ Rate Constant $k=1.85\times10^{-5}\exp\left(12.88-\frac{1855.5}{T}\right)\,\mathrm{mol/g_{cat}\,min}$ r is the rate of the reaction and P_i is the partial pressure of the species

Girdler (G3-b) (Elnashale and Elshishini) [73]	Particle size 0.62 cm equivalent diameter	$r = k\psi \dfrac{X_A X_B - \dfrac{X_C - X_D}{K}}{379\rho_b}$ $k = \exp(12.88 - 2002.6/T)$ $\psi = 0.86 + 0.14P \quad \text{for} P \leq 24.8$ $\psi = 4.0 \quad \text{for} P > 24.8$ r is the rate of reaction
CuO-ZnO-Al_2O_3 (Amadeo and Laborde) [51]	1 atm, 453-503 K	$r = \dfrac{0.92 \times e\left(\dfrac{-454.3}{T}\right) P_{CO} P_{H_2O}(1-\beta)}{\left[1 + 2.2e\left(\frac{101.5}{T}\right) P_{CO} + 0.4e\left(\frac{158.3}{T}\right) P_{H_2O} + 0.0047e\left(\frac{2737.9}{T}\right) P_{CO_2} + 0.05e\left(\frac{1596.1}{T}\right) P_{H_2}\right]^2}$
Sud Chemie Cu/ZnO/Al_2O_3 (Choi and Stenger) [52]	1 atm, 120-250 °C, particle size 200-250 μm, CO:H_2-1:2	$r = 2.96 \times 10^5 \exp - \left(\frac{-47,400}{RT}\right)\left(P_{CO} P_{H_2O} - \frac{P_{CO_2} P_{H_2}}{K_{eq}}\right)$
Girdler/Sud Chemie (G66b). Unreduced wt% – CuO (32.2), ZnO (61.8) and Fe_2O_3 (1.6) (VanHerwijnen and de Jong) [25]	0.35-0.42 mm particle size, surface area 22.3 (m^2/g), pore diameter 12.5 mn	$r = 25.9 \times 10^3 \exp\left(-\frac{-16,000}{RT}\right) \frac{P_{CO} P_{H_2O}}{(1 + 127 P_{CO} P_{H_2O} + 26 P_{CO})}$
Cu (33%) Zn (66 %) (Singh and saraf) [78]	5.09 g/cm^3	$r = E_{ff} \times 2.955 \times 10^{13} (X_{CO} - X^*_{CO}) \exp(-20,960/RT) \times A_{gf} \times P_t$ $\text{Log}\, A_{gf} = (14.66 \times 10^{-4} - 2 \times 10^{-6})\tau$ $X^*_{CO} = X_{H_2} X_{CO_2} / X_{H_2O} k_{eq}$
Noble metal catalysts		
Precious metals (Sun *et al.*) [55]		$r = A\exp\left(\dfrac{-E_a}{RT}\right) \dfrac{P_{CO} P_{H_2O}(1-\beta)}{\left[1 + A_1 \exp\left(\dfrac{-\Delta H_1}{RT}\right) P_{CO}\right]\left[1 + A_2 \exp\left(\dfrac{-\Delta H_2}{RT}\right) P_{H_2}\right]}$

Adapted from [Smith author] *et al.*, Ref [64]

REFERENCES

[1] (a) D. S. Newsome, The water gas shift reaction, Catal. Rev. Sci. Eng. 21 (1980) 275–318. (b) K. Kochloefl, Water gas shift reaction, in: G. Ertl, H. Knozinger, J. Weitkamp (Eds.), Handbook of Heterogeneous Catalysis, vol. 4, Wiley VHS Publisher, Weinheim, Germany, 1997, pp. 1831–1840.

[2] (a) N. V. Kulkova, M. I. J. Temkin, Zh. Fiz. Khim. 23 (1949) 695–713. (b) G. G. Shchibrya, N. M. Morozov, M. I. J. Temkin, Kinet. Katal. 6 (1965) 1057.

[3] G. K. Boreskov, T. M. Yureva, N. M. Morozov, A. S. Sergeeva, Kinet. Katal. 11 (1970) 1476.

[4] S. Oki, J. Happel, M. Hinatow, Y. Kancko, J. W. Hightower (Ed.), Proceedings of the 5th International Congress on Catalysis, vol. 1, North Holland Publ. Co/Elsevier, New York/Amsterdam, 1973, pp. 173–183.

[5] D. G. Rethwisch, J. A. Dumesic, Adsorptive and catalytic properties of supported metal oxides: 1. Mössbauer spectroscopy of supported iron oxides, J. Phys. Chem. 90 (1986) 1863–1871.

[6] C. R. F. Lund, J. A. Dumesic, Strong oxide-oxide interactions in silica-supported magnetite catalysts: IV. Catalytic consequences of the interaction in water-gas shift, J. Catal. 76 (1982) 93–100.

[7] M. Tinkle, J. A. Dumesic, Isotopic exchange measurements of the rates of adsorption/desorption and inter conversion of CO and CO_2 over chromia-promoted magnetite: implications for water-gas shift, J. Catal. 103 (1987) 65–78.

[8] J. O. Edstrom, The mechanism of reduction of iron oxide, J. Iron Steel Inst. 175 (1953) 289–304.

[9] M. V. C. Sastri, R. P. Vishwanath, B. Vishwanath, Studies on the reduction of iron oxide with hydrogen, Int. J. Hydrogen Energy 7 (1982) 951–955.

[10] H. Topsoe, J. A. Dumesic, M. Boudart, Alumina as a textural promoter of iron synthetic ammonia catalysts, J. Catal. 28 (1973) 477–488.

[11] D. G. Reithwisch, J. A. Dumesic, The effects of metal-oxygen bond strength on properties of oxides: II. Water-gas shift over bulk oxides, Appl. Catal. 21 (1986) 97–109.

[12] G. K. Boreskov, Kinet. Katal. 11 (1970) 374.

[13] A. Khan, P. Chen, P. Boolchand, P. G. Simrnitios, Modified nano-crystalline ferrites for high-temperature WGS membrane reactor applications, J. Catal. 253 (2008) 91–104.

[14] L. Neel, Théorie du traînage magnétique des ferromagnétiques en grains fins avec applications aux terres cuites, Ann. Geophys. 5 (1949) 99–136.

[15] G. C. de Araujo, M.do.C. Rangel, An environmental friendly dopant for the high-temperature shift catalysts, Catal. Today 62 (2000) 201–207.

[16] G. K. Reddy, P. Boolcahnd, P. G. Simrnitios, Unexpected behavior of copper in modified ferrites during high temperature WGS reaction – aspects of $Fe^{3+} \leftrightarrow Fe^{2+}$ redox chemistry from Mössbauer and XPS studies, J. Phys. Chem. C 116 (2012) 11019–11031.

[17] L. Neel, Ann. Phys. 3 (1948) 137.

[18] E. J. W. Verwey, P. W. Haayman, F. C. Romeijn, Physical properties and cation arrangement of oxides with spinel structures: II. Electronic conductivity, J. Phys. Chem. 15 (1947) 181–187.

[19] R. Bauminger, S. G. Cohin, A. Marinov, S. Offer, E. Segal, Study of the low-temperature transition in magnetite and the internal fields acting on iron nuclei in some spinel ferrites, using Mössbauer absorption, Phys. Rev. 122 (1961) 1447–1450.

[20] A. Ito, K. Ono, Y. Iskawa, A study of the low temperature transition in magnetite, J. Phys. Soc. Jpn. 18 (1963) 1465–1472.

[21] M. Gotic, T. Jurkin, S. Music, From iron(III) precursor to magnetite and vice versa, Mater. Res. Bull. 44 (2009) 2014–2021.

[22] M. Sorescu, L. Dia mandescu, R. A. Brand, T. Mihaila, Mössbauer study of manganese-doped magnetite below the Verwey transition, Mater. Lett. 58 (2004) 885–888.
[23] E. F. Armstrong, T. P. Hilditch, Proc. R. Soc. A97 (1920) 265.
[24] A. V. Davydov, G. K. Boreskov, T. M. Yurieva, N. A. Rubene, Associative mechanisms of the water gas shift reactions, Dokl. Akad. Nauk USSR 236 (1977) 1402.
[25] T. Van Herwijnen, W. A. de Jong, Kinetics and mechanism of the CO shift on Cu/ZnO: 1. Kinetics of the forward and reverse CO shift reactions, J. Catal. 63 (1980) 83–93.
[26] D. C. Grenoble, M. M. Estadt, D. F. Ollis, The chemistry and catalysis of the water gas shift reaction: 1. The kinetics over supported metal catalysts, J. Catal. 67 (1981) 90–102.
[27] T. Shido, Y. Iwasawa, Regulation of reaction intermediate by reactant in the water gas shift reaction on CeO_2 in relation to reactant promoted mechanism, J. Catal. 136 (1992) 493–503.
[28] J. M. Pigos, C. J. Brooks, G. Jacobs, B. H. Davis, Low temperature water gas shift: characterization of Pt-based ZrO_2 catalyst promoted with Na discovered by combinatorial methods, Appl. Catal. A Gen. 319 (2007) 47–57.
[29] S. Sato, J. M. White, Photocatalytic water decomposition and water gas shift reaction over NaOH-coated, platinized TiO_2, J. Catal. 69 (1981) 128–139.
[30] K. Klier, Preparation of bifunctional catalysts, Catal. Today 15 (1992) 361–382.
[31] J. M. Campbell, J. Nakamura, C. T. Campbell, Model studies of cesium promoters in water gas shift catalysts: Cs/Cu(1 1 0), J. Catal. 136 (1992) 24–42.
[32] A. Hagemeyer, R.E. Carhart, K. Yaccato, P. Strasser, M. Herrmann, R.K. Grasselli, C.J. Brooks, C.B. Phillips, U.S. Patent (2007) 7,179,442 B2.
[33] F. C. Meunier, D. Reid, A. Goguey, S. Shekhtman, C. Hardacre, R. Burch, W. Deng, M. F. Stephanopoulos, Quantitative analysis of the reactivity of formate species seen by DRIFTS over a Au/Ce(La)O_2 water gas shift catalyst: first unambiguous evidence of the minority role of formates as reaction intermediates, J. Catal. 247 (2007) 277–287.
[34] A. A. Gokhale, J. A. Dumesic, M. M. Mavrikakis, On the mechanism of low temperature water gas shift reaction on copper, J. Am. Chem. Soc. 130 (2008) 1402–1414.
[35] L. C. Grabow, A. A. Gokhale, S. T. Evans, J. A. Dumesic, M. Mavrikakis, Mechanism of the water gas shift reaction on Pt: first principles, experiments, and microkinetic modeling, J. Phys. Chem. C 112 (2008) 4608–4617.
[36] S. Hilaire, X. Wang, T. Luo, R. T. Gorte, J. P. Wagner, A comparative study of water gas shift reaction over ceria-supported metallic catalysts, Appl. Catal. A Gen. 215 (2001) 271–278.
[37] J. A. Rodriguez, P. Liu, J. Hrbek, J. Evans, M. Perez, Water gas shift reaction on Cu and Au nanoparticles supported on CeO_2(1 1 1) and Zn(00 0 1): intrinsic activity and importance of support interactions, Angew. Chem. Int. Ed. 46 (8) (2007) 1329–1332.
[38] G. G. Olympiou, C. M. Kalamaras, C. D. Z. Yazdi, A. M. Efstathiou, Mechanistic aspects of the water-gas shift reaction on alumina-supported noble metal catalysts: in situ DRIFTS and SSITKA – mass spectrometry studies, Catal. Today 127 (2007) 304–318.
[39] R. Burch, Gold catalysts for pure hydrogen production in the water-gas shift reaction: activity, structure and reaction mechanism, Phys. Chem. Chem. Phys. 8 (2006) 5483–5500.
[40] H. Bohlbro, Water gas shift reaction, Chem. Eng. World 46 (1970) 5–8.
[41] R. L. Keiski, T. Salmi, P. Niemisto, J. Ainassaari, V. J. Pohjola, Stationary and transient kinetics of the high temperature water-gas shift reaction, Appl. Catal. A Gen. 137 (1996) 349–370.
[42] S. S. Hla, G. J. Duffy, L. D. Morpeth, A. Cousins, D. G. Roberts, J. H. Edwards, Investigation of the effect of total pressure on performance of the catalytic water-gas shift reaction using simulated coal-derived syngases, Catal. Commun. 11 (2009) 272–275.
[43] Y. Lei, N. W. Cant, D. L. Trimm, Kinetics of the water-gas shift reaction over a rhodium-promoted iron-chromium oxide catalyst, Chem. Eng. J. 114 (2005) 81–85.

[44] G. C. Chinchen, R. H. Logan, M. S. Spencer, Water-gas shift reaction over an iron oxide/chromium oxide catalyst: III. Kinetics of reaction, Appl. Catal. 12 (1984) 97–103.

[45] H. Bohlbro, An Investigation on the Kinetics of the Conversion of Carbon Oxide with Water Vapor over Iron Oxide Based Catalysts, second ed., Giellerup, Copenhagen, 1969.

[46] H. Bohlbro, The kinetics of the water gas conversion at atmospheric pressure, Acta Chem. Scand. 15 (1961) 502–520.

[47] V. I. Atroshchenko, A. Y. Loboiko, Izv. Vyssh. Uchenbn. Zaved. Khim. Khim. Tekhnol. 14 (1971) 957.

[48] H. Bohlbro, The kinetics of the water gas conversion: IV. Influence of the alkali on the rate, J. Catal. 3 (1964) 207–215.

[49] H. Uchida, N. Isogai, M. Oba, T. Hasegawa, The zinc oxide-copper catalyst for carbon monoxide-shift conversion: I. The dependency of the catalytic activity on the chemical composition of the catalyst, Bull. Chem. Soc. Jpn. 40 (1967) 1981–1986.

[50] E. M. Cherednik, N. M. Moroaov, M. I. Temkin, Kinet. Katal. 10 (1969) 603.

[51] N. E. Amadeo, M. A. Laborde, Hydrogen production from the low-temperature water-gas shift reaction: kinetics and simulation of the industrial reactor, Int. J. Hydrogen Energy 20 (1995) 949–956.

[52] Y. Choi, G. Stenger, Water gas shift reaction kinetics and reactor modeling for fuel cell grade hydrogen, J. Power Sources 124 (2003) 432–439.

[53] C. V. Ovesen, B. S. Clausen, B. S. Hammershøi, G. Steffensen, T. Askgaard, I. Chorkendorff, J. K. Nørskov, P. B. Rasmussen, P. Stoltze, P. Taylor, A microkinetic analysis of the water-gas shift reaction under industrial conditions, J. Catal. 158 (1996) 170–180.

[54] N. A. Koryabkina, A. A. Phatak, W. F. Ruettinger, R. J. Farrauto, F. H. Ribeiro, Determination of kinetic parameters for the water-gas shift reaction on copper catalysts under realistic conditions for fuel cell applications, J. Catal. 217 (2003) 233–239.

[55] J. Sun, J. DesJardins, J. Buglass, K. Liu, Noble metal water gas shift catalysis: kinetics study and reactor design, Int. J. Hydrogen Energy 30 (2005) 1259–1264.

[56] O. L. Ding, S. H. Chan, Water-gas shift reaction – a 2-D modeling approach, Int. J. Hydrogen Energy 33 (2008) 4325–4336.

[57] R. Radhakrishnan, R. R. Willigan, Z. Dardas, T. H. VandersPurt, Water gas shift activity and kinetics of Pt/Re catalysts supported on ceria-zirconia oxides, Appl. Catal. B Environ. 66 (2006) 23–28.

[58] R. Leppelt, B. Schumacher, V. Plzak, M. Kinnea, R. J. Behm, Kinetics and mechanism of the low-temperature water-gas shift reaction on Au/CeO_2 catalysts in an idealized reaction atmosphere, J. Catal. 244 (2006) 137–152.

[59] Y. Denkwitz, A. Karpenko, V. Plzak, R. Leppelt, B. Schumacher, R. J. Behm, Influence of CO_2 and H_2 on the low-temperature water-gas shift reaction on Au/CeO_2 catalysts in idealized and realistic reformate, J. Catal. 246 (2006) 74–90.

[60] A. A. Phatak, N. Koryabkina, S. Rai, J. L. Ratts, W. Ruettinger, R. J. Farrauto, G. E. Blau, W. N. Delgass, F. H. Ribeiro, Kinetics of the water-gas shift reaction on Pt catalysts supported on alumina and ceria, Catal. Today 123 (2007) 224–234.

[61] K. G. Azzam, I. V. Babich, K. Seshan, L. Leffert, Role of Re in $Pt-Re/TiO_2$ catalyst for water gas shift reaction: a mechanistic and kinetic study, Appl. Catal. B Environ. 80 (2008) 129–140.

[62] C. Kalamaras, P. Panagiotopoulou, D. L. Kondarides, A. M. Efstathiou, Kinetic and mechanistic studies of the water-gas shift reaction on Pt/TiO_2 catalyst, J. Catal. 264 (2009) 117–129.

[63] O. Thinon, K. Rachedi, F. Diehl, P. Avenier, Y. Schuurman, Kinetics and mechanism of the water-gas shift reaction over platinum supported catalysts, Top. Catal. 52 (2009) 1940–1945.

[64] R. J. B. Smith, M. Loganathan, M. S. Shantha, A review of the water gas shift reaction kinetics, Int. J. Chem. React. Eng. 8 (2010) 1–32, R4.

[65] C. Rhodes, G. J. Hutchings, Studies of the role of the copper promoter in the iron oxide-chromia, low temperature water gas shift catalyst, Phys. Chem. Chem. Phys. 5 (2003) 2719–2723.

[66] R. L. Keiski, T. Salmi, V. J. Pohjola, Development and verification of a simulation model for a non-isothermal water-gas shift reactor, Chem. Eng. J. 48 (1992) 17–29.

[67] S. S. Hla, D. Park, G. J. Duffy, J. H. Edwards, D. G. Roberts, A. Ilyushechkin, L. D. Morpeth, T. Nguyen, Kinetics of high-temperature water-gas shift reaction over two iron-based commercial catalysts using simulated coal-derived syngases, Chem. Eng. J. 146 (2009) 148–154.

[68] H. F. Rase, Chemical Reactor Design for Process Plants: Volume Two – Case Studies and Design Data, John Wiley and Sons, New York, 1977.

[69] T. Salmi, R. Hakkarainen, Kinetic study of the low-temperature water-gas shift reaction over a Cu-Zno catalyst, Appl. Catal. 49 (1989) 285–306.

[70] H. Kusar, S. Hocevar, J. Levec, Kinetics of the water-gas shift reaction over nanostructured copper-ceria catalysts, Appl. Catal. B Environ. 63 (2006) 194–200.

[71] C. Wheeler, A. Jhalani, E. J. Klein, S. Tummala, L. D. Schmidt, The water-gas-shift reaction at short contact times, J. Catal. 223 (2004) 191–199.

[72] Y. Li, Q. Fu, M. F. Stephanopoulos, Low-temperature water-gas shift reaction over Cu- and Ni-loaded cerium oxide catalysts, Appl. Catal. B Environ. 27 (3) (2000) 179–191.

[73] S. S. E. H. Elnashaie, S. S. Elshishini, Modeling, Simulation and Optimization of Industrial Fixed Bed Catalytic Reactors, Gordon and Breach Science Publishers, London, 1993.

[74] C. P. P. Singh, D. N. Saraf, Simulation of high-temperature water-gas shift reactors, Ind. Eng. Chem. Process Des. Dev. 16 (1977) 313–319.

[75] W.-H. Chen, M.-R. Lin, T. L. Jiang, M.-H. Chen, Modeling and simulation of hydrogen generation from high-temperature and low-temperature water gas shift reactions, Int. J. Hydrogen Energy 33 (2008) 6644–6665.

[76] A. Criscuoli, A. Basile, E. Drioli, An analysis of the performance of membrane reactors for the water-gas shift reaction using gas feed mixtures, Catal. Today 56 (2000) 53–64.

[77] Y. S. Seo, D.-J. Seo, Y.-T. Seo, W.-L. Yoon, Investigation of the characteristics of a compact steam reformer integrated with a water-gas shift reactor, J. Power Sources 161 (2006) 1208–1216.

[78] C. P. P. Singh, D. N. Saraf, Simulation of low-temperature water-gas shift reactor, Ind. Eng. Chem. Process Des. Dev. 19 (1980) 393–396.

[79] San Shaw Hla, D. Park, G.J. Duffy, J.H. Edwards, D.G. Roberts, A. Ilyushechkin, L.D. Morpeth, T. Nguyen, Kinetics of high temperatures water gas shift reaction over two iron based commercial catalysts using simulated coal derived syngases:, Chem. Eng. J. 146 (2009) 148–154.

[80] Kusar Henrik, S. Hocevar, J. Leve, Kinetics of the water gas shift reaction over nanostructured copper-ceria catalysts, Appl. Catal. B: Envi 63 (2006) 194–200.

[81] Wei-Hsin Chen, M.-R. Lin, T.L. Jiang, M.-H. Chen, Modeling and simulation of hydrogen generation from high temperatureand low temperature water gas shift reaction, Inter. J. Hydro. Ener. 33 (2008) 6644–6656.

[82] M. V. Twigg, Catalyst Handbook, second edition, Wolfe Publishing Ltd., 1989.

Index

Note: Page numbers followed by *f* indicate figures, *t* indicate tables and *s* indicate schemes.

H

I

J

L

R

FSC
www.fsc.org
MIX
Papier | Fördert
gute Waldnutzung
FSC® C083411

Kolibri 360 GmbH
Ferdinand-Jühlke-Straße 7
99095 Erfurt, Deutschland
produktsicherheit@kolibri360.de